Advanced and Intelligent Manufacturing in China

Series Editor

This is a set of high-level and original academic monographs. This series focuses on the two fields of intelligent manufacturing and equipment, control and information technology, covering a range of core technologies such as Internet of Things, 3D printing, robotics, intelligent equipment, and epitomizing the achievements of technological development in China's manufacturing sector. With Prof. Jie Chen, a member of the Chinese Academy of Engineering and a control engineering expert in China, as the Editorial in Chief, this series is organized and written by more than 30 young experts and scholars from more than 10 universities and institutes. It typically embodies the technological development achievements of China's manufacturing industry. It will promote the research and development and innovation of advanced intelligent manufacturing technologies, and promote the technological transformation and upgrading of the equipment manufacturing industry.

Minggang Xu · Xueke Luo · Congpeng Zhang · Wen Li

Design and Case of Intelligent Electromechanical Equipment System

Chemical Industry Press Co., Ltd.

Springer

Minggang Xu
School of Mechanical and Materials Engineering
North China University of Technology
Beijing, China

Congpeng Zhang
School of Mechanical and Materials Engineering
North China University of Technology
Beijing, China

Xueke Luo
Beijing Institute of Petrochemical Technology
Beijing, China

Wen Li
School of Mechanical and Materials Engineering Technology
North China University of Technology
Beijing, China

ISSN 2731-5983 ISSN 2731-5991 (electronic)
Advanced and Intelligent Manufacturing in China
ISBN 978-981-95-3832-4 ISBN 978-981-95-3833-1 (eBook)
https://doi.org/10.1007/978-981-95-3833-1

Jointly published with Chemical Industry Press
The print edition is not for sale in China (Mainland). Customers from China (Mainland) please order the print book from: Chemical Industry Press.
ISBN of the Co-Publisher's edition: 978-7-122-39356-2

Translation from the Chinese Simplified language edition: "智能机电装备系统设计与实例" by Minggang Xu et al., © Chemical Industry Press 2022. Published by Chemical Industry Press. All Rights Reserved.

This Springer imprint is published by the registered company Springer Nature Singapore Pte Ltd.
The registered company address is: 152 Beach Road, #21-01/04 Gateway East, Singapore 189721, Singapore

Introduction

Intelligent manufacturing equipment refers to manufacturing equipment with perception, analysis, reasoning, decision-making, control functions, which is a high degree of integration and deep integration of advanced manufacturing technology, information technology and artificial intelligence technology. The manufacturing field focuses on the development of intelligent manufacturing equipment, the core of which is intelligent complete production lines with high-end numerical control technology as the core, intelligent control systems, intelligent sensing and testing instruments and key basic parts, components and general and special intelligent manufacturing equipment.

This book is about intelligent manufacturing equipment system, the book is divided into 7 chapters. Chapter 1 introduces the concept and development trend of intelligent manufacturing equipment. Chapter 2 introduces the design method and digital design of intelligent manufacturing equipment machinery. Chapter 3 introduces the design and intelligent drive system of intelligent manufacturing equipment. Chapter 4 mainly introduces the design of intelligent manufacturing equipment sensing system and the development trend of intelligent sensing technology. Chapter 5 describes the design and implementation of intelligent manufacturing equipment control system; Chap. 6 introduces the concept and application of intelligent IoT system. In Chap. 7, taking the automatic gluing machine developed by the author's research group as an example, the molding process of insulating glass and the design and development process of high-performance motion control equipment system are introduced.

The book is rich in content and detailed materials, which can be used by equipment design and control system research and development personnel and also can be read by teachers and students of related majors in colleges and universities.

This book was written by Minggang Xu, Congpeng Zhang, and others from North China University of Technology. During the writing process, they received assistance from Honghai Xu, Tan Mao, Kai Li, Ying Liu, and Leyi Li. The book was planned and reviewed by Xueke Luo from Beijing Institute of Petrochemical Technology, who also provided numerous suggestions for revisions. We would like to express our gratitude for this!

Due to the limited level of the author, the book will inevitably have shortcomings, please criticize the experts and scholars.

Contents

Chapter 1
Overview of Intelligent Manufacturing Equipment

Intelligent manufacturing (IM) is composed of intelligent machine and human experts intelligent man–machine integration system, through the analysis of man–machine cooperation to the production process, reasoning, judgment, and decision-making activities such as intelligence, can expand, extension and partially replace human experts in the manufacturing process of mental work. It updates the concept of manufacturing automation, expands it to flexibility, intelligence and highly integrated, with self-sensing, self-learning, self-decision-making, self-execution, self-adaptation and other functions. Intelligent manufacturing integrates automation, flexibility, integration and intelligence, with self-organizing ability, self-discipline ability, self-learning and self-maintenance ability, intelligent integration of the entire environment, man–machine integration and other features, involving robotics technology, 3D printing technology, virtual reality technology and other ten technical bases [1]. Its main feature is that as the manufacturing industry has once again become the driving force for the stable development of the global economy, all major industrial countries in the world have accelerated the pace of industrial development, and the manufacturing industry is gradually becoming the top priority for the development of all countries, and the direction leading the future manufacturing industry has also become a strategic commanding point for the competition of manufacturing powers [2]. China is a big manufacturing country, intelligent manufacturing is an important development trend of manufacturing industry. Manufacturing equipment is a basic and strategic industry for the national economy and the development of national science and technology, and it is an industry that countries around the world have always attached great importance to. The Outline of the 13th Five-Year Plan for National Economic and Social Development of the People's Republic of China, Made in China 2025 and the Guiding Opinions of The State Council on Deepening the Integrated Development of Manufacturing and the Internet all put forward higher requirements for China's future manufacturing industry.

"Made in China 2025" proposed that intelligent production is the main battlefield of "intelligent manufacturing engineering"; Production mode change is the

M. Xu et al., *Design and Case of Intelligent Electromechanical Equipment System*, Advanced and Intelligent Manufacturing in China,
https://doi.org/10.1007/978-981-95-3833-1_1

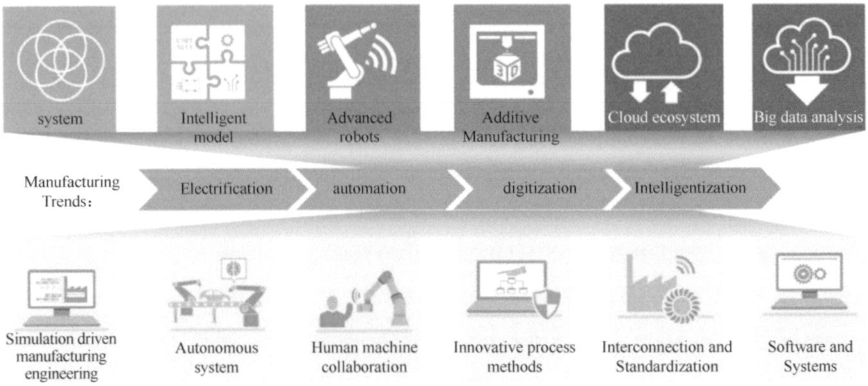

Fig. 1.1 The development trend of manufacturing industry

main battlefield of the "manufacturing service action plan", and intelligent manufacturing equipment is the main battlefield of "equipment innovation project". By 2025, China will enter the ranks of manufacturing powers, and by 2035, the manufacturing industry as a whole will reach the middle level of the world's manufacturing powers. Promote the innovation of high-end equipment manufacturing industry, promote the innovation of high-end equipment on the basis of the implementation of the Internet, high-end CNC equipment, large aircraft and other special projects, and drive the improvement of the digital level and intelligent level of the industry through intelligent manufacturing [3]. Figure 1.1 shows the development trend of the manufacturing industry.

Intelligent manufacturing equipment is the key development direction of high-end equipment manufacturing industry and an important embodiment of the deep integration of information technology and industrialization, vigorously cultivating and developing intelligent manufacturing equipment industry is of great practical significance for accelerating the transformation and upgrading of manufacturing industry, improving production efficiency, technical level and product quality, reducing energy and resource consumption, and realizing the intelligent and green development of manufacturing process. In today's world, the wave of the fourth industrial revolution has arrived [4]. Intelligent manufacturing equipment is not only the basic support for high-end equipment such as ocean engineering, high-speed rail, large aircraft, and satellites, but also can realize industrial upgrading by integrating measurement and control systems, automated complete production lines, robots and other technologies into manufacturing equipment. For China, the development bottleneck of intelligent manufacturing mainly includes weak innovation ability, insufficient core competitiveness, low added value of products, weak brand competitiveness, high energy consumption and low efficiency [5].

China's intelligent manufacturing equipment focuses on the development of high-grade CNC machine tools and basic manufacturing equipment, automated complete production lines, intelligent control systems, precision and intelligent instrumentation and testing equipment, key basic parts, components and general components, intelligent special equipment, to achieve automation of the production process, intelligent, information, precision, green, and drive the overall technical level of the industry to improve.

1.1 The Concept, Composition and Characteristics of Intelligent Manufacturing Equipment

1.1.1 Concept of Intelligent Manufacturing Equipment

Manufacturing includes intelligent manufacturing technology and intelligent manufacturing system. Intelligent manufacturing technology refers to the use of computer simulation manufacturing experts analysis, judgment, reasoning, thinking and decision-making and other intelligent activities, and organic integration with intelligent machines, throughout the entire manufacturing enterprise business decision-making, procurement, product design, production planning, manufacturing, assembly, quality assurance and marketing and other subsystems, to achieve the entire manufacturing enterprise highly flexible. It is an advanced manufacturing technology that collects, stores, improves, shares, inherits and develops intelligent information of manufacturing experts, and greatly improves production efficiency. Intelligent manufacturing system is based on intelligent manufacturing technology, the use of computer integrated application of artificial intelligence technology, intelligent manufacturing machines, material technology, modern management technology, manufacturing technology, information technology, automation technology, parallel engineering, life science and system engineering theories and methods, on the basis of international standardization and interchangeability, the entire enterprise manufacturing system in each subsystem intelligent, and make the manufacturing system form a network integrated, highly automated manufacturing system, the purpose is to automatically complete the design, processing, control and management process through the flexibility of equipment and computer artificial intelligence control, 3 improve the effectiveness of manufacturing in a highly changing environment.

The implementation of "Industry 4.0" and "Made in China 2025" strategy has put forward higher requirements for manufacturing equipment. Industry 4.0 is proposed by the German BITKOM Association, which represents the continuous progress of industrial production automation achieved through self-configuring machine automation systems [6]. The proposal and evolution process of Industry 4.0 is shown in Fig. 1.2. Its essence is the combination of information technology and traditional manufacturing, the interconnection between people and equipment and products through the industrial Internet of Things, the construction of digital intelligent manufacturing mode, and the realization of intelligent production.

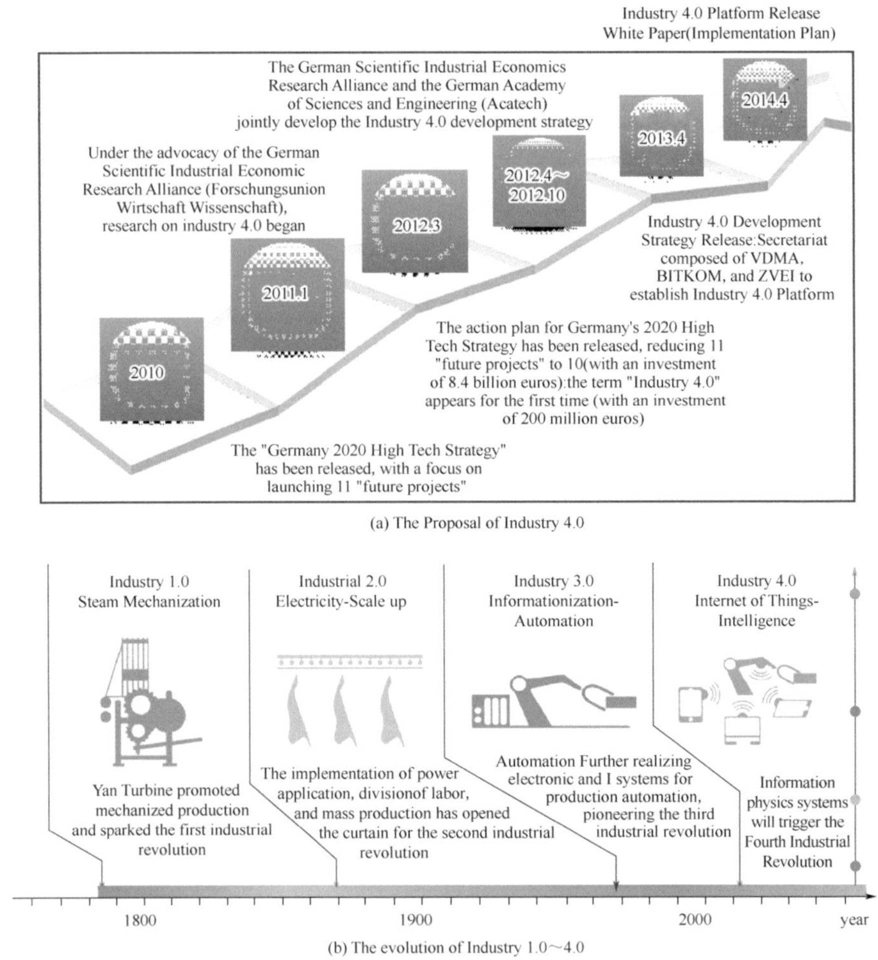

Fig. 1.2 The proposal and development of Industry 4.0

Its core is to help or even replace workers through intelligent machines and big data analysis to achieve comprehensive intelligence in the manufacturing industry [7]. Intelligent manufacturing equipment refers to manufacturing equipment with the functions of perception, analysis, reasoning, decision-making and control, which is the product of the integration and deep integration of advanced manufacturing technology, information technology and intelligent technology, and its position is becoming more and more important in the construction of manufacturing power in China.

Intelligent manufacturing equipment is mainly divided into intelligent machine tools and intelligent basic manufacturing equipment, which is a more advanced equipment with higher production efficiency and manufacturing accuracy improved on the basis of digital equipment. Unlike traditional equipment, intelligent manufacturing equipment has sensing, analysis, reasoning, decision-making and control functions, and can integrate sensors and intelligent diagnosis and decision-making software into the equipment to adapt the manufacturing process to changes in the manufacturing environment and manufacturing process. It focuses on promoting the development of high-grade CNC machine tools and basic manufacturing equipment, automated complete production lines, intelligent control systems, precision and intelligent instrumentation and test equipment, key basic parts, components and general components, intelligent special equipment, to achieve automation, intelligence, precision and green production process, and to promote the overall technical level of the industry.

Intelligent manufacturing equipment is the continuation and development of CNC manufacturing equipment, is a huge leap in equipment performance, can be more effective and more economical to achieve ordinary CNC equipment in the fields of aerospace, nuclear power, aviation, laser nuclear fusion is difficult to achieve the extraordinary manufacturing tasks, in order to enhance China's core competitiveness. China is a manufacturing country, in the process of moving from a manufacturing country to a powerful country, intelligent manufacturing equipment is the direction of manufacturing equipment development, is a major opportunity for China's manufacturing equipment products to high-end and enhance the added value of its technology, is the cutting-edge technology field of our economic strategy, but also the direction of our manufacturing industry for strategic adjustment. The category of intelligent manufacturing is shown in Fig. 1.3.

China's "12th Five-Year Plan" proposed the strategy of "adjustment, transformation, innovation and upgrading" of the equipment manufacturing industry, and formulated the goal of "promoting the equipment manufacturing industry from large to strong". As the core and key of the equipment manufacturing industry, high-end intelligent equipment manufacturing industry is listed as one of the strategic emerging industries. Its development goal is to form a complete intelligent manufacturing equipment industrial system after 10 years of efforts, the overall technical level has entered the international advanced ranks, some products have made original innovation breakthroughs, and basically meet the needs of key areas of the national economy and national defense construction. By 2020, China's intelligent manufacturing equipment industry will be cultivated into a leading industry with international competitiveness, a sound intelligent manufacturing equipment industry system will be established, equipment intelligence and manufacturing process automation will be realized, industrial production efficiency, product technology level and quality will be significantly improved, and energy, resource consumption and pollutant emissions will be significantly reduced.

With the intensive introduction of policies in the field of intelligent manufacturing, especially the proposal of the "Made in China 2025" plan, the transformation of China's manufacturing industry to the direction of intelligent manufacturing has been the general trend. At present, the key technical equipment represented by high-end CNC machine tools, advanced industrial robots, intelligent instruments and meters has made positive progress and rich achievements; Intelligent manufacturing

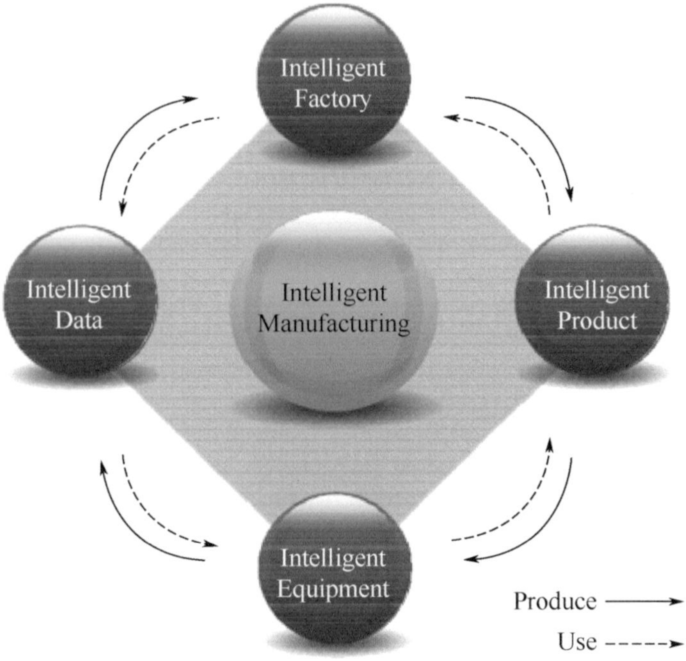

Fig. 1.3 Category of intelligent manufacturing

equipment and advanced processes are gaining popularity in key industries, and the pace of digitization, networking and intelligence of manufacturing equipment is accelerating. The Made in China vision is shown in Fig. 1.4.

1.1.2 Composition of Intelligent Manufacturing Equipment

The composition of intelligent manufacturing equipment is shown in Fig. 1.5. Mechanical system is the basis of intelligent manufacturing equipment. In addition to mechanical systems, the key technologies of intelligent manufacturing equipment involve intelligent sensing systems, mechanical execution systems, motion control systems and intelligent decision-making systems, and each part has key technologies involving hardware and software, and has a broad space for development.

(1) Intelligent perception system

Intelligent sensing system is a direction of rapid development of intelligent manufacturing equipment. The perception system simulates the vision, hearing, and touch of the intelligent agent, which is the input and starting point of intelligent manufacturing equipment, and also the basis for intelligent manufacturing equipment to play a role. Intelligent manufacturing equipment is not a complete simulation of human

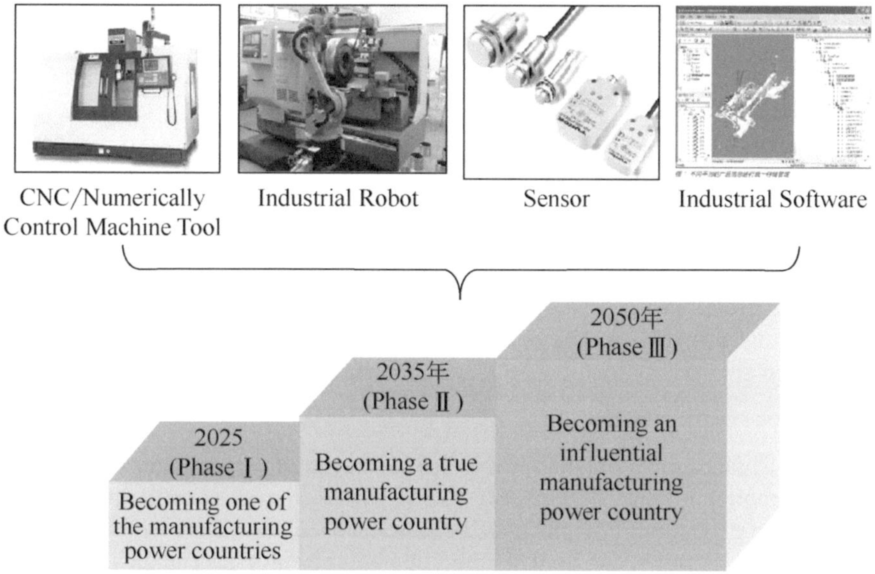

Fig. 1.4 Prospect of Chinese manufacturing

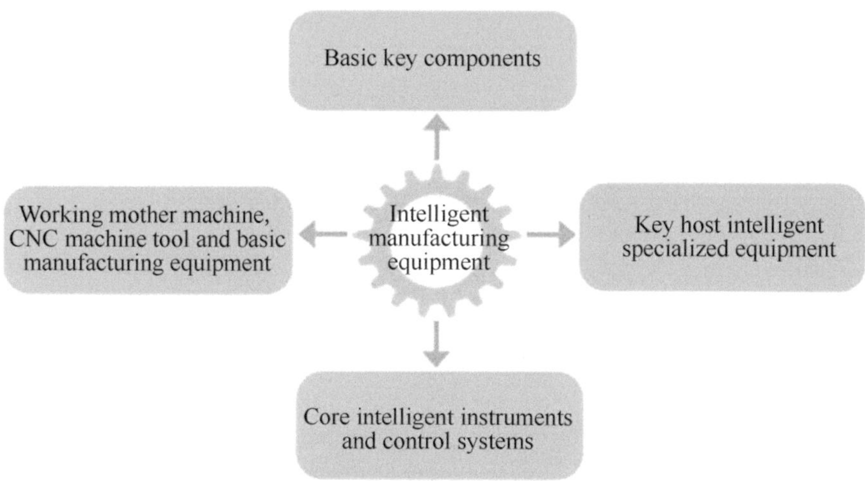

Fig. 1.5 The compose of intelligent manufacturing equipment

intelligence, but a reference to human intelligence, the use of a variety of sensors beyond human perception, such as ultrasonic sensors, infrared sensors, etc., to solve practical problems. The sensors commonly used in intelligent manufacturing equipment are vision sensors (cameras, etc.), distance sensors (laser rangefinder, infrared rangefinder, etc.), radio frequency identification (RFID) sensors, sound sensors, tactile sensors, etc. High precision, high sensitivity sensor is the goal of human unremitting pursuit.

(2) Mechanical execution system

The mechanical executive system is the mechanical executive mechanism, which is the subsystem that directly contacts and interacts with the working object in the intelligent manufacturing equipment, and interconnects with the transmission system and the support system, and is the part of the mechanical system that directly completes the expected function. The modern mechanical execution system includes the mechanism that produces the actual motion law and the prime mover that drives the motion of the mechanism. The commonly used prime movers are motors, internal combustion engines, hydraulic motors, air motors, etc.

(3) Motion control system

Control is a branch of automation that is an integral part of modern industrial production. The motion control system controls the position or speed of the machine by controlling the execution equipment of the servo mechanism, such as hydraulic pump, linear actuator or motor, etc., so that the equipment can carry out complex operations according to the given instructions, and is a bridge to realize the intelligence of the equipment.

(4) Intelligent decision system

Based on the information collected by various sensing systems, the intelligent decision system makes complex decision calculations, optimizes reasonable instructions, and the command and control system drives the execution system, thus ultimately realizing complex intelligent behaviors. Intelligent decision system is the bottleneck of the development of intelligent manufacturing equipment, and there is no good general solution in the industrial field at present.

1.1.3 Characteristics and Key Technologies of Intelligent Manufacturing Equipment

Modern production has put forward higher requirements for the equipment manufacturing industry, and the traditional equipment manufacturing industry needs to be transformed and upgraded. The traditional equipment manufacturing industry needs to be improved in terms of output and quality, and the production cycle is long, and the time for product design, production, market launch and product replacement needs to be improved. The low production efficiency of traditional equipment manufacturing industry has been unable to meet the needs of modern production. With the continuous growth of industrial economic benefits, enterprises are committed to expanding production scale, and the number of manufactured products has increased significantly [8]. At the same time, traditional equipment energy saving and consumption reduction can not meet the requirements of modern environmental protection, some traditional equipment manufacturing enterprises are produced on the basis of a large amount of energy and material consumption, and pay little attention to the problem of energy saving and consumption reduction. Therefore, the market competitiveness of traditional manufacturing equipment is weak, the product replacement is slow, the production process is poor, the production efficiency is low, and it is difficult to meet the individual needs of users. Therefore, higher requirements are put forward for traditional manufacturing equipment, and the development of advanced manufacturing technology with intelligent manufacturing equipment as the carrier is driven. The advantages of advanced manufacturing technology are mainly reflected in the following aspects:

① High precision. Speed, precision and efficiency are the key performance indexes of equipment manufacturing technology. Intelligent manufacturing equipment adopts advanced control system, high-resolution detection components, high-performance AC digital servo system, and takes effective measures to improve the dynamic and static characteristics of the machine tool, so that the production speed, accuracy and efficiency are greatly improved.

② Automation. The development of advanced manufacturing technology is inseparable from the development of automation technology. Automation technology, especially intelligent control technology, is mostly first applied to the development of advanced manufacturing technology. Modern intelligent manufacturing equipment adopts industrial robots and intelligent machine tools on a large scale to achieve a high degree of automation in the processing process.

③ Integration. The core part of intelligent manufacturing equipment such as modern numerical control system is getting smaller and smaller, and the function is getting more and more powerful. The reason is that highly integrated chips and large-scale programmable integrated circuits are used to improve the integration of numerical control system and the speed of software and hardware operation. By increasing the density of integrated circuits, it can also improve the cost performance of products, reduce the size of components, and improve the reliability of the system.

④ Informatization. The development of advanced manufacturing technology cannot be separated from the development of information technology, and the development of information technology cannot be separated from the development of manufacturing technology. Manufacturing is the basic industry for the development of information industry. The development of manufacturing technology is closely related to the development of information technology.

⑤ Flexibility. Flexible manufacturing system (FMS) refers to the integrated information control system, material storage and transportation system and a set of numerical control processing equipment, able to adapt to automation machinery manufacturing processing object transform system. The key of flexible manufacturing lies in the flexibility of equipment, including the flexibility of CNC system itself and the flexibility of group control system. The flexibility of the numerical control system itself is that the exponential control system adopts modular design, has wide functional coverage, and the system can be tailored strongly, which is convenient to meet the needs of different users. The flexibility of the group control system means that the same group control system can dynamically adjust the material flow and information flow automatically according to the requirements of different production processes, so as to maximize the efficiency of the group control system.

⑥ Graphical visualization. Man–machine interface is the dialogue interface between the CNC system and people, and a good graphical user interface can make the operator operate through the window and menu, which is convenient to use. Visual HMI uses graphics, images, animations and other visual information to efficiently process and interpret data. The "fool" man–machine interface combined with virtual reality technology has greatly expanded its application field, which can shorten the product design cycle, improve product quality and reduce product cost.

⑦ Intelligent. Today, with the rapid development of science and technology, artificial intelligence tries to realize various intelligent behaviors of human beings with computational models, so that it is developing towards a more realistic field with real-time response, and real-time systems are also developing towards more complex applications with intelligent behaviors. The combination of artificial intelligence and real-time systems has created a new field of real-time intelligent control.

⑧ Multimedia. Multimedia technology integrates computer, audio-visual and communication technology, so that the computer has the ability to process sound, text, image and video information comprehensively. In the field of advanced manufacturing technology, the application of multimedia technology can realize the integration and intelligence of information processing, and has great application value in real-time monitoring system and fault diagnosis of production site equipment, and production process parameter monitoring.

⑨ Networking. With the rapid development of the Internet of Things, manufacturing equipment can be remotely controlled and unmanned operation, and other equipment can be programmed and operated on any manufacturing equipment. Images of different devices can be displayed simultaneously on the screen of any device.

The key to intelligent manufacturing equipment is the "intelligence" of equipment. Its core technology is inseparable from key technologies such as computer technology, control systems, Industrial Internet of Things (IIoT), big data analysis and artificial intelligence.

(1) Artificial intelligence

The development of artificial intelligence technology has promoted the development of intelligent manufacturing equipment. Artificial intelligence (AI) is a research, development, used to simulate, extend and expand the theory, method, technology and the application of the intelligent system, in an attempt to understand the essence of intelligence, and can produce a kind of new to react in ways similar to human intelligence of intelligent machines, research in this field including robotics, language recognition, image recognition, natural language processing and expert systems. The equipment itself has the intelligent and knowledgeable function of process optimization, and can independently plan, control and make decisions in real time according to changes in operating status.

(2) Intelligent perception

Perception refers to the real-time perception, analysis and processing ability of equipment operating status and environment. The research content includes sensing technology with high sensitivity, high precision, high reliability and strong environmental adaptability, sensing technology with new principles, new materials and new processes, weak sensing signal extraction and processing technology, optical precision measurement and analysis instrumentation technology, real-time environment modeling, image understanding and multi-source information fusion navigation technology, force or load real-time sensing and identification technology, and multi-sensing. The technology of sensor optimization layout and sensing system network configuration.

(3) Intelligent programming and process planning

The integration technology of expert system and computational intelligence is used to improve the ability of intelligent planning and process decision-making, establish intelligent reasoning and decision-making methods of planning and programming, and realize trajectory planning and numerical control programming based on geometric and physical constraints. Establish process database and knowledge base for typical industries, improve the model base of machine tools, robots and their production lines, and realize multi-objective optimization of process parameters and job tasks according to the monitoring information during operation. By deeply studying the complex interface behaviors and coupling relationships between subsystems, the process system model and optimization method oriented to optimization goals (efficiency, quality, cost, etc.) are established to realize the analysis, prediction and simulation of processing and operation processes, so that the equipment can independently plan, control and make decisions according to the running state. The equipment itself has intelligent and knowledge-based function of process optimization, and adopts software and network tools to realize intelligent design and real-time planning of manufacturing process.

(4) Intelligent CNC system and intelligent servo drive technology

Research on intelligent servo control technology, automatic identification technology of moving shaft load and running process, automatic matching of control parameters, online compensation of various errors, intelligent processing and forming oriented to shape control and control, and research on active control technology based on intelligent materials and servo intelligent control. The single-machine system and cluster control system are seamlessly linked, and the cluster has perfect information communication function, resource optimization configuration function and scheduling function, and the cluster can efficiently cooperate in construction to achieve system optimization. Improve the robot's vision, perception and servo functions and intelligent diagnosis technology in non-structural environment to realize intelligent control and optimization of production line. Intelligent technologies such as artificial intelligence and virtual reality are used to realize voice control and intelligent operation based on virtual reality technology, and intelligent human–computer interaction technology is developed.

(5) Internet of Things + industrial Internet

The Internet of Things (IoT) is the Internet of Things connected, refers to a variety of information sensing devices, real-time collection of any need to monitor, connect, interact with the object or process of various information, combined with the Internet to form a huge network. Its purpose is to realize the connection between things and things, things and people, all items and the network, to facilitate identification, management and control. The development of the Internet of Things has promoted the development of intelligent manufacturing equipment "senses", which are the basis of manufacturing automation. The essence of the industrial Internet is to realize the transformation of production and operation organization through the depth perception, real-time transmission, rapid calculation and advanced modeling analysis of industrial data. The industrial Internet integrates networking devices, sensors, automation equipment, data storage, big data analysis, artificial intelligence, efficient computing, 4G/5G/Internet of Things and other emerging technologies, covering many industries such as computers, communications, and machinery.

(6) Fault self-diagnosis and intelligent maintenance

Prediction and Health Management (Prognostics and Health Management, PHM) is a new solution for managing health state, which is put forward by making comprehensive use of the latest research achievements of modern information technology and artificial intelligence technology. Generally speaking, the system is mainly composed of six parts: data collection, information induction and processing, condition monitoring, health assessment, fault prediction and decision-making, and guarantee decision-making. On-line and remote condition monitoring, fault diagnosis and intelligent maintenance, establish the parameter representation system of equipment condition in manufacturing process and the mapping relationship with equipment performance characterization index, realize intelligent identification, carry out active analysis and maintenance of self-performance deterioration, self-healing regulation and intelligent maintenance, and realize self-diagnosis and self-repair.

(7) Industrial big data, cloud computing

With the arrival of the "cloud" era, big data has achieved rapid development. Industrial big data applies big data's concept to the industrial field to connect equipment data, operation data, environmental data, service data, business data, market data and upstream and downstream industrial chain data. Realize the connection between people and people, things and things, people and things. Industrial big data has more professional, relevance, process, timing and analytical characteristics, especially to realize the connection between the end user and the manufacturing and service process, through the new processing mode, according to the real-time requirements of the business scene, realize the conversion of data, information and knowledge, so that it has stronger decision-making ability, insight and process optimization ability.

Cloud computing is a kind of distributed computing, which refers to decomposing huge data processing programs into countless Mini Program through the network "cloud", and then processing and analyzing these Mini Program through a system composed of multiple servers. Get the results and return them to the user. In the early days, cloud computing was a simple distributed computing, which solved the task distribution and merged the computing results, also known as grid computing. Cloud computing can process tens of thousands of data in a very short time, so as to achieve powerful network services.

(8) Network integration and network collaboration

The network-based intelligent perception technology of the environment inside and outside the factory, including the information of logistics, environment and energy flow, as well as the related information in the Internet and enterprise information system. Networked integrated manufacturing integrates the information, process, organization and knowledge of partners under the support of computer network and distributed database by comprehensively using modern design technology, manufacturing automation technology, system engineering method, dynamic alliance method, concurrent engineering method, supply chain management technology, Agent technology, knowledge management technology, distributed database management technology, Internet and Web technology and network communication technology. And achieve the comprehensive optimization of the whole system, so as to achieve the goals of fast product listing, high quality, low cost, good service and little environmental impact, so that the system can win the competition and achieve good economic and social benefits.

(9) Information security

Information security is to apply the concept of information security to the industrial field to protect the systems and terminals covered by the use and maintenance of factories and products. The terminal equipment and systems involved include industrial Ethernet, data acquisition and monitoring, distributed control system, process control system, programmable logic controller, remote monitoring system and other network equipment and industrial control system, to ensure that the industrial Internet and industrial systems cannot be accessed, used, leaked, interrupted, modified and destroyed without authorization, so as to provide information services for the normal production of enterprises and the normal use of products.

1.1.4 Intelligent Equipment Application Fields

Intelligent equipment is mainly distributed in manufacturing and agriculture. Intelligent equipment is widely used in almost all aspects of production and life, but the concept of intelligent equipment and intelligent equipment is becoming more and more vague. If divided according to product form, intelligent equipment can be roughly divided into three categories: process automation production line focusing on process processing, assembly and sub-assembly in a manufacturing industry, with the main functions of efficient production and manual substitution; compared with the intelligent terminal equipment sold in the form of standardized products, it has anthropomorphic characteristics and can complete special environment operations or high-frequency repetitive operations. Focus on detection and data collection, often gather a variety of detection components of equipment, detect the target project data and give a comprehensive evaluation feedback.

(1) Intelligent manufacturing

Intelligence is the development direction of manufacturing automation. Intelligent manufacturing technology is based on modern sensing technology, network technology, automation technology, anthropomorphic intelligent technology and other advanced technologies, through intelligent perception, human–computer interaction, decision-making and execution technology, to realize the intelligence of design process, manufacturing process and manufacturing equipment, is the deep integration and integration of information technology, intelligent technology and equipment manufacturing technology, and is oriented to the whole life cycle of products [9]. The realization of information-based manufacturing under ubiquitous perception is a major trend of the deep integration of informatization and industrialization. Intelligent manufacturing is increasingly becoming the major trend and core content of the development of the manufacturing industry in the future. It is also an important measure to speed up the transformation of the mode of development, promote the industry to the middle and high end, and build a manufacturing power. It is also the inevitable choice to create a new international competitive advantage under the new normal. In 2011, Germany put forward the concept of "Industrial 4.0", that is, to improve the level of manufacturing through digitization and intelligence. China has also put forward the concept of "made in China 2025". Its core is to help or even replace workers through intelligent machines and big data analysis, so as to realize the overall intelligence of the manufacturing industry. In the United States, Tesla Automobile Company has tried to use robots to assemble cars, which is one of the models of "intelligent manufacturing", which not only liberates human labor, but also improves the performance and quality of products [10].

(2) Intelligent sensors

Traditional sensors can no longer meet the needs of intelligent equipment. Intelligent Sensor is a sensor with information processing function and a "sensory organ" of intelligent equipment. Intelligent sensor is embodied in "intelligence". The sensor itself has a microprocessor (chip) and has the ability of data acquisition, processing and analysis. It is the product of modern microelectronics technology, information technology, material technology and processing technology, and has been widely used in intelligent equipment. The function of intelligent sensor is put forward by simulating the coordinated action of human senses and brain, combined with long-term research and practical experience of testing technology. Intelligent sensor is a relatively independent intelligent unit. Its appearance reduces the requirements for hardware performance, and the performance of sensor is greatly improved by software. Compared with general sensors, intelligent sensors have the characteristics of high precision, high reliability, high cost performance and diversified functions, which can realize multi-sensor and multi-parameter comprehensive measurement and expand the measurement and application scope through programming. It also has a certain adaptive ability, and can change the form of the range reverse output data according to the detected object or condition. The intelligent sensor has the function of digital communication interface, which can be directly sent to a remote computer for processing. It has a variety of data output forms and is suitable for various application systems. As the front-end sensing device of a wide range of intelligent equipment systems, intelligent sensors can promote the upgrading of traditional industries, and have been widely used in many fields such as industrial automation, astronomical exploration, earth-sea exploration, environmental protection and energy saving, medical health, national defense, biopharmaceuticals and so on.

(3) Intelligent control software

The value of intelligent manufacturing equipment is not only embodied in hardware such as mechanical ontology and sensors, but also in various intelligent software. The highest level is the software that realizes optimal control, linear programming and other scheduling and decision-making layers. Intelligent software has the ability of on-site induction, self-organization and self-adaptability, and can perceive, learn, reason, judge and make corresponding actions to the environment. Through the prediction, judgment and adaptive adjustment functions of intelligent control software, intelligent processing of equipment can be realized. Intelligent control software provides intelligent process optimization and technical support for manufacturing equipment enterprises, and powerful enterprises can develop the original enterprise system of manufacturing equipment into a system integration enterprise.

(4) Industrial robots

Industrial robot is a multi-joint manipulator or multi-degree-of-freedom mechanical device widely used in industrial field, which has certain automation and can realize various industrial processing and manufacturing functions by relying on its own power source and control ability. Industrial robots are widely used in electronics, logistics, chemical industry and other industrial fields. The popularization of industrial robots is a typical representative of intelligent manufacturing, which greatly reduces the labor intensity of workers and improves the production environment. Domestic industrial robots are mainly low-end, and need to be driven by industrial robots applied in automobiles, airplanes, military industries and other industries, focusing on overcoming the common technologies of core components such as controllers, reducers, servos and motors, and breaking through six subsystems: mechanical structure system, drive system, sensing system, robot-environment interaction system, man–machine interaction system and control system. Solve the problems of design technology, reliability and testing standards of industrial robot automatic production line and core components, improve the market retention rate of domestic industrial robots, break foreign monopoly, strive to demonstrate the application of industrial robot production line in typical industries such as automobile manufacturing, numerical control processing, aircraft assembly and shipbuilding, and form China's independent industrial robot standards.

(5) Additive manufacturing

Traditional manufacturing thinking is to form a three-dimensional concept according to the purpose of use, transform it into a two-dimensional drawing, and then manufacture it into a three-dimensional entity. Traditional machining is turning, milling, planning and grinding to remove materials. Different from traditional material removal processing, additive manufacturing (AM), also known as 3D printing technology, is a kind of "bottom-up" manufacturing technology, which is based on computer-aided design, material processing and molding, and uses software and numerical control system to stack special non-metallic materials, metal materials and medical biomaterials layer by layer according to extrusion, sintering, melting, light curing and spraying, etc.

The feature of 3D printing is the rapid manufacturing of single piece or small batch, which can realize the rapid development and innovative design of products. The method of directly manufacturing parts by heating materials with laser beam, electron beam, plasma or ion beam as heat source is called high energy beam rapid manufacturing. High-energy beam rapid manufacturing is an important branch of additive manufacturing field, which is the most common in industrial field. High-energy beam rapid manufacturing of metal, non-metal or metal-based composites is the fastest-growing research direction at present, which has a broad application prospect in aerospace and industrial fields, and it should focus on developing high-end application industries of 3D printing in aerospace, biomedical and other fields, such as manufacturing complex and precise metal parts in control. Based on the application of 3D printing technology in personalized consumer products, cultural and creative industries and other fields, we will strengthen its technical research and popularization and application in product development. Relying on 3D printing technology, we will establish a technical service network, form a complete industrial chain such as product design, raw materials, key components, equipment and industrial applications, form a "cloud manufacturing" model based on Internet technology, and integrate with traditional manufacturing industries to carry out systematic innovation and development. Give full play to the innovative ability of all social groups and promote the realization of an innovative society.

(6) Smart agriculture

China is a big agricultural country, not a strong agricultural country, and agricultural production is still dominated by traditional production mode. Traditional farming can only be irrigated by fertilization based on experience, which not only wastes a lot of manpower and material resources, but also poses a serious threat to environmental protection and soil and water conservation, and poses a severe challenge to the sustainable development of agriculture. The integration of intelligent agriculture

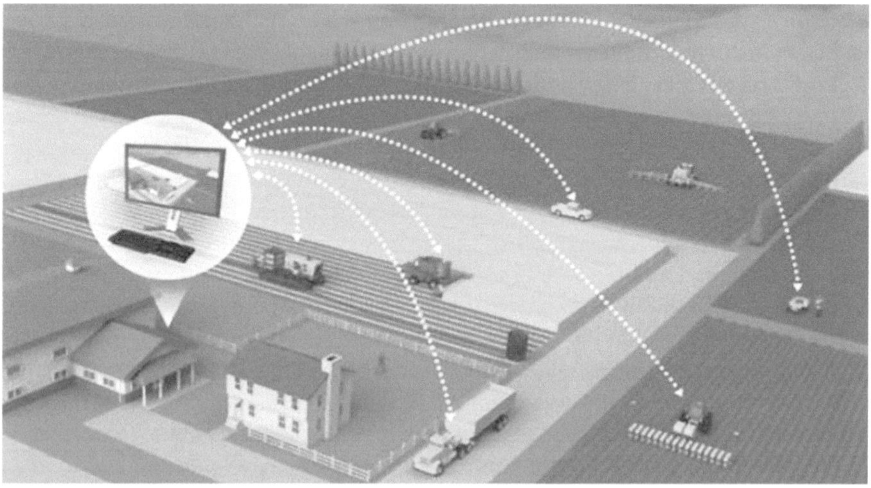

Fig. 1.6 Intelligent agriculture

(Fig. 1.6) with modern biotechnology, planting technology and other high and new technologies is of great significance to building world-class agriculture.

Smart agriculture is to apply the Internet of Things technology to traditional agriculture, fully apply the achievements of modern information technology, integrate computer and network technology, Internet of Things technology, audio and video technology, 3S technology, wireless communication technology and expert wisdom and knowledge, and control agricultural production through mobile terminals or computer platforms such as mobile phones and tablets, making traditional agriculture more "smart". Realize intelligent perception, intelligent early warning, intelligent decision-making, intelligent analysis and expert online guidance of agricultural production environment, realize intelligent management of agricultural visual remote diagnosis, remote control and disaster early warning, and realize modern advanced agricultural production mode with intensive, efficient and sustainable development. In the process of transforming from traditional agriculture to modern agriculture, the development of agricultural informatization has roughly experienced four processes: computer agriculture, digital agriculture, precision agriculture and intelligent agriculture.

(7) Smart healthcare

In 2020, the global spread of novel coronavirus will put forward a severe test for the medical systems of all countries in the world, and at the same time, it will also bring great opportunities for the transformation and upgrading of the medical system. In the future, intelligent equipment in the medical field will be put into use on a large scale. When faced with unpredictable epidemic viruses or other diseases that endanger human health, it can perceive, handle and analyze major medical events in real time, so as to respond quickly and effectively.

Smart medical care is the development trend. The integration of Internet and medical industry has produced Internet medical treatment, that is, using the Internet as a technical means and platform to provide users with various forms of health management services such as medical consultation, health index monitoring, health education, electronic health records, remote diagnosis and treatment, electronic prescription and remote rehabilitation guidance [11]. Through the wireless network, the handheld PDA is used to conveniently communicate with various diagnosis and treatment instruments, so that medical staff can grasp the medical record information and the latest diagnosis and treatment report of each patient at any time and make a diagnosis and treatment plan quickly at any time and anywhere; In any place in the hospital, medical staff can log in to the nearest system to inquire about medical imaging data and doctor's advice; The patient's referral information and medical records can be accessed in any hospital through medical networking. The doctor-patient information warehouse becomes a shareable record, and medical information and records are integrated and shared in order to build a comprehensive professional medical network. With the help of big data or blockchain technology, authorized doctors can check the patient's medical records, patient history, treatment measures and insurance rules at any time, and patients can also choose to change doctors or hospitals. The establishment of medical database enables practitioners to search, analyze and cite a large number of scientific evidence to support their diagnosis.

Village and smart medical care support township hospitals and community hospitals to seamlessly link to central hospitals, so as to obtain expert advice, arrange referrals and receive training in real time, improve the professional knowledge and process handling ability of township doctors, and further promote clinical innovation and research. Intelligent robots can not only help diagnosis, but also perform surgery, as shown in Fig. 1.7. The mechanical arm is far more flexible than human beings, and it is equipped with a camera, which can enter the human body for surgery, with small surgical wounds, and can perform difficult operations, which is equivalent to the realization of "doctors entering the patient's body for treatment". Take thyroid surgery as an example. Before that, a knife needs to be cut horizontally below the throat, and the patient will leave a scar. Now robot surgery is very mature, which can open the armpit and breast to let the micro manipulator in and perform thyroid resection accurately. There is no scar on the exposed part of the patient after operation, which greatly improves the surgical effect and postoperative state of the patient. Laparoscopic robot can perform most abdominal surgery, which greatly improves the efficiency and accuracy of surgery and reduces the amount of bleeding of patients. On the control terminal, the computer can use two-dimensional images taken by several cameras to restore high-definition three-dimensional images in human body, so as to monitor the whole operation process. Doctors can also manually intervene in the operation process remotely.

(8) Smart environment

Environmental monitoring is the activity of environmental monitoring institutions to monitor the environmental quality. By monitoring and measuring the indicators reflecting the environmental quality, the environmental pollution and environmental

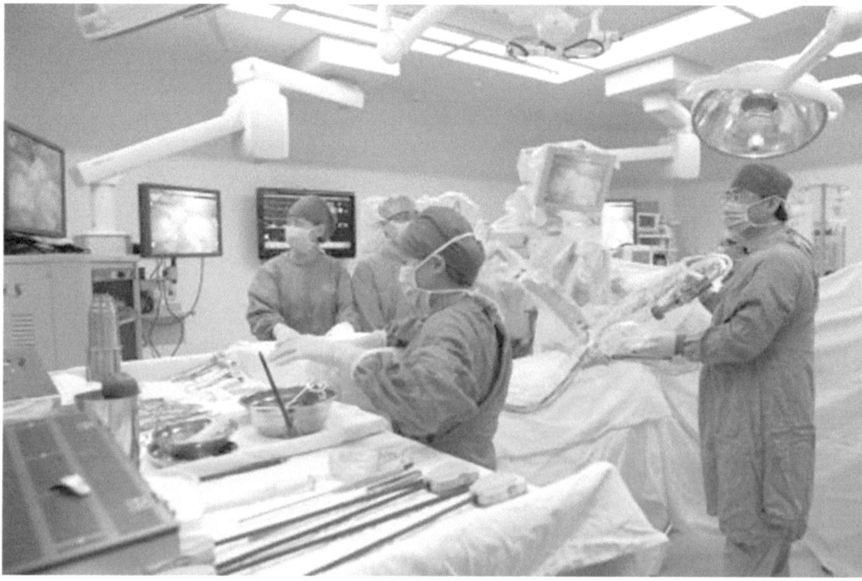

Fig. 1.7 Robotic surgery

quality can be determined. The content of environmental monitoring mainly includes the monitoring of physical indicators, chemical indicators and ecological system, which is the basis of scientific management of the environment and environmental law enforcement supervision and the essential basic work of environmental protection. The core goal of environmental monitoring is to provide data on the present situation and changing trend of environmental quality, judge environmental quality, evaluate current major environmental problems and serve environmental management [12]. The online environmental monitoring system is shown in Fig. 1.8. Intelligent environmental monitoring is the development trend. With the development of Internet of Things technology, traditional environmental monitoring methods have been gradually replaced by remote monitoring. Environmental monitoring covers many online environmental monitoring applications such as water quality monitoring, ambient air quality monitoring, stationary pollution source monitoring (CEMS), atmospheric environmental monitoring and video monitoring. Based on on-line monitoring of pollutants, the system fully implements the principles of total amount management and total amount control, and contains many important functions of environmental management information system, which fully meets the requirements of environmental information network construction of environmental protection departments at all levels and provides reliable data support for decision-making of relevant functional departments. Support environmental supervision and monitoring work of environmental protection departments at all levels to meet the management needs of users at different levels.

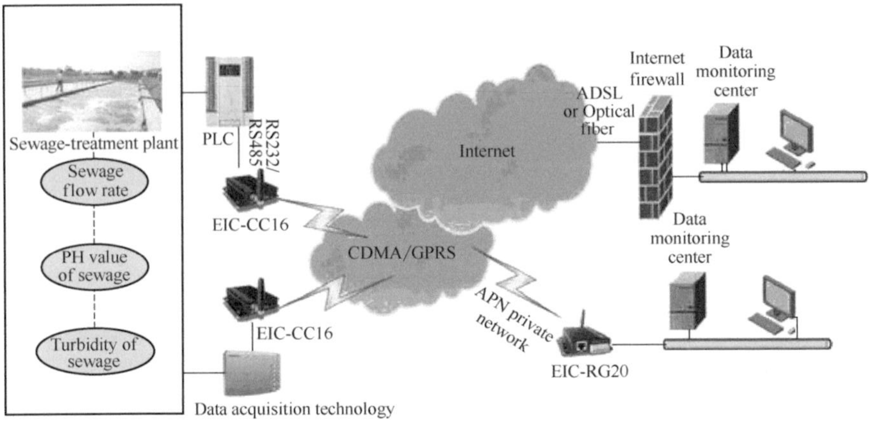

Fig. 1.8 The environment online monitoring system

In terms of automatic monitoring, some developed countries have mature technologies and products, such as mature automatic continuous monitoring systems in atmosphere, surface water, enterprise waste gas and wastewater and urban sewage. After years of efforts, the research team of the author has developed intelligent monitoring equipment for groundwater level and surface water quality, which has been widely used in groundwater level and surface water quality monitoring in some provinces and in several irrigation areas.

1.2 Development Overview of Intelligent Manufacturing Equipment

The United States, Germany, Japan is the current world CNC machine tool strength of the country, is the world's CNC machine tool technology pioneers, intelligent manufacturing equipment multinational enterprises are mainly concentrated in these countries. The development process of intelligent manufacturing is shown in Fig. 1.9. Intelligent machine tools and basic manufacturing equipment mainly by the United States, Germany, Japan and other countries of multinational companies to improve research and development, the current world's four major international machine tool manufacturers CNC machine tool technology innovation, mainly reflects the progress of these companies in intelligent manufacturing equipment, CNC equipment performance and efficiency has been greatly improved. The development of intelligent manufacturing equipment is highly dependent on the accumulation of engineering manufacturing science, technology foundation and development experience, which leads to a generally strong monopoly in this industry, and the monopoly power is mainly from developed countries and leading multinational enterprises.

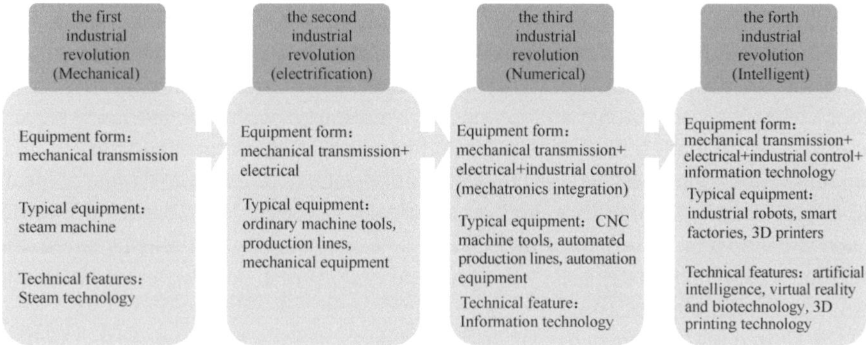

Fig. 1.9 The development of intelligent manufacturing

China has become the world's second largest economy and manufacturing power, intelligent equipment manufacturing industry is still in the initial stage of automation to intelligent development, intelligent manufacturing equipment industry as a whole is not high level, some industries even basic equipment automation has not been completed. The degree of CNC in China's equipment manufacturing industry is not high, and the performance of most high-precision and ultra-high precision machine tools cannot meet the requirements, and the equipment of CNC machine tools is basically medium sized lathes, milling machines and machining centers [13]. The technical development direction of China's development of intelligent manufacturing equipment should be the combination of hardware and software, and focus on the development of networking and modularization of control software.

Intelligent manufacturing technology is one of the important directions for the future development of the world manufacturing industry. In order to promote the transformation and upgrading of China's traditional manufacturing industry structure, the State Council issued the "decision on speeding up the cultivation and Development of Strategic emerging Industries". Including high-end equipment manufacturing industry, comprehensively carrying out intelligent manufacturing technology research will be the core content of the development of high-end equipment manufacturing industry. It is also a necessity to improve China's international competitiveness and promote the transformation of our country from a big manufacturing country to a manufacturing power.

1.2.1 Development Status of CNC Machine Tools

Digitization of processing machine tools is the development trend of manufacturing industry. CNC machine tool, as the mainstream representative product of machine tool industry, has become the key equipment to realize the modernization of equipment manufacturing industry, and is a strategic material for the development of national defense military equipment. Numerical control machine tool is the abbreviation of digital control machine tool, is an automatic machine tool equipped with a program control system, can logically process the program with control code or other symbolic instructions, and decode it, represent it with a coded number, and input the numerical control device through the information carrier. After calculation and processing, the CNC device sends out a variety of control signals, controls the action of the machine tool, and automatically processes parts according to the shape and size required by the drawing. It is a flexible and high-efficiency automatic machine tool, which better solves the problem of complex, precise, small-batch and multi-variety parts processing.

The United States, Germany and Japan are the countries with advanced technology and rich experience in the research, design, manufacturing and use of CNC machine tools in the world today. In recent years, China's numerical control technology has had a certain development, every year the international machine tool exhibition will have domestic new numerical control machine tools come out, but compared with other advanced countries, there is still a certain gap. Domestic CNC machine tools have experienced more than 30 years of development, quality and reliability can meet the needs of most users. China can design its own system supporting, can design and manufacture high-speed, high-performance, multi-faceted, multi-axis linkage CNC machine tools, has developed a vertical machining center, horizontal machining center, as well as CNC lathes, CNC milling machines and other CNC machine tools, in heavy machine tools, high-precision machine tools, special processing machine tools, forging equipment, cutting-edge high-tech machine tools and other fields, especially in the five-axis linkage CNC machine tools, CNC super heavy machine tools, vertical horizontal machining center, CNC lathes, CNC gear machining machine tools, some technologies have reached the advanced level.

60 to 70% of domestic CNC machine tools are domestic machine tools, but the proportion of these domestic CNC machine tools with foreign CNC systems is very high, and the domestic production rate of high-end CNC machine tools is low. In terms of market demand, low-grade machine tools and mid-grade machine tools account for about 50% and 40% respectively, and the demand for high-grade CNC machine tools is about 10%, so the development of CNC machine tools in China has a long way to go and broad prospects for development.

At present, the development of CNC machine tools with each passing day, high-speed, high reliability, high precision, compound, intelligent, open, networked, new functional components, green has become the trend and direction of the development of CNC machine tools.

(1) High speed

The high speed is mainly reflected in the spindle speed, feed rate, numerical control system operation speed, tool change speed and so on. For example, modern CNC machine tools generally use electric spindle, the maximum speed of the spindle up to 200,000 r/min; German Chiron Company through a special design, the tool change time is only 0.9 s.

(2) High reliability

Compared with traditional machine tools, CNC machine tools are complex, the processing cycle is long, the parts processed are complex, and the application of various devices is more, which leads to the increase of the probability of machine tool failure. At the same time, the stability of the numerical control system, the fluctuation and interference of the industrial power grid voltage, and the factors of the workshop environment may also affect the reliability of the numerical control machine tool. Therefore, reliability is one of the development directions of CNC machine tools.

(3) High precision

CNC machine tool accuracy requirements are now not limited to static geometric accuracy, machine tool motion accuracy, thermal deformation and vibration monitoring and compensation more and more attention. High speed interpolation technology and high resolution position detection device are used to improve the accuracy. The thermal deformation error and space error of the equipment are compensated by reverse clearance compensation, lead screw pitch error compensation and tool error compensation.

(4) Functional compounding

A variety of elements from blank to finished product can be processed or completed as far as possible on a single machine. According to its structural characteristics, it can be divided into two categories: process compound and process compound. Process complex machine tools such as boring, milling and drilling complex machining center; Process compound machine tools, such as multi-sided multi-axis joint machining of composite machine tools and double spindle turning centers.

(5) Intelligent control

With the development of artificial intelligence technology, in order to meet the development needs of manufacturing production flexibility and manufacturing automation, the intelligence level of CNC machine tools is constantly improving. It is embodied in the process adaptive control technology, intelligent optimization and selection of processing parameters, intelligent fault diagnosis and self-repair technology, intelligent fault playback and fault simulation technology, intelligent AC servo drive device, intelligent 4M numerical control system, measurement (Meas_urement), modeling (Modelling), processing (Manufac)turing ("Turing") and machine operation ("Manipula_tor") are integrated in one system to realize information sharing and promote the integration of measurement, modeling, processing, clamping, and operation.

(6) Open system

The hardware and software interfaces of modern CNC machine tools follow the recognized standard protocols, and the new generation of general hardware and software resources are easier to be adopted, absorbed and compatible with the existing system, while the international community is committed to the standardization of product information throughout the manufacturing process and even various industrial fields. Standardized programming languages and standardized interfaces are not only convenient for users to use, but also can reduce labor consumption directly related to operation efficiency.

(7) Networking of information exchange

With the development of the Internet of Things technology, CNC machine tools with two-way, high-speed networking communication functions can ensure the smooth flow of information, enterprises can realize network resource sharing, but also realize the remote monitoring and control of CNC machine tools, but also realize the digital service of CNC equipment. For example, the new generation of machining center launched by Japan Mazak Company is equipped with information tower, which can realize voice, graphics, video and text communication fault alarm display, online help troubleshooting and other functions.

(8) New functional components

New functional parts with high precision and high reliability can improve the performance of CNC machine tools in all aspects, and the representative new functional parts include high-frequency electric spindle, linear motor, ball screw and so on. The use of new functional components can simplify the machine structure and improve the performance of the machine.

(9) Greenization of the processing process

With the increasingly prominent resources and environmental problems, green manufacturing has been paid more and more attention. In recent years, without or less use of coolant to achieve dry cutting, semi-dry cutting energy saving and environmental protection machine tools continue to appear, and in further development, is the mainstream of future machine tool development.

1.2.2 Development Status of Industrial Robots

The aging of population and the shortage of labor force is a problem facing mankind. The increase in labor costs forces people to use robots to replace people in production, and it is a general trend for machines to replace people.

Industrial robots are the most representative equipment of intelligent manufacturing industry. According to data released by the International Federation of Robotics, global industrial robot sales continued to maintain rapid growth in 2016, with sales of about 290,000 units, of which 90,000 industrial robots were sold in China, an increase of 31% year-on-year. The International Federation of Robotics predicts that in the next decade, the average annual growth rate of global industrial robot sales will remain at about 12%. At present, industrial robots are basically classified according to their uses, such as welding robots, handling robots, processing robots, and equipment robots. In the application of industrial robots in major industries, the automotive industry accounts for 38%, and most of the production lines of automation equipment in the automotive industry use industrial robots. Domestic independent brands of industrial robots should most appear in the field of automotive, but the data shows that domestic auto factories basically do not use domestic independent brands of industrial robots. At present, automobile manufacturers, especially Chinese automobile manufacturers, basically use the international robot four masters (ABB, KUKA, FANUC, YASKAWA) industrial robots.

ABB, KUKA, FANUC, YASKAWA in the industrial robot industry known as the international four. ABB is mainly a joint venture between Switzerland and Sweden, headquartered in Europe; KUKA is a German company that was acquired by Midea for $30 billion in 2015. FANUC and YASKAWA are both Japanese companies, which also reflect Japan's unique advantages in the field of industrial robots. In terms of humanoid robots, the most popular ones include Aldebaran's Nao, Honda's ASIMO, Kawada's HRP-4, SONY's AIBO and QRIO [14].

China's industrial robot market has broad prospects for development, and the competitiveness and profitability of relevant domestic robot manufacturers are gradually strengthening. As China's demographic dividend continues to fade, the transformation and upgrading of industrial development around the country is accelerating, the "machine replacement" tide pushed by the government and implemented by enterprises is accelerating the deployment, and the "black light factories" that are completely replaced by robots for manual production continue to emerge. Since 2013, China has become the world's largest industrial robot market. In 2013, the Ministry of Industry and Information Technology's "Guiding Opinions on Promoting the development of industrial robot Industry" pointed out that the installed capacity of industrial robots will reach 1 million units in 2020, and about 200,000 employees related to industrial robot applications will be needed. In addition, deep underground, deep sea, deep space and other extremely dangerous environment operations also need to use robots to achieve less human and unmanned.

Some of the core components of our robot (such as RV reducer, servo controller, controller, etc.) have gradually been localized, for low-end robots have realized the localization of core components, but high-end is not yet available; With the emergence of new technologies, it will promote the further development of robotics. With China's industrial transformation and upgrading, labor costs continue to rise and robot production costs decline, during the "13th Five-Year Plan" and "14th Five-Year Plan" period, robots have become one of the key development objects, industrial robots have a certain foundation for development, and are now entering the stage of full popularity. In the future, China's industrial robot industry demand will continue to increase, and the scale will be further expanded [15].

"Robot Industry Development Plan (2016–2020)" proposed that in 2020, the annual output of China's independent brand industrial robots will reach 100,000 units, and the annual output of six-axis and above industrial robots will reach more than 50,000 units; The speed, load, accuracy, weight ratio and other main technical indicators of industrial robots have reached the level of similar foreign products, and the average trouble free time has reached 80,000 h; The performance, accuracy and reliability of precision reducer, servo motor, drive and controller for robots have reached the level of similar foreign products, and they have achieved batch application in six-axis and above industrial robots, with a market share of more than 50%. Complete comprehensive robot application solutions in more than 30 typical fields, and form corresponding standards and norms to achieve large-scale application of robots in key industries.

However, it should be seen that China's domestic robot market share is low. China's robot industry has achieved great development in recent years, and a number of high-tech industries have emerged. However, in general, it is still mainly in the low-end market, the largest field of domestic application of independent brand industrial robots is the electronics industry, and the high-end market is still mainly imported. Taking the automobile industry as an example, the domestic automobile production line is basically the use of the international four industrial robots mentioned earlier.

The performance of core components such as reducer and servo controller needs to be improved. High-end gear reducer and other core components import proportion is too high. The precision reducer needs to be further improved in terms of rated torque and transmission efficiency. The motor dynamic response, overload capacity and efficiency of the servo system have great room for improvement.

There is a certain gap between the proportion of domestic robot high-end technical talents and the international developed countries. Robot engineering is an applied interdisciplinary and technology that combines perception, decision making, calculation and executive drive in the real world environment. It is a multi-field interdisciplinary subject that studies intelligent perception, optimal control and system design of robots, human–computer interaction mode, etc. With the promotion of "Made in China 2025", domestic universities are vigorously developing robotics majors. From 2015 to 2019, 249 universities across the country successfully applied for the major of "robot engineering". Robot education has shown strong vitality, and it is believed that in the near future, a large number of robot professionals will come out of the campus and shoulder the heavy responsibility of the rise of China's robots [16].

1.2.3 Future Development of Intelligent Manufacturing Equipment

In the future, China's intelligent manufacturing equipment industry will focus on developing manufacturing industries related to major national needs and strategic security, such as aerospace, national defense industry urgently needed difficult-to-process materials and equipment in the field of new materials application. These areas are supported by the state, with high technical requirements and great difficulties, mainly through planned and continuous support and research by the state to large backbone equipment manufacturing enterprises, institutions of higher learning and scientific research institutes with technological advantages, and state-level research bases. In order to overcome difficulties, solve core technologies, and expand application and market promotion. Taking the manufacturing industry related to national major demand and strategic security as the object, this paper focuses on the development of several kinds of manufacturing equipment with good automation foundation and urgent intelligent requirements. These manufacturing equipment mainly include high-speed machining center, ultra-precision machine tool, impeller blade machine tool, aircraft large flexible structure machine tool, difficult-to-process material processing machine tool in aerospace field, five-axis curved surface milling machine and so on. It lays the foundation for the development of intelligent manufacturing equipment in our country. Intelligent special equipment mainly includes large-scale intelligent construction machinery, high-efficiency agricultural machinery, environmental protection machinery, automatic textile machinery, intelligent printing machinery, coal machinery, metallurgical machinery and other special equipment, to achieve various manufacturing process automation, intelligence, lean, promote the overall level of intelligent manufacturing equipment.

Vigorously carry out process optimization research and intelligent sensor development, develop home-made sensor network systems, develop sensor manufacturing enterprises, support interdisciplinary research, develop practical machine tool parameter measurement, focusing on the development of MEMS and wireless sensors. Develop the intelligent numerical control system suitable for manufacturing equipment, lay the technical foundation of intelligent manufacturing equipment, and form the industrialization chain of intelligent manufacturing equipment. While promoting large-scale manufacturing equipment enterprises to develop intelligent equipment, support a number of specialized scientific and technological enterprises to carry out intelligent transformation of CNC machine tools, form their own brands, and open up and occupy the international market.

1.3 Research and Development Content of Intelligent Manufacturing Equipment

Intelligent manufacturing equipment is the general name of all kinds of manufacturing equipment with the functions of perception, analysis, reasoning, decision-making and execution. It is the integration and deep integration of advanced manufacturing technology, information technology and intelligent technology.

According to the 12th five-year Development Roadmap of Intelligent Manufacturing equipment Industry formulated and issued by the Ministry of Industry and Information Technology, the development of intelligent manufacturing equipment includes the following contents:

(1) Key intelligent basic common technologies. Vigorously develop intelligent common technologies, including intelligent sensing technology, modular embedded control system design technology, advanced control and optimization technology, system collaboration technology, fault diagnosis and health

maintenance technology, highly reliable real-time communication network technology, functional safety technology, special technology and precision manufacturing technology and identification technology, etc., is the basis of intelligent equipment.

(2) The core intelligent measurement and control devices and components. It includes new sensors and their systems, intelligent control system fieldbus, intelligent instruments, precision instruments, industrial robots and special robots, precision transmission devices, servo control mechanisms, hydraulic and pneumatic sealing components and systems.

(3) Major intelligent manufacturing complete sets of equipment. Including petroleum and petrochemical intelligent equipment integration, metallurgical intelligent equipment integration, intelligent forming and processing equipment integration, automatic logistics equipment integration, building materials manufacturing equipment integration, intelligent food manufacturing production line integration, intelligent textile equipment integration, intelligent printing equipment integration.

(4) Focus on the areas of application, demonstration and promotion. Including electric power, energy saving and environmental protection, agricultural equipment, resource exploitation, national defense and military industry, infrastructure construction and other fields.

1.3.1 Intelligent Machine Tools and Basic Manufacturing Equipment

Machine tool has gone through three stages from birth to intelligence. The first stage was from the 1930s to the 1960s, with the development of manual bed machines, electrical and hydraulic high-efficiency automatic machine tools and automatic lines, liberating workers from manual labor. The second stage is the birth of CNC machine tools in the 1960s to the beginning of the twenty-first century, the rapid development of CNC machine tools has further solved the problem of reducing physical strength and part of mental work. The third stage is intelligent machine tools. The accelerated development of intelligent machine tools will further solve the problem of reducing mental work.

The current intelligent machine tool (Fig. 1.10) has the function of automatic vibration suppression; automatic measurement and automatic compensation to reduce the influence of thermal deformation of high-speed spindle, column and bed; automatic knife-proof function; automatic replenishment of lubricating oil and noise suppression; voice information system has man–machine dialogue function and remote fault diagnosis function. The functions of intelligent CNC machine tools are shown in Fig. 1.11.

Fig. 1.10 A machining center

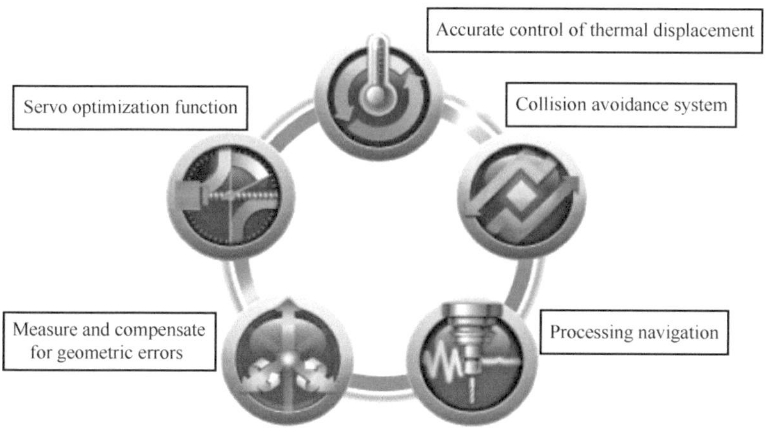

Fig. 1.11 The function of intelligent CNC

1.3.2 Industrial Robots

Industrial robot is an important branch in the field of robotics, which is a highly automatic equipment that integrates many advanced technologies, such as machinery, electronics, control, computer, sensor, artificial intelligence and so on. It plays an important role in modern industry, national defense and other industries. The structural block diagram of the industrial robot is shown in Fig. 1.12. The robot application of the automobile production line is shown in Fig. 1.13.

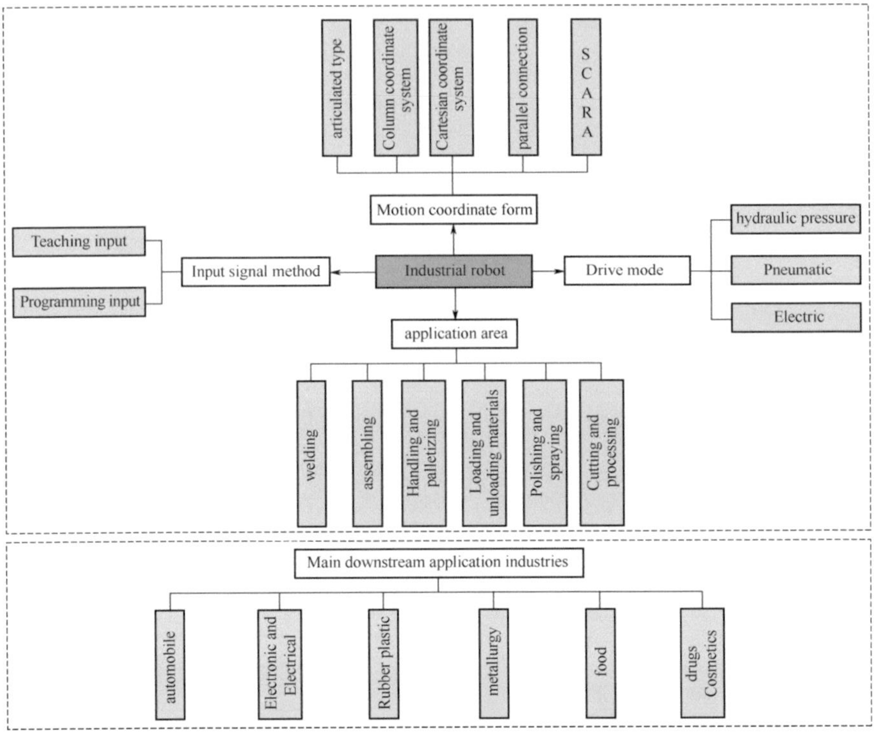

Fig. 1.12 The frame diagram of industry robot

Fig. 1.13 The application of robot in automobile production line

The research of robot technology in manufacturing industry is constantly developing in the direction of intelligence, cooperation, modularization, multi-function, high performance, self-diagnosis and self-repair, in order to meet the needs of agile manufacturing, diversification and personalized manufacturing. So that the robot can adapt to the changeable working environment and is widely used in the mechanical manufacturing industry.

The main contents are as follows: (1) according to the given grinding wheel diameter, feed rate and rotation speed, the relevant foreign scholars measure the grinding depth that can be obtained by different grinding normal forces according to the given grinding wheel diameter, feed rate and rotation speed. Through this measurement curve and according to the initial grinding depth, the desired grinding force is obtained, and the model of grinding depth and grinding force is established. Based on the feedback of this force, the robot independently controls the grinding parameters to improve the grinding effect. In addition to grinding, the same is true for turning and milling. The system can adjust the cutting parameters in real time according to the cutting force to achieve the optimal machining effect. (2) Autonomous welding robot welding based on machine vision is one of the important application fields of robot. The robot welding system based on vision makes the robot grow "eyes". The robot can independently make up for the welding error caused by the position deviation of the welding torch, the clamping position of the fixture and the workpiece itself, and improve the welding accuracy and quality. Some famous international welding equipment manufacturers have developed intelligent welding systems based on vision. (3) Vision-based workpiece inspection machine vision refers to using machines instead of human eyes to measure and judge, which belongs to the category of artificial intelligence and has excellent performance in the task of perceiving and understanding the working environment and workpiece information. It has been widely used in the field of robot. Through the machine vision information, the industrial robot can automatically identify and locate the workpiece, which can be applied to many processes of the automobile production line. Many companies at home and abroad have developed visual inspection equipment for part size measurement, which has been more and more widely used in the automobile industry and other fields of mechanical manufacturing. (4) Self-assembly based on force sensing (torque sensing) robot control technology is one of the research focuses of well-known universities and major robot companies at home and abroad. Based on the force sensor, the industrial robot can sense the contact force between the manipulator and the workpiece and realize high-precision assembly. ABB has developed an industrial robot platform based on force or position hybrid control. By controlling the contact force and torque of the end manipulator of the industrial robot, the industrial robot has the ability to respond to the contact information. The industrial robot assembly system based on force control has been successfully applied to the automobile assembly line. FANUC Company of Japan mainly studies the three-dimensional assembly technology based on force information, and the industrial robot assembles the parts under the control of force or torque controller.

(5) The traditional multi-variety and variable batch production technology of man-machine coordinated unit production is mainly completed by manual, which not only

requires high cost of time and material to train skilled operators, but also is difficult to carry out quality management. By adding robot and manual coordination to the unit production mode, the autonomy of the machine and the subjective initiative of the human are brought into full play, and the production efficiency is greatly improved. ABB and other major international industrial robot enterprises have developed and launched industrial dual-arm robots to coordinate production with people, which can cooperate with workers to complete assembly work under visual guidance. (6) The development of robot itself the development of industrial robot technology also includes the development of the mechanical structure of industrial robot to modularization and reconfiguration. For example, the servo motor, reducer and detection system in the joint module are integrated, and the whole robot machine is constructed by recombination with the joint module and connecting rod module. The industrial robot control system develops to the open controller based on personal computer (PC), which makes it develop to the standardized and networked mode. The integration of robot devices is improved, the controller is becoming smaller and smaller, and the modular structure is adopted, which greatly improves the availability, ease of operation and maintainability of the system [17].

1.3.3 Additive Manufacturing

(1) Overview of additive manufacturing

Manufacturing (Additive Manufacturing, AM) technology is based on CAD design data, using the material layer-by-layer accumulation method to manufacture solid parts. Compared with the traditional material removal technology, it is a bottom-up, material accumulation manufacturing method, also known as "rapid prototyping" technology. Material augmentation manufacturing is a process of connecting materials to make objects based on three-dimensional CAD data. Without traditional tools, fixtures and multiple processing procedures, parts of arbitrary complex shape can be quickly and accurately manufactured on one equipment, which shortens the production cycle and has been widely used.

3D printing technology is the representative form of incremental manufacturing technology, and its development research has the following aspects:

① The study of original technology, common technology and standards. 3D printing technology is a manufacturing technology which is widely used in many disciplines and new technologies, such as computer graphics, computer-aided design, material science and so on. With the help of high-end talents and scientific research teams, relying on universities and scientific research institutes to study original technology, common technology and standards, and using dynamic R&D and operation mode for development research.

② Principle innovation and its related supporting technology. Establish a research system for the integration of mathematics, physics, chemistry, materials, life and other disciplines, study new materials, new devices, new software, new forming

Fig. 1.14 A 3D printer

principles, new equipment and technology, vigorously develop supporting technologies and explore new application fields. Carry out research and development of key devices, intelligent control software, raw materials and forming process, and build 3D printing intelligent manufacturing equipment industry chain.

③ To promote the industrialization of 3D printing technology with major projects and industry applications as traction. Carry out application technology research in key fields such as aerospace, biology, medical treatment, automobile, home appliances, micro-nano sensors, etc., and study the combination and process optimization of 3D printing technology and traditional manufacturing technology. A certain type of 3D printer is shown in Fig. 1.14.

(2) Development trend of increasing timber manufacturing

① Develop in the direction of daily consumer goods manufacturing. 3D printing technology has a wide application prospect and commercial value in science education, industrial modeling, product creativity, arts and crafts and other fields. With the development of high-precision, low-cost and high-performance materials, 3D printing is no longer a laboratory product. It is developing in the direction of daily consumer goods manufacturing, as shown in Fig. 1.15. Both the 3D printer itself and the products relying on 3D printing technology are gradually moving towards the family, and a huge consumer market is coming into being.

② Develop to the manufacture of functional parts. 3D printing is not only limited to resin, plastic and other materials, but also can be directly melted metal powder by laser or electron beam, and metal direct forming technology can be formed by stacking metal layer by layer. Figure 1.16 shows a 3D printed metal part. This technology can directly manufacture metal functional parts with complex structure, and the mechanical properties of the parts can reach the performance index of forgings. The future development direction is to further improve the accuracy and performance, and to develop the manufacturing technology of ceramic parts and composites at the same time.

③ The development of intelligent equipment. Intelligent equipment is the development trend, and 3D printing is no exception. At present, there are still many problems to be optimized in the software function and post-processing of the

Fig. 1.15 3D printed automobile

Fig. 1.16 3D printed metal parts

adding material manufacturing equipment, for example, supporting parts need to be added in the forming process; software intelligence and automation need to be further improved; the matching of process parameters and materials needs to be intelligentized; the powder or supporting parts need to be removed after processing. Intelligent 3D printing equipment will solve these problems in the future.

④ To the integration of organization and structure manufacturing. Achieve controlled 3D manufacturing from micro-structure to macro-structure. For example, in the manufacture of composite materials, the design and manufacture of composite materials are synchronized with the design and manufacture of external structures, and synchronous manufacturing is achieved from micro to macro scales to achieve the integration of "design-material-manufacturing" of structures. It supports the manufacturing of complex structural parts such as biological tissue manufacturing and composite materials, bringing revolutionary development to manufacturing technology. In 2020, the market value of 3D printing technology in healthcare will reach $2.13 billion. For example, the commercial operation of the dental industry has been so successful that one company estimates that 50,000 customised orthodontics will be printed using 3D printers every day. Experts predict that a truly fully functional 3D printed heart will be available within 20 years (Fig. 1.17). Now the difficulty of 3D printing technology is to print complex blood vessels. In the future, 3D printing technology has unlimited prospects, and printing complex organs such as hearts, livers and kidneys will no longer be a dream. 3D printing is also developing in two extreme directions, and can be large-scale, such as printing houses; It can also be miniaturized, such as using a miniaturized low-cost desktop printer to print living cells. The British Economist magazine believes that additive manufacturing, together with other digital production modes, will promote the third industrial revolution, change the future production and life mode, realize social manufacturing, change the way goods are manufactured and the way people live. The United States has made additive manufacturing technology a top strategic priority for national manufacturing development and has supported it. In the future, 3D printing will enter thousands of households, people may be able to manufacture their own products at any time, these products may involve all aspects of life, 3D printing development prospects.

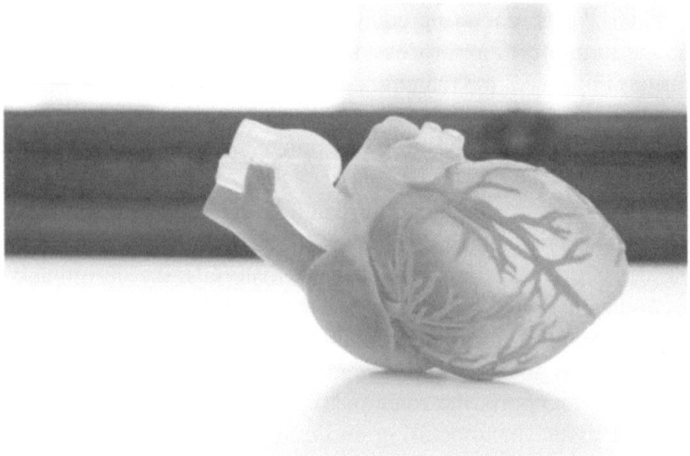

Fig. 1.17 3D printed 'heart'

1.4 Development Trends of Intelligent Manufacturing Equipment

1.4.1 Intelligent Machine Tool Equipment

The development of CNC machine tools is shown in Table 1.1. The prospect of the development of machine tool technology is to realize the automation of equipment manufacturing industry, from single machine automation to FMC, CIM, CIMS development, improve processing accuracy, efficiency, and reduce manufacturing costs. The emergence of intelligent machine tools has created conditions for the future equipment manufacturing industry to achieve total production automation. By self-suppressing vibration, reducing thermal deformation, preventing interference, self-adjusting lubricating oil amount, reducing noise, etc., the machining accuracy and efficiency of the machine tool can be improved. After the automation level of machine tools is improved, the workload of people in the management of machine tools can be reduced. The development and innovation of CNC system plays a very important role in the intelligence of machine tools.

With the development of artificial intelligence technology, in order to meet the development needs of manufacturing production flexibility and manufacturing automation, the intelligence degree of CNC machine tools is constantly improving, and intelligence and networking are the general trend. Intelligent machine tool refers to the machine tool that can make intelligent decision, automatic perception, intelligent monitoring, intelligent regulation and intelligent maintenance in its processing and manufacturing process, and can realize the efficient, high-quality and green multi-objective optimization operation of the processing and manufacturing process. The functions of an intelligent machine include intelligent operation, intelligent

Table 1.1 The development of CNC machine

Stage	Years	Development history
Stage 1	1930–1960	Developing from manual machine tools to highly efficient semi automated, automated machine tools, and automated production lines, addressing the issue of reducing manual labor for workers
Stage 2	1960–2006	The development of digital control machine tools has solved the problem of further reducing physical and mental labor
Stage 3	2006–	The stage of machine tool intelligence. The accelerated development of intelligent machine tools will further solve the problem of reducing mental labor

management, and intelligent maintenance (Fig. 1.18). It is reflected in the following aspects:

① Process adaptive control technology. Through real-time monitoring of the cutting force, spindle and feed motor power, current, voltage and other information during the machining process, the system independently identifies the force, wear, damage state of the tool and the stability state of the machine tool processing, and real-time adjustment of the processing parameters (spindle speed, feed speed) and processing instructions, so that the equipment is in the best running state, improve the processing accuracy, reduce the rough processing surface roughness, improve the safety of equipment operation.

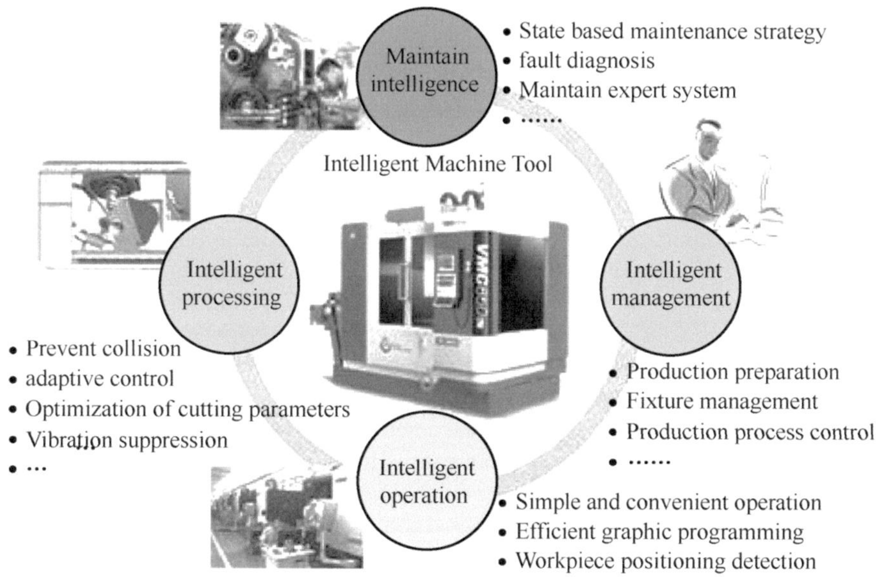

Fig. 1.18 The demand for intelligent of CNC machine tools

② Intelligent optimization and selection of processing parameters. Construct "intelligent optimization and selector of machining parameters" based on expert system or model to obtain optimized machining parameters, improve programming efficiency and machining technology level, and shorten production preparation time.
③ Intelligent fault diagnosis and self-repair technology. According to the existing fault information, the modern intelligent method is used to locate the fault quickly and accurately. Complete record all kinds of information of the system, playback and simulation of various errors and accidents of CNC machine tools, and timely repair and solve problems.
④ Intelligent AC servo drive device. The intelligent machine tool control system can automatically identify the load and automatically adjust the parameters, the intelligent spindle AC drive device and the intelligent feed servo device can automatically identify the moment of inertia of the motor and the load, and automatically optimize and adjust the control system parameters.
⑤ Intelligent 4M numerical control system. With the development of technologies such as big data and cloud storage, intelligent machine tools can integrate Measurement, modeling, Manufacturing and machine operation in a system to achieve information sharing and promote measurement, modeling, processing and clamping integration of operations.

For example, Okuma's intelligent digital control system is named "THINK" and has the ability to think. Okuma believes that the design, implementation and use of the current classic CNC system are outdated, and the time has come to fundamentally change it. "THINK" can not only make smart decisions about changed situations without human intervention, but also enable the machine to grow its functions in the application in an incremental way after the user plant, and be more adaptive to new situations and needs, more fault-tolerant, easier to program and use. In short, without human intervention, the machine will bring users higher productivity. OSP suite, an open architecture CNC numerical control system, as shown in Fig. 1.19, is based on the OP-P300 control platform and combines a visually pleasing and easy-to-use operating interface with customizable layouts, machine applications, widgets, shortcuts and touch screen technology to improve shop floor productivity.

It is worth mentioning that in addition to foreign CNC products, many domestic CNC systems and CNC machine tools are also excellent, such as Huazhong CNC (Fig. 1.20). The comprehensive performance of the domestic high-end numerical control system represented by the Central 8 can be comparable to that of foreign numerical control systems, reaching the international advanced level. Huazhong CNC has been widely used in key enterprises in aircraft manufacturing, automobile manufacturing and other fields, such as Chengdu Aircraft, Shenyang Aircraft, etc. Vigorously develop domestic numerical control, as soon as possible to fill the gap in domestic high-end numerical control, occupy the numerical control market, and lay the foundation for the overall improvement of China's equipment manufacturing industry.

Fig. 1.19 OSP suite numerical system

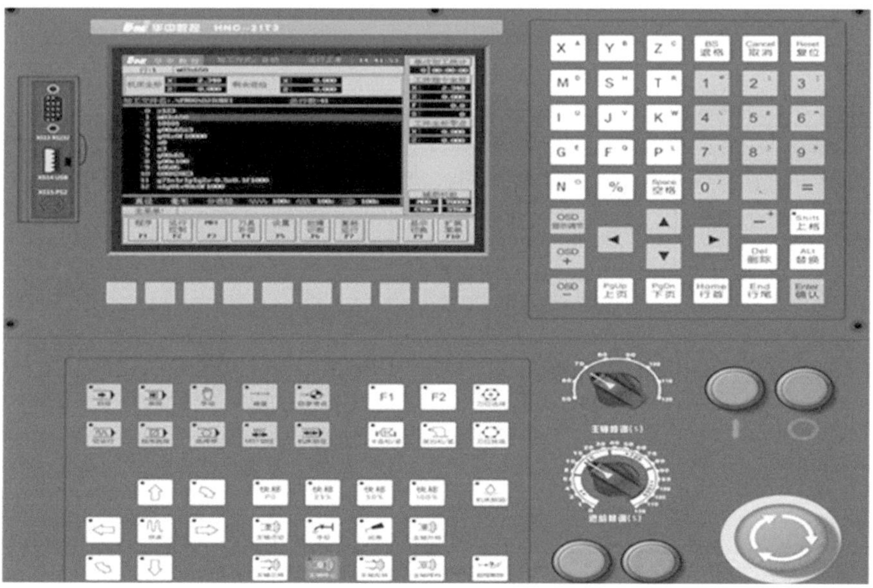

Fig. 1.20 China Huazhong CNC

1.4.2 Intelligent Construction Machinery

The intelligence of the construction machinery industry and the development of intelligent construction machinery are the most direct embodiment of the introduction of robots and the application of robot technology. Nowadays, many enterprises use robots for welding, painting, assembly and other work on a large scale in the production process. The intelligence of construction machinery is not only reflected in the intelligent work, but also in the fault diagnosis, product life prediction and other links. With the help of big data, the Internet of Things, deep learning and other technologies, enterprises can grasp the basic status of construction machinery products in real time, find faults in time, and respond in time. At the same time, enterprises can also obtain real-time operation data of intelligent construction machinery through wireless or wired networks, and carry out real-time data analysis and research and judgment to further play the value of these data. In the near future, under the influence of multiple factors such as changes in the development of science and technology and changes in the requirements of construction projects, the demand for integrated operation and intelligent control of construction machinery applications will be increasingly high. Under the promotion of information technology and intelligent upsurge, the development goal of a new generation of construction machinery will become increasingly clear, and the pace will continue to accelerate.

1.4.3 Intelligent Power Equipment

Power equipment can not only provide power, but also a "smart" machine. By collecting the data of the working process of the power equipment in real time through the sensor, the whole life cycle of the intelligent power equipment can be monitored and supported, and the whole life cycle fault diagnosis and the whole life cycle performance optimization of the power equipment can be carried out to achieve the dynamic adaptive monitoring of the equipment. Through data analysis and research and judgment, the health state prediction and evaluation, fault warning and rapid intelligent diagnosis, remote and remote communication can be achieved. The key technology of fast dynamic balance maintenance and intelligent maintenance decision.

1.4.4 Intelligent Robots

Intelligent robot is characterized by "intelligence", it has a very developed brain (central processor) and nerve center (intelligent sensor), can judge, logical analysis, understanding and other intellectual activities, can understand human language, use human language to talk to the operator, can analyze the situation, can adjust their actions to achieve all the requirements of the operator and work out desired actions and the ability to perform these actions autonomously with insufficient information and under rapidly changing environmental conditions. Intelligent robots also have "five features", which are all kinds of internal information sensors and external information sensors, such as visual sensors, auditory sensors, tactile sensors, smell sensors and so on. In addition, intelligent robots also have perceptual functions [18, 19]. The Ministry of Industry and Information Technology, the National Development and Reform Commission, and the Ministry of Finance jointly issued the "Robot Industry Development Plan (2016–2020)", pointing out that the development of the robot industry should promote major landmark products to take the lead in breakthroughs. Intelligent robots include industrial robots and service robots. In the field of industrial robots, focus on intelligent production, intelligent logistics, to overcome the key technologies of industrial robots, improve operability and maintainability, focus on the development of arc welding robots, vacuum (clean) robots, fully autonomous programming intelligent industrial robots, man–machine collaborative robots, two-arm robots, heavy-duty AGV6 iconic industrial robot products, to guide China's industrial robots to medium–high end development. In the field of service robots, focus on the development of fire rescue robots, surgical robots, intelligent public service robots, intelligent nursing robots four iconic products, promote the serialization of professional service robots, personal/family service robots to achieve commercialization.

1.4.5 Intelligent Terminal Product

Now we live in the world of intelligent terminal products, daily life contact with smart phones, smart electronic products, vehicle navigation, intelligent navigation robots and so on. In the future, the machines, equipment, tools, etc. used by anyone or associated with people may be an "intelligent terminal", which has the function of information collection, analysis and processing.

1.5 Summary of This Chapter

This chapter mainly introduces the concept, composition and characteristics of intelligent manufacturing equipment, the development overview of intelligent manufacturing equipment, the research and development content of intelligent manufacturing equipment, and the development trend of intelligent manufacturing equipment, so that readers have a full understanding of intelligent manufacturing equipment, so as to facilitate the understanding of the content of subsequent chapters.

References

1. Tan Jianrong, Liu Zhenyu. Key Technologies and Enterprise Applications of Intelligent Manufacturing [M]. Beijing: China Machine Press, 2017.
2. Siemens Industry Software. Practice of Industry 4.0 - The Digital Path for Equipment Manufacturing Industry [M]. Beijing: China Machine Press, 2016.
3. Advisory Committee for the Construction of a Powerful Manufacturing Country. Made in China 2025 Blue Book [M]. Beijing: Electronic Industry Press, 2018.
4. Zhang Ce. History of Mechanical Engineering [M]. Beijing: Tsinghua University Press, 2015.
5. Wan Rong, Zhang Zegong, Gao Qian, et al. Internet + Intelligent Manufacturing [M]. Beijing: Science Press, 2016.
6. Brauckmann, O. Intelligent Manufacturing: Subversion and Reconstruction of Future Industrial Modes and Formats [M]. Translated by Zhang Xiao, Yu Ji. Beijing: China Machine Press, 2015.
7. Wu Jun. The Age of Intelligence [M]. Beijing: CITIC Press, 2016.
8. Sun Lei, Sun Jinan. Opportunities and Challenges Faced by the Development of Intelligent Manufacturing in China's Industrial Economy [J]. Finance and Economics, 2020(20):19–20.
9. Serope Kalpakjian, Steven R. Schmid; Translated by Wang Xiankui. Manufacturing Engineering and Technology - Machining [M]. Beijing: China Machine Press, 2012.
10. Jiao Bo. Development Status and Trend of Intelligent Manufacturing Equipment [J]. Internal Combustion Engines and Parts, 2020(09):214–215.
11. Wen Danfeng, Wei Shaofeng. Internet + Medical Care - Medical and Health Revolution in the Mobile Internet Era [M]. Beijing: China Economic Publishing House, 2015.
12. Xiao Xin. Environmental Monitoring [M]. Beijing: Science Press, 2017.
13. Zhang Fu, Wang Xiaofang. Design of Mechanical Manufacturing Equipment [M]. Wuhan: Huazhong University of Science and Technology Press, 2017.
14. Kalma Toth. The Age of Artificial Intelligence [M]. Translated by Zhao Li. Beijing: Posts & Telecom Press, 2017.
15. Qiao Xuetao, Zhang Libin, Yan Cunfu, et al. Analysis on the Development Status of RV Reducers for Industrial Robots in China [J]. Journal of Mechanical Strength, 2019, 41(06):1486–1492.
16. "Made in China 2025" Again Supports the Development of Industrial Robots [J]. Tool Engineering, 2015, 49(08):58.
17. Zhao Jie. Development Status and Challenges of Industrial Robots in China [J]. Aeronautical Manufacturing Technology, 2012(12):26–29.
18. Lei Zongyou. High-end Equipment Manufacturing Industry [M]. Shanghai: Shanghai Scientific and Technical Literature Press, 2014.
19. Meng Qingchun, Qi Yong, et al. Intelligent Robots and Their Development [J]. Periodical of Ocean University of China (Natural Science Edition), 2004, 34(5):831–838.

Chapter 2
Intelligent Manufacturing Equipment Mechanical Body Design

The design of mechanical body of intelligent manufacturing equipment is mainly the design of mechanical part. Mechanical design is to determine the function that the product should have according to the requirements of use, conceive the working principle of the product, structural shape, mode of movement, force and energy transfer and materials used, and convert it into a specific description, such as drawings and design documents, etc., as the basis for manufacturing. Mechanical design is an important link in the product life cycle from demand analysis, design, manufacturing, sales, use to recycling, and the impact on product cost accounts for 80%.

2.1 Basic Requirements for Mechanical Design

Mechanical design is an important part of product design, and its main tasks are shown in Fig. 2.1. The basic requirements of mechanical design are: the mechanical products designed under the premise of completing the specified functions, and strive to produce beautiful appearance, excellent performance, high production efficiency and low use cost; Within the specified service cycle, the product should be safe and reliable, easy to operate, and simple to maintain. In the long-term engineering design and manufacturing use process, people have summed up nine word evaluation methods: production (productivity), rely (reliability), can (good performance, energy saving), repair (easy maintenance), guarantee (environmental protection), use (easy to use), into (cost, including manufacturing cost and use and maintenance cost), spirit (flexibility), beauty (beauty) [1].

M. Xu et al., *Design and Case of Intelligent Electromechanical Equipment System*, Advanced and Intelligent Manufacturing in China,
https://doi.org/10.1007/978-981-95-3833-1_2

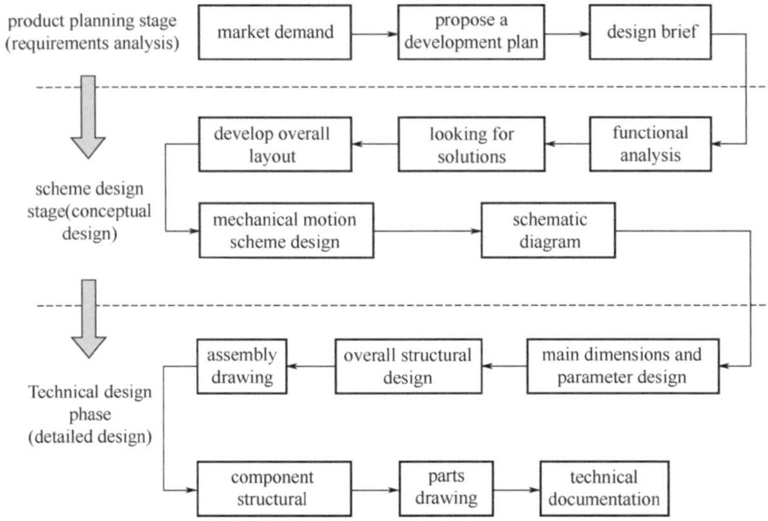

Fig. 2.1 The main task of mechanical design

Mechanical design should generally meet the following requirements:

(1) Use requirements

Use requirements are the primary requirements for mechanical products, but also the most basic requirements. The use requirements of machinery refer to the mechanical products must meet the specific requirements of users for the required product functions, which is the most fundamental starting point of mechanical design. If you can't meet the customer's use requirements, the design will lose its meaning.

(2) Reliability and safety requirements

The reliability and safety of mechanical products refers to the ability of mechanical products to complete the specified functions under the specified conditions of use and life cycle. Safe and reliable is a necessary condition for mechanical products, the safe operation of machinery is the premise of safe production, and the protection of the operator's personal safety is an important embodiment of people-oriented.

(3) Economic and social requirements

Economic requirements refer to the design of mechanical products in the design, manufacturing cycle is short, low cost; High efficiency in use, low energy consumption, high productivity, low maintenance and management costs. In addition, mechanical products should be easy to operate, safe and reliable, comfortable appearance, pleasant color, product production process and use process must meet the requirements of national environmental protection and labor regulations.

(4) Other requirements

Some mechanical products due to the different working environment and requirements, the design will have some special requirements. For example, the design of aircraft has the requirements of small mass, small flight resistance and large carrying capacity; Mobile machinery (such as tower cranes, drilling machines, etc.) should be easy to install, disassemble and transport; When designing the machine tool, the requirement of maintaining accuracy for a long time should be considered; Food, printing, textile, papermaking machinery, etc. should be kept clean and should not pollute products.

According to the basic requirements of mechanical design, the design of mechanical products should follow the following basic principles:

(1) Market-oriented

Mechanical design and the market is closely linked. Good design can quickly capture the market and gain profits. From determining the design project, use requirements, technical indicators, design and manufacturing period, to coming up with the overall plan, feasibility demonstration, comprehensive utility analysis, profit and loss analysis, to the specific design, trial production, identification, and information feedback after the product is put on the market are closely around the market demand to operate. How to design to make the product competitive and win the market is the problem that the mechanical designer should think about at all times.

(2) Creative principle

Creation is the production of things that did not exist before, is a typical human autonomous behavior. Creation is unique to human beings and is the labor of conscious exploration of the world. Design has strong vitality only as a creative activity. Conformism, dare not innovate, can only be "beaten" behind. With the rapid development of science and technology in the world today, the creative principle of mechanical design is particularly important.

(3) The "three" principle

Standardization, serialization, generalization referred to as "three", is a very important technical policy in our country. Standardization refers to the product quality, specifications, performance, structure and other aspects of the technical indicators to be unified and implemented as standards. Common standard codes are GB (National standard of the People's Republic of China), JB (Machinery industry standard) and ISO (International Organization for Standardization Standard). Serialization refers to the same product, under the same basic structure or basic conditions to specify a number of different size series. Generalization means that different kinds of products or similar products of different specifications use the same structure and size parts as far as possible.

The implementation of "three" can reduce the design workload, improve the design quality, shorten the production cycle; Reduce tool and measuring tool specifications, easy to design and manufacture, reduce costs; It is convenient to organize the large-scale and specialized production of standard parts; Easy to ensure product quality, save materials, reduce costs; Improve interchangeability, easy maintenance; Facilitate the country's macro-management and regulation as well as internal and external trade; Easy to evaluate product quality and resolve economic disputes.

(4) Overall optimization principle

Mechanical designers should have the idea of systematization and optimization, and consider the product as a whole. The best performing machines do not necessarily have the best internal parts or the best benefits. Designers should put the design scheme in the large system to consider, from the economic, technical, social benefits and other aspects to analyze, calculate, weigh the advantages and disadvantages, and seek the best scheme, so that the design effect is the best, the best economic benefits.

(5) Practical principle

Design should be used for it, and all design can not be divorced from reality. Mechanical designers, in the design of a product, to consider the current material supply situation, the enterprise's own production conditions, the user's use conditions and requirements, so that the design of the product is in line with the actual needs.

(6) Ergonomics principle

Machine is to serve people, but the machine also needs people to operate and use in the working process, people are always leading, is the most active, the most vulnerable, good product design needs to meet the ergonomic principle. Ergonomics emphasizes the coordination between human, machine and environment in different operations. The research methods and evaluation methods involve many fields such as psychology, physiology, medicine, anthropometry, aesthetics, design and engineering technology. Through multidisciplinary knowledge, the design and transformation of working instruments, working methods and working environment are guided to improve product efficiency, safety, health and comfort the characteristics of aspects. How to make the machine adapt to the operation requirements of people, the input-output ratio is high, and the overall effect is the best, which is the problem that designers should consider. The design should reasonably allocate man–machine functions to minimize the chance of operator intervention or intervention danger. When determining the relevant size of the machine, human parameters should be considered to make the machine equipment adapt to the characteristics of the human body. It is necessary to have friendly man–machine interface design and reasonable working space layout.

The mechanical system of intelligent manufacturing equipment is mainly divided into five parts: power system, executive system, transmission system, support system and operation control system, and the interaction relationship is shown in Fig. 2.2. This chapter will focus on the mechanical body design of intelligent manufacturing equipment, that is, transmission system, support system and execution system design.

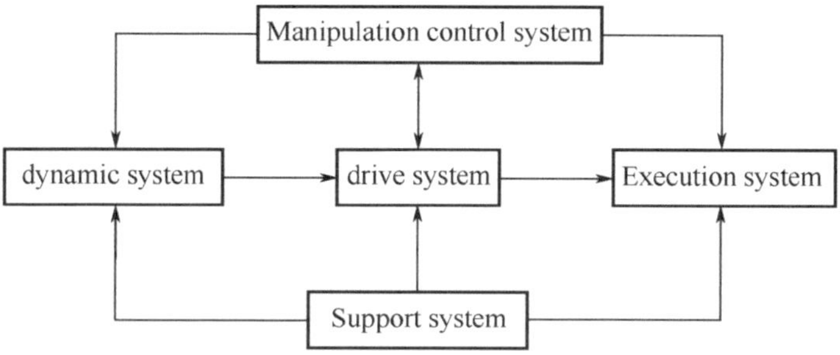

Fig. 2.2 The relationship of mechanical parts

2.2 Overall Design of Intelligent Manufacturing Equipment Mechanical Body

2.2.1 *Task of Mechanical Structure Design*

The task of structural design is to determine and draw a specific mechanical structure diagram on the basis of the overall design and according to the mechanical principle scheme to meet the actual requirements of the function. The process of structural design is to convert the abstract working principle into a certain type of component or part. While determining the material, shape, size, tolerance, heat treatment and surface condition of the structural part, it is also necessary to comprehensively consider the processing technology, strength, stiffness, accuracy and assembly relationship and exchange between the parts. The direct product of mechanical structure design is technical drawing, including 3D model, 2D engineering drawing, simulation analysis and so on. Although the product of mechanical structure design is the technical drawing, the structural design work is not only a simple mechanical drawing, the drawing is only the language to express the design scheme, and the embodiment of comprehensive technology is the basic content of structural design.

2.2.2 *Characteristics of Mechanical Structure Design*

The product structure design stage includes appearance modeling and review, appearance prototype production, and product internal structure design Review and other contents are important stages in product structure design [2]. Mechanical structure design is mechanical design main tasks. Its main characteristics include:

① Mechanical structure design plays a pivotal role in the success or failure of mechanical design, and is the stage involving the most problems, the most specific and the largest workload in mechanical design. Structural design is a design process integrating thinking, discussion, calculation, drawing (including

3D modeling and 2D engineering drawing), simulation, experiment and rapid prototyping. About 80% of the time in the mechanical design process is spent on structural design, and the engineers engaged in mechanical structure design are also the most.

② The mechanical structure design problem has multiple solutions, that is, the mechanical structure that meets the same design requirements is not the only one. Therefore, it is necessary for mechanical structure design engineers to have complete professional knowledge and rich practical experience, and to choose the most suitable design scheme according to the actual requirements in many design schemes, in order to achieve the best working performance, the lowest manufacturing cost, the smallest size and quality, the highest reliability in use, the lowest consumption and the least environmental pollution. These requirements are often contradictory, and their relative importance varies depending on the type and use of the machine. The designer's task is to weigh the weight according to the specific situation, overall consideration, so that the designed machinery has the best comprehensive technical and economic effects.

③ Mechanical structure design stage is a more active part of the mechanical design process, in the mechanical structure design, we must start from the machine as a whole to analyze the basic requirements of the mechanical structure, often need to repeat, spiral upward.

2.2.3 Structural Elements and Design Methods of Mechanical Structural Components

(1) Geometric elements of structural components

The function of mechanical structure is mainly realized by the geometric shape of mechanical parts and the relative position relationship and relative motion relationship between each part. Each geometric surface constitutes the geometric shape of the parts, a part usually has multiple surfaces, some of these surfaces are in direct contact with the surface of other parts, this part of the surface is the functional surface, the connecting part is called the connecting surface. The functional surface of a part is an important factor to determine the mechanical function, and the design of the functional surface is the core of the structural design of the part. The main geometric parameters of the functional surface are the geometric shape, scale, number, position and sequence of the surface. Different designs of functional surfaces can realize the design of multiple structural schemes with the same technical requirements.

(2) Connection between structural components

No part of a device or product exists in isolation, but is interrelated. Therefore, in the structural design, in addition to studying the function and other features of the parts themselves, it is also necessary to study the relationship between the parts. The correlation of parts is divided into direct correlation and indirect correlation. If the two parts have a direct assembly relationship, it is called direct correlation, otherwise it is indirect correlation. Indirect correlation can be divided into two categories: position correlation and motion correlation. Position correlation means that the two parts have requirements in the mutual position, such as the two adjacent drive shafts in the reducer, the center distance must ensure a certain accuracy, and the two axes must be parallel to ensure the normal meshing of the gear. Motion correlation means that the motion trajectory of a part is related to another part, such as the motion trajectory of the lathe tool holder must be parallel to the center line of the spindle, which is guaranteed by the parallel of the bed guide rail and the spindle axis. Therefore, there is a position correlation between the spindle and the guide rail, and a motion correlation between the tool holder and the spindle. Most parts have two or more directly related parts, and each part has two or more parts that are structurally related to other parts. In the structural design, the directly related parts of the two parts must be considered at the same time to reasonably select the heat treatment method, shape, size, precision and surface quality of the parts. At the same time, indirect related conditions must be considered, such as dimensional chain and precision calculation. In general, if a part is directly related to more parts, its structure is more complex; The more indirectly related parts of a part, the higher its accuracy requirements.

(3) Consider the material and heat treatment of structural components

Materials are the material basis on which human beings depend for survival and development. Information, materials and energy are important components of contemporary civilization, and many aspects of the international community's competition are the competition of materials. There are many materials that can be selected in mechanical design, and different materials have different properties and correspond to different processing technologies. In the structural design, it is necessary to reasonably select the appropriate material according to the functional requirements, and determine the appropriate processing technology according to the type of material, and determine the appropriate structure according to the requirements of the processing technology, and only through the appropriate structural design can the selected material give full play to the advantages. In order to select materials correctly, designers must fully understand the mechanical properties, processing properties, use costs and other information of the selected materials. In the structural design, different design principles should be followed according to the characteristics of the selected materials and the corresponding processing technology.

Metal heat treatment is one of the important processes in mechanical manufacturing, and it is an important means to improve the performance of mechanical products. Compared with other processing processes, heat treatment generally does not change the geometry of the workpiece and the overall chemical composition, but improves the performance of the workpiece by changing the internal microstructure of the workpiece, or changing the chemical composition of the workpiece surface. Heat treatment process generally includes quenching, normalizing, tempering and so on. The structural design requirements of the parts requiring heat treatment are: the geometry of the parts should be simple and symmetrical; For parts with unequal cross sections, the change of size section should be gentle to avoid mutation (if the change of adjacent parts is too large, the cooling of size section is uneven, and internal stress will inevitably be formed); Avoid sharp edge structure (in order to prevent the sharp edge from melting or overheating, generally cut 2 ~ 3 mm chamfering on the edge of the groove or hole); Avoid a cross-section with a wide thickness (because it is easy to deform during quenching cooling and has a larger tendency to crack).

2.3 Main Content of Intelligent Manufacturing Equipment Ontology Design

2.3.1 Functional Principle Design

Scheme design refers to the functional principle design of products according to actual needs. This stage is a very important stage in the design process, which mainly carries out the abstraction of design tasks, functional decomposition, establishment of functional structure, search for solution methods, formation of schemes and evaluation.

(1) Functional decomposition

In the process of mechanical design, the equipment that needs to be designed is often more complex, and it is difficult to directly find the optimal principle scheme that meets the total function. In the design process, functional analysis method can be used to decompose the total function into multiple functional elements, seize the main requirements, and take into account the secondary requirements. Then by solving and combining the function elements, various solutions of the principle scheme are obtained. For example, automatic beverage filling machines are used to achieve automatic beverage filling. Its sub-functions include beverage bottle, cap, beverage storage, transportation, filling, capping, sealing, jet coding, labeling, finished product transportation, packaging. For a machine, it can be decomposed according to prime mover, working machine, transmission motor, controller, supporting parts, etc. The functional elements commonly used in mechanical design include physical functional elements, mathematical functional elements and logical functional elements. The physical function element reflects the basic physical relationship of matter, energy and information in the transfer and transformation of technical system, including transformation, amplification and reduction, separation and merger, conduction and blocking, storage, etc. Mathematical functional elements include addition, subtraction, multiplication, division, multiplication, differentiation and integration, etc. Logical function elements include three kinds of logical relations: with, or, and not.

(2) Draw functional structure

The decomposition and combination of function elements are called functional structure. The decomposition and abstraction of tasks can clarify the total function of the product. The functional structure intuitively reflects the transfer and conversion process of matter, energy and information in the working process of the system. Functional structure is mainly divided into chain structure, parallel structure and cyclic structure, as shown in Fig. 2.3. F_1, F_2 and F_3 in the figure represent different functions.

Figure 2.4 shows the relationship between functional layers. The functional structure of different functional layers is connected and passed layer by layer until the total functional requirements are met.

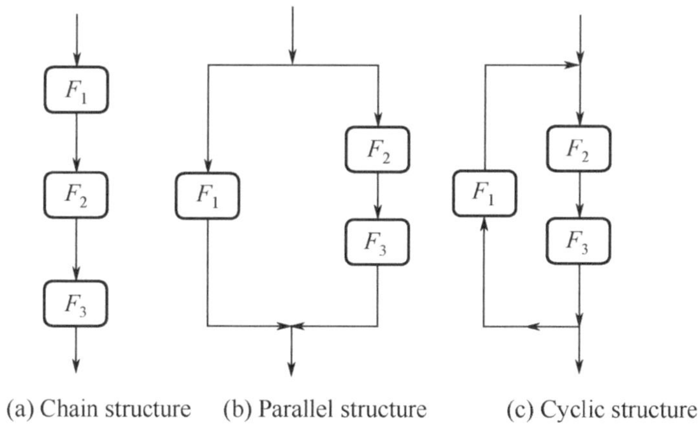

Fig. 2.3 Function structure

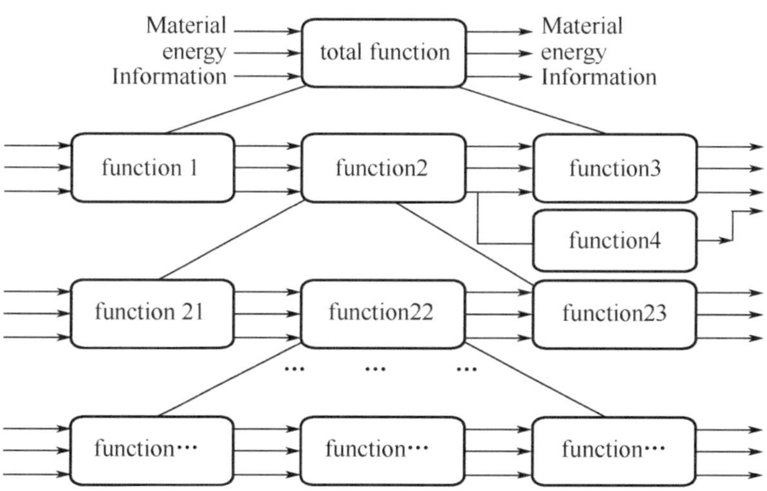

Fig. 2.4 Functional layer relationship

(3) Functional element solving

Functional metasolution methods generally include design catalog solution method, innovative solution method, etc. The design catalog is a carrier of design information, which can classify, sort and store a lot of information required in the design process according to certain rules, so as to facilitate designers to find and use it. The design catalog is divided into three categories: object catalog, operation catalog and method catalog. Figure 2.5 shows the solution directory for the motion transformation of the four-bar mechanism, ○ indicates yes and ⊗ indicates no.

	four-bar mechanism diagram	motion sub-conversion				four-bar mechanism diagram	motion sub-conversion		
		rotation / rotation	rotation / pan	pan / pan			rotation / rotation	rotation / pan	pan / pan
1	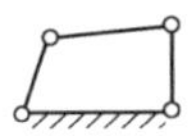	○	⊗	⊗	9	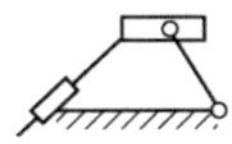	⊗	○	⊗
2	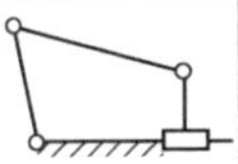	⊗	○	⊗	10	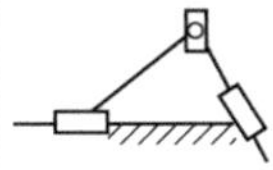	⊗	⊗	○
3	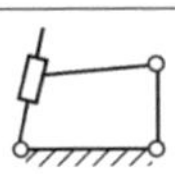	○	⊗	⊗	11	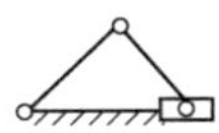	○	○	⊗
4	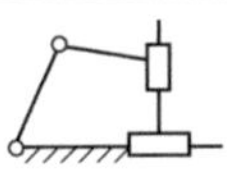	⊗	○	⊗	12	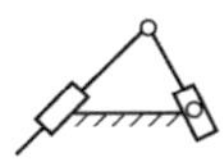	⊗	○	○
5	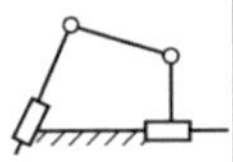	⊗	⊗	○	13	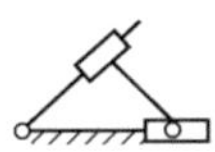	○	○	⊗
6	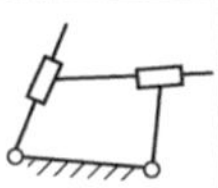	○	⊗	⊗	14	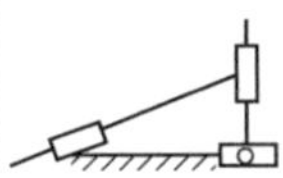	⊗	⊗	○
7	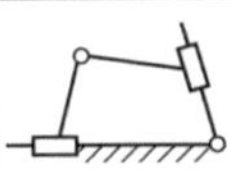	⊗	○	⊗	15	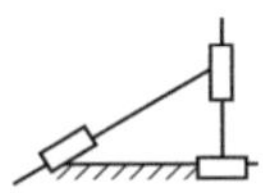	⊗	⊗	○
8	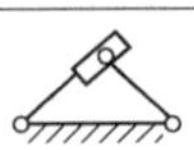	○	⊗	⊗	16		⊗	⊗	○

Fig. 2.5 Contents for solving the motion conversion of four-bar mechanism

(4) Principle solution

The original understanding is the working principle that can realize a certain function and the technical means and structural principle that can realize the working principle. The analysis and solution of principle schemes are generally based on morphological matrix. The system function unit and its corresponding solutions are taken as coordinates respectively, the "function solution matrix" of the system is listed, and then a corresponding solution is selected from each function unit and organically combined to form a system solution [3].

(5) The preliminary design took shape

Combine all the sub-functional principles to form the total function.

Multiple design schemes can be obtained by the combination of the original understanding, and the ideal preliminary design scheme can be obtained by system association or mathematical method association. Then we can evaluate the scheme, arrange the overall structure and calculate the parameters.

(6) General structural arrangement

After selecting the preliminary plan, the scheme can be concretized. For example, the space layout, quality, technical parameters, materials, performance, process, cost, maintenance and so on are quantified. The overall layout of mechanical system is an important part of structural design, which should meet the requirements of reasonable function and performance, compact structure, clear hierarchy, coordinated proportion and expandability. The design sequence of the overall layout should be repeated many times from simple to complex. According to the layout direction of the actuator, it is divided into horizontal type, inclined type, upright type, etc. According to the movement mode of the actuator, it is divided into rotary type, linear type, vibration type, etc. According to the relative position of the prime mover, it can be divided into front type, middle type and rear type.

(7) Main parameter calculation

The main parameters of the mechanical system are divided into performance parameters, size parameters, dynamic parameters and motion parameters. During the design, the overall parameters should be selected for structural design according to the actual requirements, and the overall parameters should be checked to determine the overall parameters, and the parameters and structure should be adjusted according to the existing problems until the overall technical parameters meet the requirements (Fig. 2.6). To do a good job of the overall parameters of the primary election requires certain design experience.

Fig. 2.6 The process of mechanical system design

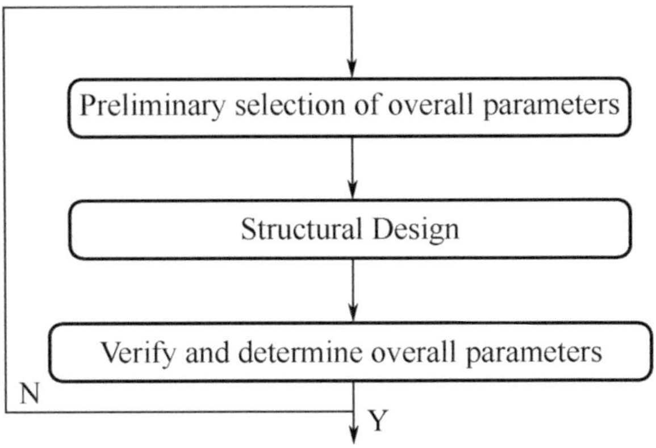

2.3.2 Scheme Evaluation and Screening

(1) Determine evaluation indicators

Overall scheme evaluation is an important step in the process of product design, and it is the key link of detailed design after determining the overall design. In the evaluation index system of total product scheme, there are both quantitative and qualitative indicators. The evaluation value of qualitative indicators can not be quantified because of incomplete information or qualitative attributes of evaluation indicators. The evaluation value of different quantitative indicators is not comparable. The evaluation index needs to establish the evaluation index value of the same dimension that can be compared with each other, so it is necessary to standardize the qualitative index and the quantitative index. For qualitative indicators, the same level of fuzzy number is usually used to represent the evaluation value of indicators. For quantitative indicators, the performance evaluation methods often used include linear transformation, standard 0–1 transformation, vector normalization and so on.

(2) Determine evaluation model

Evaluation index is the basis of evaluation. Evaluation indicators include technical evaluation indicators (that is, technical possibility and advancement, including work performance indicators, reliability, maintainability, etc.), economic evaluation indicators (including cost, profit, implementation costs and investment payback period, etc.), and social evaluation indicators (whether it conforms to the national science and technology development policies and plans, whether it is conducive to improving the environment, and whether it is conducive to resource development and conservation energy, etc.).

Through the analysis of the overall design objectives, the most important requirements and constraints are selected as evaluation indicators. At the same time, the weighting coefficient is set according to the importance of each evaluation objective. Generally, the weighted coefficient of each evaluation index is Wi $\in [0, 1]$, and $\sum$Wi

= 1. The commonly used evaluation model building methods include effective value scoring method, fuzzy evaluation method, analytic hierarchy process, etc.

There are many contents of design evaluation. For the evaluation of intelligent mechanical and electrical equipment, there are mainly technical and economic evaluation, reliability evaluation, structural technology evaluation, ergonomics evaluation, product modeling evaluation, standardization evaluation and so on.

The manufacturability evaluation of structure is introduced as an example. Structural manufacturability evaluation can reduce production cost, shorten production time and improve product quality. The evaluation contents of structure manufacturability include manufacturability, assembly, maintenance and so on.

The processability can be evaluated from two aspects: the reasonable combination of product structure and the processability of parts. A product is made up of components, assemblies and parts. The fewer parts that make up the product, the simpler the structure, and the lighter the weight may be, but the shape of the part may be complex and the processing process is poor. According to the process requirements, the design should reasonably consider the structural combination of the product, the process is not good or large parts into a number of small parts with good process, so that the size of the parts and the size of the enterprise production equipment, but the increase of processing costs and assembly costs, the increase of the connection surface will also reduce the stiffness, shock resistance and sealing performance. The reasonable combination of product structure also includes the design of a number of simple structure, small size parts into one part, to reduce weight, reduce the number of connecting surfaces, save processing and assembly costs, improve the mechanical properties of the structure.

The structural shape, material, size, surface quality, tolerance and fit of the part determine its processability. There is no unified standard for the evaluation of processing technology, which can be determined according to the existing production conditions of the workshop, such as traditional process habits, the processing capacity and tooling conditions of the existing processing equipment of the workshop, external processing conditions and capabilities, the supply of materials, blanks and semi-finished products and the possibility of quality inspection. Casting parts, forging parts, cold pressing parts, turning parts, special processing parts, milling parts, grinding parts, etc., have their own process requirements, the length of the relationship here is no longer detailed, when necessary, you can refer to the metal processing manual for reference in the design.

Product design not only determines the cost and quality of parts processing, but also determines the cost and quality of assembly. The cost and quality of assembly depend on the type and number of assembly operations, which in turn are related to the product structure, the structure of the parts and their combined parts and the type of production.

The product structure that is easy to assemble will decompose the product into parts reasonably, the parts into components, and the components into parts, so as to realize parallel assembly, shorten the assembly cycle, and ensure the assembly quality. Parts with simple structure are combined into one part to reduce assembly workload; Under the premise of meeting the function, the number of parts, joint parts and joint surfaces should be reduced as much as possible; As far as possible, the assembly should adopt unified tools, unified assembly direction and methods. The rationality of the structure of the joint parts can improve the assembly process and maintenance process.

Product design should fully consider the entire product life cycle, and the maintenance of the product should be considered at the beginning of the design: the average repair time should be short; Good interchangeability of components or parts required for maintenance and easy to find; Sufficient operating space between equipment components; The number and type of maintenance tools, accessories and auxiliary maintenance equipment are small; Low maintenance cost and less working hours.

Other evaluations are not detailed here. In the case of a large number of preliminary schemes, the schemes can be selected. For example, the schemes that cannot be realized in the schemes can be eliminated first through observation and comparison, and each scheme can be quantified by index scoring, and the final scheme can be determined by scoring.

2.3.3 Basic Requirements for Mechanical Structure Design

Since industrial civilization, human beings have created a variety of mechanical products and applied them to production and life, changing the way of human life. The content and requirements of mechanical structure design vary greatly, but different machines have common parts. The requirements of mechanical structure design can be explained from three different levels of mechanical structure design.

(1) Functional principle design

Function is an abstract description of the working function of a product, and the basic function is the efficiency of the product to meet certain needs of users, that is, the use or use value of the product. Product design is to give a machine or device from scratch to give value, that is, to be useful. The functional design should meet the main mechanical functional requirements, be technically specific, and be expressed in the form of engineering drawings, such as the realization of the working principle, the reliability of the work, the process, materials and assembly requirements. Functional principle design must first determine the functional objectives that meet customer requirements through investigation and research, and then carry out innovation, carry out functional design and principle verification, determine the program and evaluation, and obtain the optimal program. Specifically, it includes function principle analysis, function decomposition, sub-function solution and function principle scheme determination.

(2) Quality design

Quality design is very important, is a feature of modern engineering design, is an important factor to improve product competitiveness. The design takes into account various requirements and restrictions (operation, aesthetics, cost, safety, environmental protection, etc.) to improve product quality and cost performance. Taking into account various requirements and improving the quality of products is the key to modern machinery design, and it is also the key to the strong competitiveness of products. Product quality is not only a matter of process and materials, improving quality should start from design. Excellent design can make the product quickly occupy the market, for the enterprise to win profits and competitiveness.

(3) Optimal design and innovative design

With the improvement of product performance requirements in the market, the role of optimized design and innovative design in modern machinery design is becoming more and more important, and has become the focus of competition in the development of future technical products. To survive, enterprises need to continuously introduce innovative products with competitiveness. Innovative design needs to adjust the product structure according to the forecast of the development of market demand, transform and upgrade traditional products with new technical means and technical principles, develop a new generation of products, enhance the added value of products, improve their functions and technical indicators, reduce production costs and energy consumption, adopt advanced production processes, and narrow the difference between advanced similar products at home and abroad distance, improve the competitiveness of products [4]. Innovative design is the design to solve the invention problem, and the core of this process is conceptual design [5]. Optimization design and innovation design require the systematic construction of optimization design space by means of structural design variables, creative design thinking and other scientific methods for optimization and innovation. The task of mechanical design is to seek the best solution among many feasible solutions. The premise of structural optimization design is to construct a large number of optimal feasible schemes, that is, to construct a large number of optimization solution space, which is also the most creative place in structural design. At present, structural optimization design is still limited to such problems described by mathematical models. The optimization design based on the structural design solution space composed of process, material, connection mode, shape, sequence, orientation, quantity, size and other structural design variables has more development potential. In general, innovation and optimized design need to start with market research and demand forecasting. On the basis of market research, clear equipment product design tasks, product planning, scheme design, technical design and construction design. Finally, it is necessary to produce a product prototype or a rapid prototype, conduct product tests, verify the performance of the new product, and then conduct pilot tests.

2.3.4 Basic Design Criteria for Mechanical Structures

The final result of mechanical design is a certain structural form and according to the designed structure for processing, assembly, debugging, the formation of new products. Structural design should meet the product in function, reliability, technology and economy and many other aspects of the requirements, but also to deal with the parts of the force balance, strength, stiffness, accuracy and life and other continuous improvement, so mechanical structure design is a comprehensive technical work. Wrong or unreasonable structural design will cause the failure of parts, so that the machine can not meet the requirements of design accuracy, and bring great inconvenience to assembly and maintenance. The structural design criteria that should be considered in the process of mechanical structure design are:

(1) Clarify expected functions

The main purpose of product design is to achieve the predetermined functional requirements, so the design criteria to achieve the expected function is the first consideration of structural design. The principle is clear, simple, safe and reliable.

① Clarify functions. The structural design must first clarify the function of the product, and the problems in the product design should be clearly reflected in the structural scheme, so that the function is clear, the working principle is clear, the working condition and the stress state are clear. Structure design is based on the function of parts in the machine and the connection relationship with other parts, design structure and size parameters. The main functions of the parts are to bear the load, transmit the motion and power, and ensure or maintain the relative position or movement trajectory between the relevant parts or components. The designed structure should first be able to meet the basic functional requirements of the machine, on this basis, and then gradually optimize.

② Reasonably allocate functions. In product design, it is sometimes necessary to reasonably allocate tasks according to specific circumstances, and decompose a function into multiple sub-functions. Each sub-function must have a certain structure, and the structure of each part should have a reasonable and coordinated connection to achieve the total function. Multiple structural parts bearing the same function can reduce the burden of a single part and extend the service life. Take V-belt with transmission as an example, the tensile layer fiber rope is used to withstand the tension; The rubber filling layer is subjected to alternating bending stress when the belt is bent. The coating layer acts with the pulley groove to produce the friction required for transmission. For example, in threaded connections that bear transverse loads, if only the friction generated by bolt pretightening is used to bear the transverse loads, the size of the bolts will be too large or the number of bolts will be too large, and the shear elements can be added to share the transverse loads in the design. In this way, the connection is mainly completed by bolts, and the shear resistance is mainly completed by shear elements, as shown in Fig. 2.7.

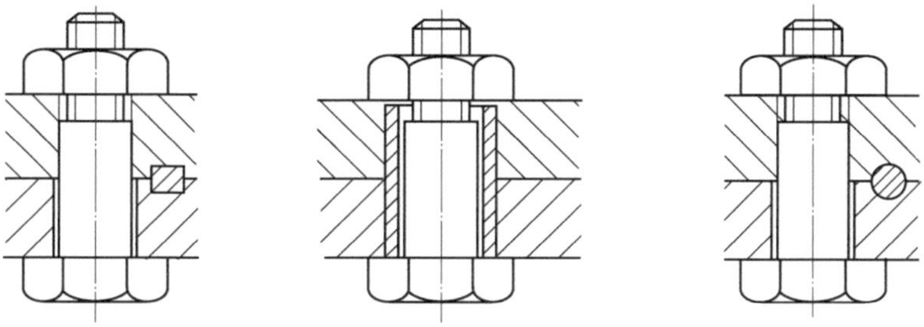

Fig. 2.7 Using shear resistant components to share the load

③ Centralized functionality. In order to simplify the structure of mechanical products, reduce processing costs, and facilitate installation, in some cases, multiple functions can be assumed by one part or component. Functional concentration has certain advantages, but excessive will make the shape of the parts more complex, but affect the processing technology, increase the processing cost, the design should be based on the specific situation.

④ Simple and reliable. When determining the structural scheme, the number of parts and processing procedures should be reduced as much as possible, the shape and structure of the parts should be simple, the processing surface should be reduced as much as possible, and the number of machining and heat treatment procedures should be reduced.

(2) Design criteria that meet strength requirements

① Equal strength criterion. The change of the section size of the part should adapt to the change of its internal stress, so that the strength of each section is equal. According to the principle of equal strength design of the structure, the material can be fully utilized, thereby reducing the weight and cost. Such as cantilever bracket, ladder shaft, aircraft wing design (Fig. 2.8).

② The force flow structure should be reasonable. Force flow is the trajectory of force in the process of transmission. The force in the mechanical system is transmitted through each connecting surface, and the direction of force transmission is the direction of force flow. In order to intuitively represent the state of force transmission in mechanical components, it can be represented by force flow. Force flow plays an important role in structural design. The force flow is not interrupted in the component, and any line of force will not suddenly disappear, and it must be introduced from one place and transmitted from another. Another property of the force flow is that it tends to travel along the shortest route, so that the force flow is dense near the shortest route, forming a high stress zone. Other parts of the force flow is sparse, and even there is no strong flow, and from the point of view of stress, the material is not fully utilized. The transmission route of force in the mechanical system under load follows the rule of the shortest transmission path. In order to improve the stiffness of the component, the shape of the part should be designed as far as possible according to the shortest route of the force

Fig. 2.8 Equal strength design of aircraft wings

flow, reducing the bearing area, so that the cumulative deformation is smaller, improving the material utilization rate and the stiffness of the entire component. Figure 2.9 shows the plate spring and shaft support structure designed according to the force flow structure.

③ Reduce stress concentration structure. The rapid change of the direction of force flow can cause stress concentration, which often occurs in structural design. In the design, measures should be taken in the structure to make the force flow smooth. Stress concentration is an important factor affecting the fatigue strength of parts, so it should be avoided or reduced as much as possible. Measures to avoid stress concentration can be consulted in mechanical design books or manuals, such as increasing excessive corners, using unloading structures, etc. Figure 2.10 shows a structure that reduces stress concentration in a threaded connection. Table 2.1 shows the measures to reduce axial stress concentration.

④ Balance the load. When the machine is working, it often produces some useless forces, such as inertial force, helical gear axial force, etc. These forces increase

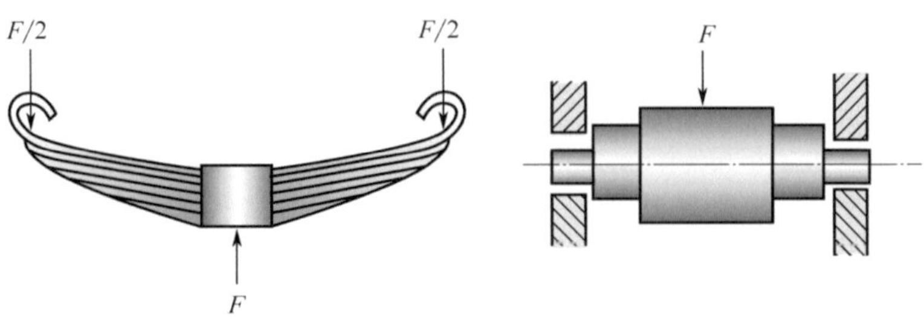

Fig. 2.9 Design of leaf spring and shaft support structure

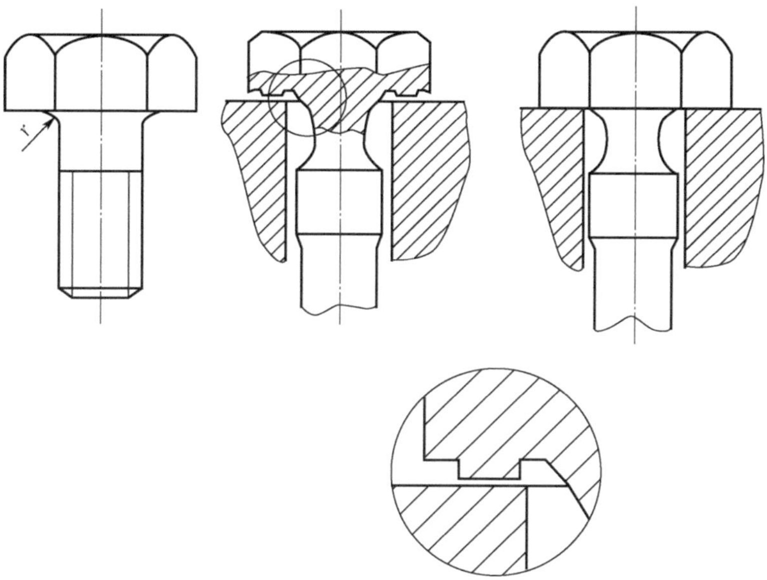

Fig. 2.10 Structure for reducing stress concentration in threaded connections

the load of parts such as shaft and bearing shell, reduce its accuracy and life, and also reduce the transmission efficiency of the machine. The so-called load balance refers to the structural measures taken to balance part or all of the non-force in order to reduce or eliminate its adverse effects. These structural measures mainly include balanced elements, symmetrical arrangement and so on.

(3) Meet structural stiffness

Stiffness refers to the ability of a material or structure to resist elastic deformation when stressed, and represents the difficulty of elastic deformation of a material or component. Component deformation often affects the work of components, for example, excessive deformation of the gear shaft will affect the gear mesh state, and the machine tool deformation will reduce the machining accuracy. The factors affecting the stiffness are elastic modulus and structural form of the material, and the change of structural form has a significant effect on the stiffness. In order to ensure that the mechanical parts normally achieve their functions during the service cycle, they must have sufficient stiffness.

(4) Consider processing technology

The main purpose of structural design of mechanical parts is to make the product achieve the required function and achieve the required performance. Structural design has a great impact on the processing technology, production cost and final product quality of the product parts, therefore, in the structural design should strive to have good processing technology of the product. Machining technology refers to

Table 2.1 Measures to reduce axial stress concentration

Structure name	Diagram	Measures
Rounded comers	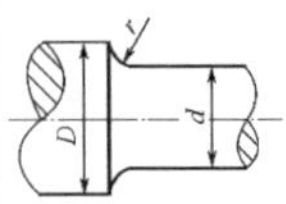	Increase the fillet radius r/d > 0.1 Reduce diameter difference D/d < 1.15 ~ 1.2
	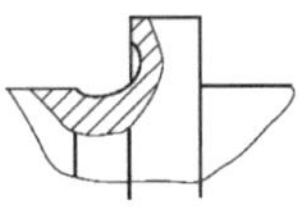	Add concave rounded corners
	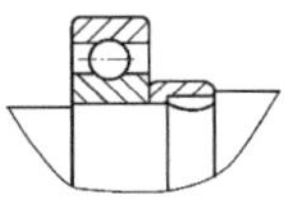	Increase the fillet radius and set a middle ring
	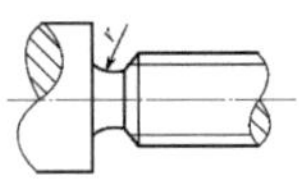	Add and remove rounded corners of the blade
Keyway	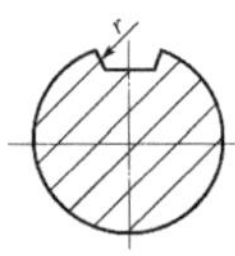	Add rounded corners at the bottom
	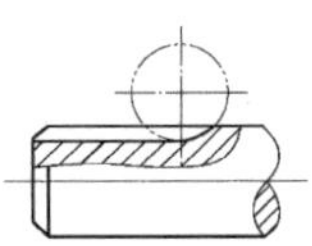	Use a disc milling cutter
Spline	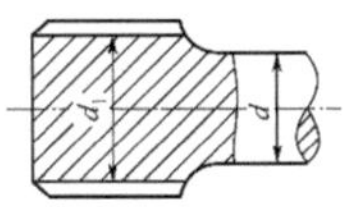	Increase the diameter of the spline $d_1 = (1.1 \sim 1.3)d$
	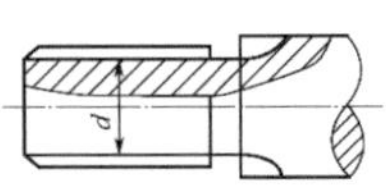	Spline with return groove

the whole process of using machining methods to make the blank into a part with qualified shape, size, relative position and nature according to the pattern and size of the drawing. Conventional processing processes have turning, milling, planning, grinding, fitters, special processing, etc., any kind of processing process has its limitations, may not be applicable to the processing of parts of certain structures or parts of a process of processing, or the production cost is high, or the quality is affected. Therefore, for mechanical designers, it is very important to be familiar with the characteristics and scope of application of conventional processing methods, and to understand the processing capacity of the unit workshop, so that the advantages and disadvantages can be avoided as much as possible when designing the structure. In the actual production, the structural manufacturability of parts is restricted by many factors, such as the size of the production batch will affect the generation method of blank parts; The conditions of the production equipment may limit the size of the workpiece; In addition, the molding, precision, heat treatment, cost and other aspects may have a restricting effect on the manufacturability of the component structure. Therefore, the influence of the above factors on the manufacturability should be fully considered in the structural design.

(5) Design criteria for considering assembly

Products are made up of several parts and components. In accordance with the specified technical requirements and the design of the assembly drawings, a number of parts are connected to the synthesis of parts or a number of parts and components assembly, and after debugging, inspection to make it a qualified product process, called assembly, is an important process in the product manufacturing process. The structure of parts has a direct impact on the quality and cost of assembly. Structural design criteria for assembly are:

① Reasonable division of assembly units. The basic task of assembly is to study how to assemble high-quality products with high efficiency and low cost under certain production conditions. Assembly can be divided into parts and final assembly. The designed machine should be able to be decomposed into several units (parts or components) that can be assembled separately to achieve parallel and professional assembly operations, shorten the assembly cycle, and facilitate step-by-step technical inspection and maintenance, and extend the service life of the product.

② Install the parts correctly. Ensure accurate alignment of parts, avoid double alignment, and prevent assembly errors. Arrange the assembly sequence and process reasonably, minimize manual labor, meet the requirements of assembly cycle, and improve assembly efficiency.

③ Ensure assembly accuracy. Assembly accuracy not only affects the working performance of mechanical parts, but also affects the service life of equipment. The main contents of assembly accuracy are: the mutual position accuracy of each part, the relative motion accuracy of each moving part, the matching accuracy and contact quality of the matching surface. Appropriate measures can be taken to ensure assembly accuracy.

④ Make the parts easy to assemble and disassemble. In the structural design, there should be enough assembly space, such as wrench space; Avoid too long to increase the difficulty of assembly, so that the matching surface scratches, such as some ladder shaft design; In order to facilitate the disassembly of parts, the location of the disassembly tool should be given, such as the disassembly of bearings.

⑤ Try to reduce assembly costs.

(6) Design criteria for considering maintenance and repair

① The configuration of the product should be based on the level of its failure rate, the difficulty of maintenance, the size and quality of the size and installation characteristics of the overall arrangement, where the parts that need to be repaired, the parts that have a high failure rate and need frequent maintenance and the emergency switch should have the best accessibility.
② The disassembly and assembly of products, especially wearing parts, often disassembled parts and additional equipment should be simple, and the parts should be soft when disassembling, and the route is best to be straight or smooth curve.
③ Maintenance points such as inspection points, test points, observation holes, and oil injection holes of the product should be arranged in a position that is convenient for the operator to access.
④ Products that need to be repaired and disassembled should be surrounded by sufficient operating space for the operator.
⑤ Consider easy maintenance. Maintenance should generally be able to see the internal operation, the channel in addition to accommodate the hand or arm of the maintenance personnel, but also should leave appropriate gaps for observation.

(7) Guidelines for considering styling design

The design of the product should not only meet the functional requirements, but also consider the industrial design, improve the aesthetic value of the product shape, and improve the market competitiveness. The social attributes of technical products are commodities. In the era of buyer's market, it is an important design requirement for products to design an appearance that can attract customers. Fashionable appearance with a sense of The Times can quickly lock the consumer group and occupy the market in time. The problems that should be paid attention to in modeling design are:

① The size proportion of the whole machine should be coordinated. In the mechanical structure design, we should fully consider the uniform and coordinated proportional relationship between the dimensions of each part of the outline, and use some popular accepted aesthetic principles as much as possible, such as the "golden section method" to determine the size, so that the product shape is more aesthetic.
② Product appearance color. Color is an important factor in product modeling elements, with the effect of preemption, can seize the attention of consumers at the first time, improve product quality and competitiveness.
③ The shape is simple and uniform. The shape of mechanical products is usually composed of cuboid, cylinder, cone and other basic geometric forms, through the combination of difference, intersection, and so on. When designing the structure, these shapes should be matched appropriately, and the basic shapes should be visually balanced, minimizing changes in shape and position, avoiding excessive clutter, and achieving simplicity and not simplicity.

(8) Design criteria considering cost

Design cost is according to a certain production conditions, according to the product design scheme, through technical analysis and economic analysis, using a certain method to determine the most reasonable processing method under the product expected cost. Product costs, while occurring primarily in the manufacturing phase, are largely dependent on the design phase. The cost waste in the design will cause the "congenital" lack of cost control. Design cost control is the key to cost control. The measures that can be taken to control the cost are:

① Functional decomposition of the product, combining the same or similar functions, removing unnecessary functions, and simplifying product use and maintenance operations as much as possible.
② Under the condition of meeting the specified functional requirements, simplify the structure as much as possible, reduce the number of product levels and component units, and simplify the shape of parts.
③ Design a simple and reliable adjustment mechanism for the product to eliminate common faults caused by wear or drift. For valuable parts that are prone to local loss, they should be designed as adjustable or removable assemblies to facilitate local replacement or repair, and to avoid or reduce repeated adjustments involving each other.
④ Reasonable arrangement of the position of each component, reduce the connection, fixed parts, detection, replacement parts easy to operate, minimize disassembly, movement.
⑤ Preference for standard parts. When designing, we should give priority to standardized equipment, components, parts and tools and other products, and minimize their varieties and specifications.
⑥ Improve the degree of interchangeability and universality.

2.3.5 Mechanical Structure Design Steps

The structural design of machinery is usually to determine the shape, size and layout of the components that complete the given function. The process of structural design is a combination of comprehensive analysis, drawing and calculation, from inside to outside, from important to minor, from local to overall, from rough to fine, weighing advantages and disadvantages, repeated checks, and gradually improving the process. The mechanical structure design process is roughly as follows:

① Clarify the primary and secondary, overall consideration. Define the main tasks and limitations of the structural parts to be designed, and decompose the functions to achieve its purpose into several sub-functions. Then from the realization of the main function of the machine (refers to the basic function of the machine to achieve energy or material conversion) parts, usually start from the structural surface to achieve the function, consider the mutual position and connection relationship with other related parts, and gradually connect with other surfaces into a part, and then connect this part with other parts into a part, and finally combine into a solid now the main function of the machine. Then, determine the secondary, complementary or supporting components of the main components, such as sealing, lubrication and maintenance parts.

② Draw a sketch. Before drawing the sketch, there should be a motion scheme design of the mechanism, on this basis, while analyzing and determining the structure, the main size of the structural parts is roughly estimated, and the sketch is drawn in a certain proportion to determine the structure of the parts. The drawing of the sketch at this stage should indicate the basic shape of the parts, the main size, the limit position of the moving components, the space limitation, the installation size, etc. At the same time, the structural design should pay full attention to the application of standard parts, common parts and universal parts, and serialize as much as possible, that is, "three", in order to reduce the design and manufacturing workload and reduce costs.

③ Comprehensive analysis to determine the structure scheme. The main work of the synthesis process is to find out various alternative structures to achieve the functional purpose of the product, and then analyze, evaluate, discuss, compare, and finally determine the structure. By changing the size, orientation and quantity of the working surface and the component materials, surface characteristics and connection modes in the structure, new schemes can be generated.

④ Calculate and improve the structural design. Carry out load analysis on the structure of the bearing parts, calculate the strength and stiffness of the structural parts under the load, and improve the design according to the calculation results until it meets the requirements, in order to improve the bearing capacity and working accuracy. A very important part of structural design is the requirements of parts assembly and disassembly, materials and processing technology, and the structure should be improved according to the actual situation.

⑤ Improve the structural design. Consider the whole life cycle of the product, constantly improve according to technical, economic and social indicators, find the defects and weak links in the selected program, and improve repeatedly against various requirements and restrictions. Consider the universality and standardization of parts, reduce the variety of parts, and reduce production costs. Note standard parts and purchased parts in the structural sketch. Pay attention to safety and labor protection, and improve the structure.

⑥ Appearance design. Comprehensively consider whether the mechanical appearance is symmetrical and beautiful. Uneven appearance will cause waste of materials or mechanisms, the occurrence of inertial forces will lose balance, a small external interference force may be destabilized, anti-stress concentration and anti-fatigue performance will be weakened. The appearance design should be completed or participated in by designers with industrial design background. In addition, market research can be done when necessary to understand the psychological needs and positioning of potential customers, so as to guide the design of products.

2.4 Design of Feed Transmission System for Intelligent Manufacturing Equipment

2.4.1 Functional Requirements for the Feed Transmission System of Intelligent Manufacturing Equipment

(1) Meet sports requirements

The feed transmission system needs to realize the transformation of the motion form and law of the actuators and the motion distribution function of different actuators, so that the actuators can meet the working requirements under different working environments. The most important thing is that the feed transmission system needs to achieve the variable speed function of the actuator, and achieve the function of raising and lowering the speed from the power source to the actuator. The system should have good response characteristics, low speed feed or micro feed does not crawl, high motion sensitivity.

(2) Meet the power requirements

The feed transmission system should have high transmission efficiency to realize the power and torque conversion from the power source to the actuator; With a wide enough speed range, it can transmit large torque to meet the needs of different working conditions.

(3) Meet performance requirements

The actuator in the feed transmission system needs to have sufficient strength, stiffness and precision, stiffness including dynamic stiffness and static stiffness, and the processing and assembly technology is good. If the transmission and the actuator are concentrated in a box, the vibration generated during the operation of the transmission will directly affect the smoothness of the operation of the actuator, and the heat generated by the transmission will also cause the thermal deformation of the actuator, affecting the processing accuracy. Therefore, the actuator should have both good shock resistance and small thermal deformation characteristics.

(4) Meet economic requirements

On the premise that the feed transmission system meets the working requirements, the number of transmission parts should be minimized to make its structure compact, reduce efficiency loss, save materials and reduce costs.

At present, the transmission devices widely used in the equipment mainly include ball screw nut pair, static pressure worm gear pair, double lead worm and so on. Taking CNC machine tool as an example, compared with the traditional feed transmission system, CNC machine tool feed system, each movement is driven by a separate servo motor, the transmission chain is greatly shortened, the transmission system compared with the traditional feed system adopts a large number of gear transmission form, not only greatly reduces the number of transmission parts, but also simplifies the structure, but also reduces the transmission error, to ensure the transmission precision degree.

2.4.2 Composition of Feed Transmission System for Intelligent Manufacturing Equipment

(1) Variable speed device

Transmission, also known as transmission, is the most common transmission device used to change the output speed and torque of the prime mover to adapt to the working requirements of the execution system, which can fix or shift the output shaft and input shaft transmission ratio, composed of variable speed transmission mechanism and control mechanism. The common transmission modes are gear train transmission, belt transmission, clutch transmission, meshing gear transmission and so on. The transmission device should meet the requirements of the transmission range and series, with high transmission efficiency and enough power or torque, simple structure, light weight and good process, lubrication and sealing.

(2) Start stop and reversing device

Start-stop and reversing are the most basic functions of the traditional feed system. The start-stop and reversing device is used to control the start-stop of the actuator and the change of the direction of motion. The commonly used start-stop and reversing devices in the equipment are generally divided into three cases: infrequent start-stop and no reversing (automatic machinery), infrequent reversing (lifting machinery), frequent start-stop and reversing (general machine tool), and the common reversing ways are engine reversing, gear-clutch reversing, sliding gear reversing and so on. The start-stop and reversing device should meet the requirements of simple structure, easy operation, safe and reliable, and able to transmit sufficient power.

(3) Feed motion device

The function of the feed motion device is to equip a moving part with linear or circumferential feed, which is also one of the most basic functions of the traditional feed system. The feed motion device is mainly composed of a ball screw nut pair and a guide rail. In some modern intelligent manufacturing equipment, such as high-speed cutting machine tools, feed transmission devices such as electric spindle are widely used. In terms of linear motion devices, linear motors have also been widely used.

(4) Braking device

The device is a device that makes the actuator stop quickly from the motion state, which is generally used for the transmission system with frequent start and stop, large inertia of the moving component or high movement speed, and can also be used for emergency stopping in the event of a safety accident or emergency. There are two commonly used braking modes: motor braking and brake braking. The motor braking has the advantages of simple structure, convenient operation and rapid braking, but the inertia impact of the transmission is large. Brake braking is usually used in the transmission system with frequent start, long transmission chain, transmission inertia and high transmission power. The brake device should have the characteristics of simple structure, easy operation, high wear resistance, easy heat dissipation and smooth and rapid braking.

(5) Safety protection device

The safety protection device is a device that plays a safety protection role in the transmission system to avoid damage to the parts due to overload. Common safety protection devices include pin type safety coupling, steel ball type safety clutch, friction type safety clutch, etc. The transmission parts should have protective devices such as shells, and should not be exposed to the environment to avoid personal injury to the operator. Equipment should be designed with an emergency stop device, emergency power failure in the event of an accident.

2.4.3 *Analysis and Calculation of Transmission System for 3 Intelligent Manufacturing Equipment*

(1) Transmission system diagram

In order to facilitate the analysis of the motion and transmission of the mechanical system, the designer usually needs to draw the transmission system diagram, which is a schematic diagram of the whole motion transmission relationship of the mechanical system. The transmission system diagram should be drawn as far as possible on a projector surface that can reflect the mutual position of the transmission parts, and the transmission parts should be drawn in the form of an expansion diagram in accordance with the transmission sequence. The transmission diagram can only represent the transmission relationship, and does not represent the actual size and spatial position of each component. The figure usually needs to mark the number of teeth of the gear (worm gear), the diameter of the belt wheel, the lead and the number of the lead screw, the speed of the motor, the number of the drive shaft and other information. Figure 2.11 shows the transmission system of a machine tool.

(2) Determination of gear tooth number

When the transmission ratio of each transmission group is determined, the number of gear teeth and pulley diameter can be determined. Generally speaking, the number of teeth of the gear is positively correlated with the center distance between the gears, and the center distance depends on the transmitted torque, the main transmission is a deceleration transmission, the more the transmission torque of the transmission group, so the center distance is also larger. However, the number of teeth and should

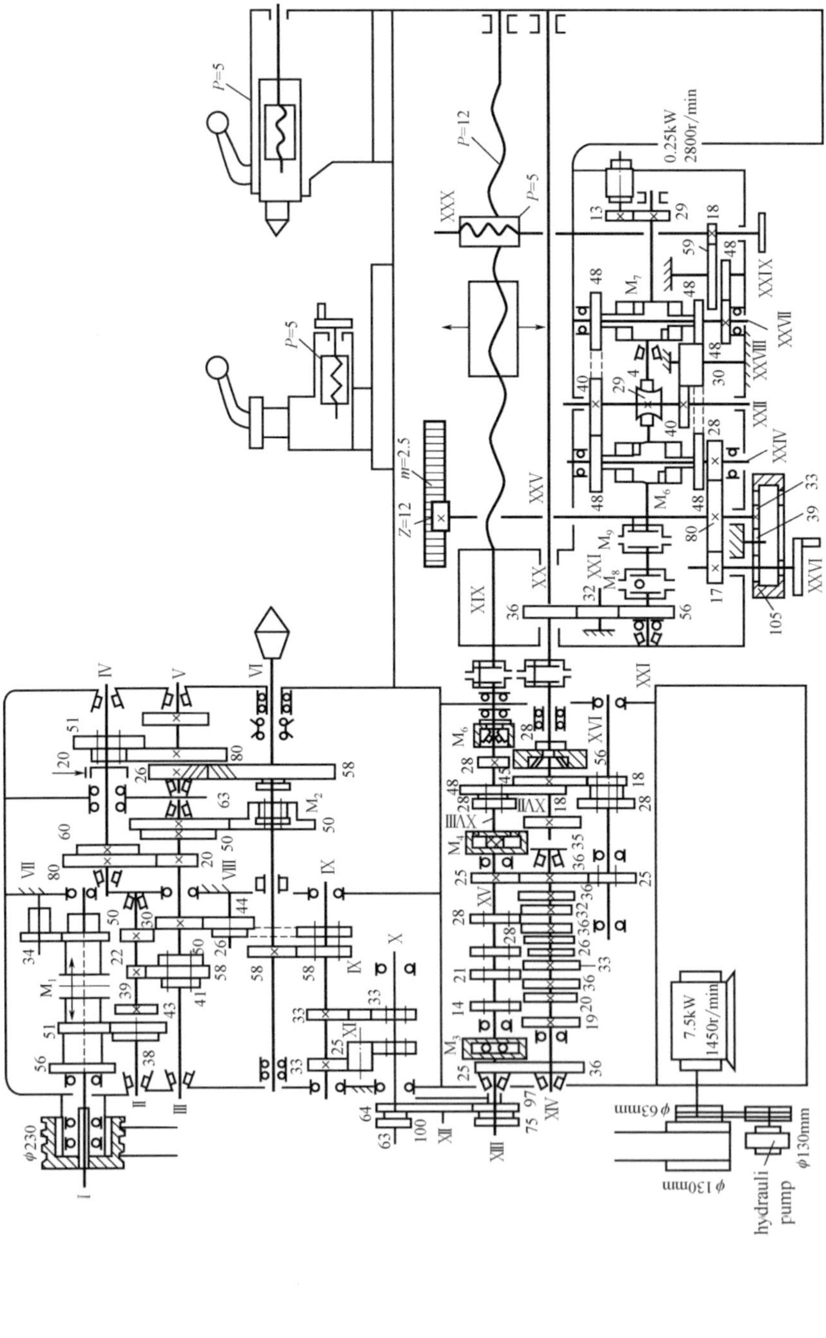

Fig. 2.11 A certain machine tool transmission system

not be too large, generally recommended Sz ≤ 100 ~ 120. The number of teeth should not be too small, the number of teeth of the minimum gear should be as small as possible, but to meet the minimum number of teeth conditions that do not occur root cutting, and to ensure that the main transmission has a better smoothness of motion. Each gear with the minimum number of teeth limited by the gear structure should be able to be reliably installed, and the wall thickness of the gear groove to the hole wall or keyway is a ≥ 2m (m is the modulus) to ensure sufficient strength to avoid deformation and fracture. Minimum number of teeth Zmin ≥ 6.5 + D/m (D is the large diameter of the gear spline hole; m is the gear module). The minimum center distance between the two axes should be appropriate, if the number of teeth and Sz is too small, it will cause the distance between the bearings and other structures of the two axes to be too close or collision. In addition, when distributing the transmission ratio, lubrication and other conditions should be considered.

The transmission ratio can be expressed as

$$i = \frac{n_1}{n_0} = \frac{Z_1}{Z_0} = \frac{d_1}{d_0}$$

In the formula, n_0 and n_1 are the rotation speed of active and passive axes respectively; Z_0 and Z_1 are respectively the number of teeth of active and passive gears (sprockets); d_0 and d_1 are active and passive pulley diameters respectively.

2.4.4 Design of Transmission System Structure for Intelligent Manufacturing Equipment

(1) Determination of transmission route

Transmission system transmission routes can generally be divided into four categories: series single transmission, parallel shunt transmission, parallel mixed transmission, and hybrid transmission, as shown in Table 2.2.

Table 2.2 Transmission route

Series single flow transmission	Parallel split transmission	Parallel mixed flow transmission	Hybrid transmission
	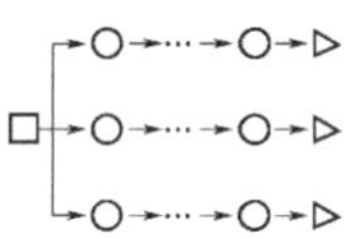	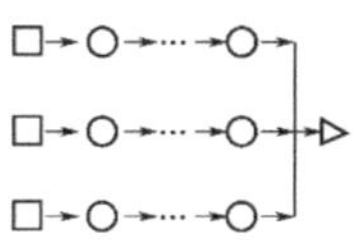	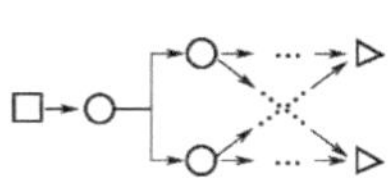

(2) Arrangement of transmission sequence

When both helical gear and spur gear drive exist, the helical gear should be placed in the high speed level; When both bevel gear and cylindrical gear drive exist, the bevel gear should be placed in the high-speed level; When both closed and open gear drives exist, the closed gear drive should be placed at the high speed level; The chain drive is placed in the low speed stage of the transmission system; The belt drive should be placed in the high-speed level of the transmission system; The transmission or mechanism that changes the form of motion, such as rack and pinion drive, screw drive, connecting rod mechanism and CAM mechanism, is generally placed at the end of the transmission chain, so that it is close to the actuator; When both the variable speed transmission and the constant transmission ratio transmission exist, the variable speed transmission should be placed in the high speed stage. When the transmission ratio is assigned, it should generally not exceed the recommended transmission ratio of various drives.

When distributing the transmission ratio, attention should be paid to the coordination of the size of each transmission component and the symmetry of the structure to avoid interference; For multi-stage deceleration drive, the transmission ratio can be distributed according to the principle of "small front and large back", and the difference between the two adjacent stages should not be too large; In multistage gear transmission, the low speed transmission ratio is relatively small, which is conducive to reducing the profile size and overall mass.

2.4.5 *Design of Rolling Guide Rail Pairs*

Guide rail is an important link of the feed system and one of the basic structural elements of intelligent manufacturing equipment. The accuracy and service life of intelligent manufacturing equipment largely depend on the quality of the guide rail [6].

In the work of the feed system, the rolling guide pair directly participates in and completes the work of determining the relative position of the workpiece and the tool, and the accuracy of the rolling guide pair directly determines the working accuracy of the whole feed transmission system. The rolling guide rail is composed of standard guide rail block, easy assembly and disassembly, simple lubrication, small difference in dynamic and static friction factor, light and flexible movement. Because the relative movement of the ball between the guide rail and the slider is rolling, the friction loss can be reduced, and the rolling friction coefficient is about 2% of the sliding friction coefficient, so the transmission mechanism using the rolling guide rail is far better than the traditional sliding guide rail.

(1) Characteristics of rolling guide pair

① High positioning accuracy. The movement of the rolling linear guide rail is realized by the steel ball rolling, the friction resistance of the guide rail pair is small, the difference between static and static friction resistance is small, and it is not easy to crawl at low speed. High repeated positioning accuracy, suitable for frequent starting or reversing moving parts. The positioning accuracy of the machine can be set to the level of ultrafine meters. The increase of preload during assembly can not only achieve smooth movement, improve transmission accuracy, but also reduce the impact and vibration of movement.

② Small friction and wear. For fluid lubrication of the sliding guide surface, direct friction caused by metal contact is unavoidable because the fluid lubrication is limited to the boundary area. Because the friction energy consumption of rolling contact is small, the friction loss of rolling surface is also reduced, so the rolling linear guide system can be in a high precision state for a long time. The lubricating oil or grease required for rolling surface lubrication is also less, simplifying the lubrication system design and easy maintenance.

③ Suitable for high-speed movement, green energy saving. Due to the small friction resistance, the required driving torque is reduced, and the power source and power transmission mechanism are miniaturized, the energy consumption of the machine tool is greatly reduced, and the energy saving effect is obvious. Because the rolling guide rail is suitable for high-speed movement, the working efficiency of the machine tool is increased by 20–30%.

④ Strong carrying capacity. The rolling linear guide pair has good load bearing performance and good load adaptability, and can withstand the force and moment load in different directions, as well as the jolting moment, shaking moment and swinging moment during the working process. By preloading, damping can be increased to improve vibration resistance and high-frequency vibration can be eliminated.

⑤ Simple installation, good interchangeability. The traditional sliding guide rail must be scraped on the surface of the guide rail, which is time-consuming and laborious, and needs to be scraped repeatedly when the accuracy of the machine tool is reduced. The requirements of scraping are very high, and there are fewer and fewer experienced senior fitters engaged in scraping work. The rolling guide rail is interchangeable, and the machine can regain high precision as long as the slide or guide rail or the entire rolling guide pair is replaced.

(2) Design and calculation of rolling guide pairs [7]

Figure 2.12 shows the structure of a rolling guide rail. Figure 2.13 shows the components of a rolling guide rail.

1) Weight estimation of the table

First determine the effective stroke of the table installed on the guide rail pair, and estimate the load weight WX and WY of the X- and Y-direction tables. The center distance of the supporting steel ball of the X-direction guide rail is x_0, and the center distance of the supporting steel ball of the Y-direction guide rail is y_0. The length of both guide rails is L and the height is h.

X-axis pallet size: $S_X = Lx_0h$;

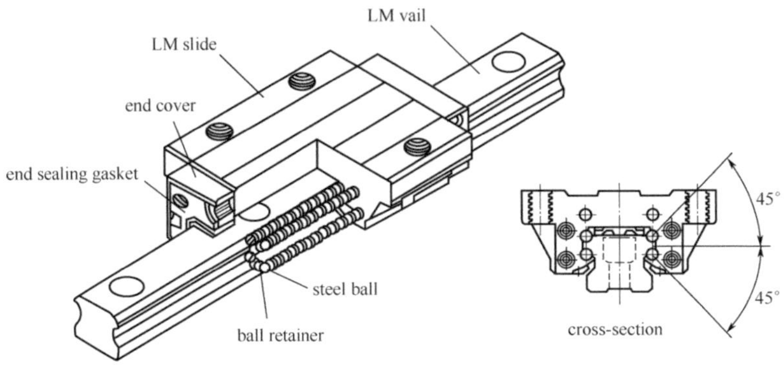

Fig. 2.12 Structure of rolling rail

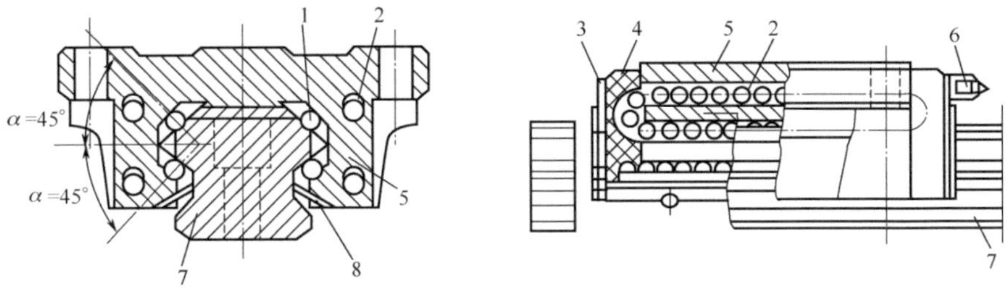

Fig. 2.13 Rolling rail component, 1 rolling element; 2 ball return hole; 3, 8 sealing gasket; 4 baffle plate; 5 sliding plate; 6 lubrication injection hole; 7 guide rail

The weight of the *X*-axis pallet is: $Wx = Sx \times$ relative density of the material, in N;

Y-axis pallet size: $S_Y = Ly_0h$;

The weight of the *Y*-axis pallet is: $Wy = Sy \times$ relative density of the material, in N;

The total weight *W* of the moving part of the workbench is:

$$W = Wx + W + \text{weight of upper guide rail (including motor)} + \text{weight of fixture and workpiece} \tag{2.1}$$

2) Design and calculation of rolling guide pair

Calculate the static safety factor of the guide according to the given working load Fz and the estimated Wx and W_Y:

$$f_{sL} = C_0/P \tag{2.2}$$

In the formula, C_0 is the basic rated static load of the guide rail, in kN; P is the working load, $P = 0.5\ (Fz + W)$.

Select the corresponding f_{sL}; according to the working conditions: $f_{sL} = 1.0 \sim 3.0$ is the general operating condition; $f_{sL} = 3.0 \sim 5.0$ is the impact and vibration condition during movement. The corresponding static safety factor can be selected according to the design requirements for the next step of design calculation.

$$C_0 = f_{\mathrm{sL}} P_{\mathrm{X,Y}} \tag{2.3}$$

$$P_{\mathrm{X,Y}} = 0.5\left(F_{\mathrm{Z}} + W_{\mathrm{X,Y}}\right) \tag{2.4}$$

$$P_{\mathrm{X}} = 0.5(F_{\mathrm{Z}} + W_{\mathrm{X}}),\ \ C_{0\mathrm{X}} = f_{\mathrm{sL}} P_{\mathrm{X}} \tag{2.5}$$

$$P_{\mathrm{Y}} = 0.5(F_{\mathrm{Z}} + W_{\mathrm{Y}}),\ \ C_{0\mathrm{Y}} = f_{\mathrm{sL}} P_{\mathrm{Y}} \tag{2.6}$$

where f_{sL} is the static safety factor of the guide rail; C_0 is the basic rated static load of the guide rail; C_{0X}, C_{0Y} are the static loads of the X and Y axes respectively; P_X is the working load of the X axis; P_Y is the working load of the Y axis; $P_{X,Y}$ is the working load of the X and Y directions; $W_{X,Y}$ is the load-bearing weight of the X and Y directions.

Calculate the service life of the guide rail based on the operating speed (m/min) and the basic rated dynamic load C_0 (kN) of the initially selected guide rail.

Rated stroke length life:

$$T_{\mathrm{s}} = K(f_{\mathrm{H}} f_{\mathrm{T}} f_{\mathrm{C}} / f_{\mathrm{W}} \times C_{\mathrm{a}} / F)^3 \tag{2.7}$$

In the formula, f_H is the hardness coefficient; f_W is the load coefficient; f_T is the temperature coefficient; f_C is the contact coefficient; C_0 is the basic rated dynamic load of the guide rail; F is the calculated load, $F = F_z/M$, F_z is the working load, M is the number of slides; K is the guide rail parameter.

Rated working life of guide rail:

$$T_{\mathrm{H}} = 10^3 T_{\mathrm{s}} / (2 l_{\mathrm{s}} n) \tag{2.8}$$

In the formula, T_H is the rated working time life of the guide rail; T_s is the rated stroke length life; l_s is the stroke length; n is the number of round trips per minute. l_s and n are the basic parameters of the pre-selected guide rail. If T_H is longer than the design required life, the pre-selected guide rail meets the design requirements and can be selected according to the design plan.

2.4.6 *Design of Ball Screw*

Ball screw is widely used in moving parts of modern CNC machine tools and high-performance intelligent manufacturing equipment. Ball screw is the ideal product to realize the mutual conversion of rotating motion and linear motion. It is composed of screw, nut and ball, which is the further extension and development of ball screws, and has the characteristics of high precision, reversibility and high efficiency. Ball screw is to change the bearing from rolling action to sliding action. Because of its small friction resistance, ball screws are widely used in various industrial equipment and precision instruments. Figure 2.14 shows the structure and physical object of the ball screw.

2.4.6.1 Ball Screw Characteristics

① High transmission efficiency. The transmission efficiency of the ball screw transmission system is as high as 90 ~ 98%, which is traditional. The sliding screw system is 2–4 times larger, so it can obtain larger thrust with smaller torque. Inversion, which can change from linear motion to rotational motion.

② Smooth movement. The ball screw transmission system is a point contact rolling motion with low frictional resistance during operationHigh sensitivity, no shaking during startup, and no crawling phenomenon at low speeds, therefore precise control of micro feed can be achieved.

③ High precision. The ball screw transmission system has a small temperature rise during movement and can be pre tightened to eliminate axial clearance and impact. The lead screw is pre stretched to compensate for thermal elongation, thus achieving high positioning accuracy and repeated positioning accuracy.

④ High durability. The rolling contact area of the ball screw rolling element and steel ball is hardened (58–63HRC). Through precision grinding, the circulation

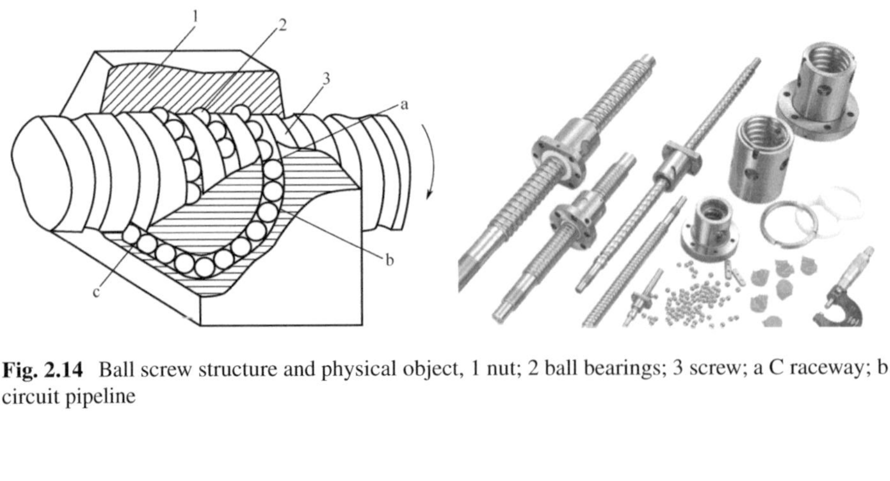

Fig. 2.14 Ball screw structure and physical object, 1 nut; 2 ball bearings; 3 screw; a C raceway; b circuit pipeline

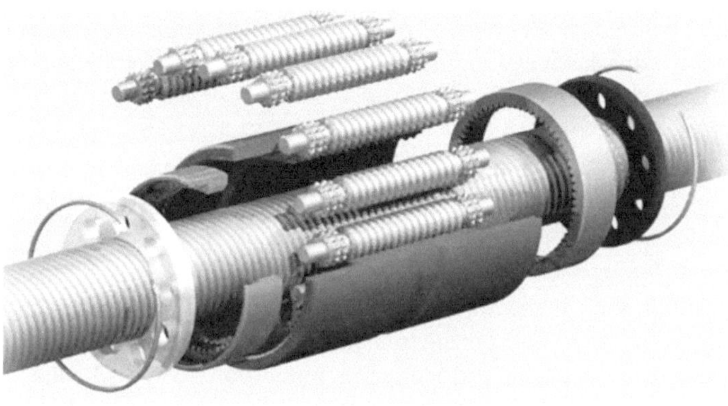

Fig. 2.15 Planetary roller screw

system process is purely rolling and has minimal relative wear, making it highly useful lifetime and accuracy retention.

⑤ Good synchronization. Due to the smooth movement, sensitive response, no obstruction, and no slip of the ball screw, it is recommended to use a few. By using the same ball screw transmission system to simultaneously drive several identical components or devices, good results can be achieved step effect.

⑥ High reliability. Compared to other transmission mechanisms, the failure rate of the ball screw transmission system is very low, and maintenance is guaranteed. Maintenance is also relatively simple, requiring only general lubrication and dust prevention, and can work in a non lubricated state for special occasions

⑦ High stiffness. The ball screw transmission system adopts a Goethe style groove shape to achieve optimal steel ball and groove performance. Contact for easy operation, by pre tightening, the ball has better stiffness, reducing the gap between the ball and the nut and screw. Elastic deformation improves the stiffness of the transmission system. Another relatively novel transmission form is the planetary roller screw (Fig. 2.15). Planetary roller screw and rolling. The structure of the ball screw is similar, but the difference is that the load transfer component of the planetary roller screw is a threaded roller, which is typical. Line contact; The load transfer component of the ball screw is a ball, which is a point contact. The main advantages of planetary roller screws. There are numerous contact points to support the load. Replacing the ball with a threaded roller will allow the load to pass quickly through numerous contact points. Release, thus enabling higher impact resistance.

2.4.6.2 Design and Calculation of Horizontal and Vertical Ball Screws

Firstly, the weight of the workbench W1, the maximum weight of the workpiece and fixture W2, the maximum stroke of the workbench LK, the friction coefficient μ of the guide rail of the workbench, the rapid feed speed vmax, the positioning accuracy, the repeated positioning accuracy, and the required life time are determined. And the coefficient values such as longitudinal cutting force Fa, speed v and time ratio q determined by the cutting mode.

(1) Determine the lead of the ball screw pair

$$P_{\mathrm{h}} = \frac{V_{\max}}{i n_{\max}} \tag{2.9}$$

In the formula, V_{max} is the maximum moving speed of the worktable; n_{max} is the maximum speed of the motor; i is the transmission ratio. Substitute V_{max}, i, and n_{max} to get P_h, and then check the "Modern Machine Tool Design Manual" to get the standard P_h.

(2) Determine the equivalent speed and equivalent load

Screw speed under various cutting methods

$$n_i = \frac{v_i}{P_h} \tag{2.10}$$

where, v_1 is the feed speed under strong cutting; v_2 is the feed speed under general cutting; v_3 is the feed speed under fast feed for fine cutting, v_4 is the feed speed for rapid feed and the standard, P_h value obtained above is substituted into the feed speed n_1, n_2, n_3, n_4.

Under various cutting methods, the axial load of the screw is

$$F_i = P_{\mathrm{x}i} + (w_1 + w_2 + P_{\mathrm{z}i})/10 \tag{2.11}$$

In the formula, F_i is the axial load of the screw; P_{xi} is the longitudinal cutting force; P_{zi} is the vertical cutting force. W_1, W_2, P_{xi}, and P_{zi} are known, and F_i can be obtained under various cutting methods by substituting them.

Equivalent speed n_m (r/min) is calculated from formula (2.12):

$$n_{\mathrm{m}} = n_1 \frac{t_1}{100} + n_2 \frac{t_2}{100} + \cdots + n_i \frac{t_i}{100} + n_n \frac{t_n}{100} \tag{2.12}$$

where t_i is the percentage of working time determined by the cutting method, and n_i is the percentage of working time determined by the cutting method, which has been obtained above. Substitute it into the formula to obtain n_m.

The equivalent load F_m is calculated by formula (2.13):

$$F_{\mathrm{m}} = \sqrt[3]{F_1^3 \times \frac{n_1 t_1}{100 n_{\mathrm{m}}} + F_2^3 \times \frac{n_2 t_2}{100 n_{\mathrm{m}}} + F_3^3 \times \frac{n_3 t_3}{100 n_{\mathrm{m}}} + F_4^3 \times \frac{n_4 t_4}{100 n_{\mathrm{m}}}} \tag{2.13}$$

(3) Preliminary selection of ball screw pair

According to the "Modern Machine Tool Design Manual", the required life of the ball screw pair is

$$L_{\mathrm{h}} = \frac{10^6}{60n_{\mathrm{m}}} \times \left(\frac{c_{\mathrm{a}}}{F_{\mathrm{m}}} \times \frac{f_t f_{\mathrm{h}} f_{\mathrm{a}} f_{\mathrm{k}}}{f_{\mathrm{w}}}\right)^3 \tag{2.14}$$

Check the "Modern Machine Tool Design Manual" to get $f_t\, f_{hr}\, f_a\, f_k\, f_w$, and L_h is the required life span. Substituting the data into the data, we can get C_a.

(4) Determine the minimum allowable thread bottom diameter

1) Calculate the maximum allowable axial deformation of the screw.

① (1/4 ~ 1/3) Repeat positioning accuracy;
② (1/5 ~ 1/4) Positioning accuracy.

δm (μm) is the maximum axial deformation. If the repeatability and positioning accuracy are known, the values of cases ① and ② can be calculated and the minimum value of the two results can be taken.

2) Estimate the minimum thread bottom diameter.

The screw requires pre-stretching and the support form is fixed at both ends

$$d_{2\mathrm{m}} = 0.039\sqrt{\frac{F_0 L}{\delta_{\mathrm{m}}}} \tag{2.15}$$

where d_{2m} is the minimum thread bottom diameter, mm.

$$L = (1.1 \sim 1.2)\text{travel} + (10 \sim 14)P_h \tag{2.16}$$

Static friction

$$F_0 = \mu_0 W_1 \tag{2.17}$$

The stroke, W_1, and μ_0 are known. Substituting the data, we get L, F_0, and d_{2m}.

(5) Determine the specification code of the ball screw pair

Select the internal circulation floating flange and the straight nut type gasket preload form. According to the calculated P_h, c_a, d_{2m}, select the ball screw pair of the corresponding specifications in the "Modern Machine Tool Design Manual".

(6) Determine the preload force of the ball screw pair

$$F_{\mathrm{P}} = \frac{1}{3}F_{\max} \tag{2.18}$$

In the formula, F_{max} is the maximum value of the screw axial load in various cutting methods considered, and F_P can be calculated by substituting it into the formula.

(7) Stroke compensation value and tensile force

1) Stroke compensation value

$$C = 11.8 \times 10^{-3} \Delta t l_{\mathrm{u}} \tag{2.19}$$

In the formula, $l_u = L_k + L_n + 2L_a$. Check the "Modern Machine Tool Design Manual" and take L_k, L_n, L_a, Δt respectively, and substitute them to get C (μm),

2) Pre-stretching force

$$F_{\mathrm{t}} = 1.95 \Delta t d_2^2 \tag{2.20}$$

In the formula, d_2 is the bottom diameter of the screw, and the values of t and d_2 can be obtained by looking up the "Modern Machine Tool Design Manual".

(8) Determine the bearing code and specifications for the ball screw pair support

1) The maximum axial load that the bearing can bear.

$$F_{\mathrm{Bmax}} = F_{\mathrm{t}} + F_{\mathrm{max}} \tag{2.21}$$

2) Bearing type: For the support form with fixed ends, choose back-to-back 60° angular contact thrust ball bearings.
3) The inner diameter d of the bearing should be slightly smaller than d_2,

$$F_{\mathrm{BP}} = \frac{1}{3} F_{\mathrm{Bmax}} \tag{2.22}$$

4) Bearing preload: preload load $\geq F_{BP}$, select bearing model specifications according to the "Modern Machine Tool Design Manual".

(9) Design of ball screw working diagram

1) Screw thread length,

$$L_{\mathrm{s}} = L_{\mathrm{u}} + 2L_{\mathrm{e}} \tag{2.23}$$

From the table, we can find the remaining length L_e.

2) Distance between two fixed supports L_1 lead screw L.
3) The distance between the starting point of the stroke and the fixed support is L_0.

(10) Transmission system stiffness

1) Screw compressive stiffness.

Minimum compressive stiffness of screw

$$k_{\text{smin}} = 6.6d_2^2/(100L_1) \tag{2.24}$$

In the formula, d_2 is the bottom diameter of the screw; L_1 is the fixed support distance. Substituting the data into k_{min} (N/μm).

Maximum compressive stiffness of screw

$$k_{\text{smax}} = 6.6d_2^2 L_1/400L_0(L_1 - L_0) \tag{2.25}$$

In the formula, L_0 is the distance from the starting point of the trip to the fixed support. Substituting the data into k_{max} (N/μm).

2) Support bearing combined stiffness,

Combined stiffness of a pair of preloaded bearings

$$K_{\text{B0}} = 2 \times 2.34 \times \sqrt[3]{d_{\text{Q}} z^2 F_{\text{amax}} \sin^5 \beta} \tag{2.26}$$

where d_Q is the ball diameter, mm; z is the number of balls; F_{amax} is the maximum axial working load, N; β is the bearing contact angle.

The bearing number, the relationship between F_{amax} and preload, and the values of k_{amax} and K_{B0} are obtained from the "Modern Machine Tool Design Manual".

From $k_b = 2K_{B0}$, the combined stiffness of the support and bearing k_b (N/μm) can be calculated.

The contact stiffness of the ball and raceway of the ball screw pair is

$$k_{\text{e}} = k'_{\text{e}} \left(\frac{F_{\text{P}}}{0.1C_{\text{a}}} \right)^{\frac{1}{3}} \tag{2.27}$$

In the formula, k_e is the stiffness in the "Modern Machine Tool Design Manual"; c_a is the ball screw pair parameter selected in the "Modern Machine Tool Design Manual"; F_p is the preload force of the ball screw pair, and substituting the data into k, we get k_e (N/μm).

(11) Stiffness verification and accuracy selection

From the "Modern Machine Tool Design Manual", we can find the bearing ball diameter d_Q (mm), the number of balls z and the bearing contact angle β.

① From the formula

$$\frac{1}{k_{\min}} = \frac{1}{k_{\text{smin}}} + \frac{1}{k_{\text{b}}} + \frac{1}{k_{\text{e}}} \tag{2.28}$$

$$\frac{1}{k_{\max}} = \frac{1}{k_{\text{smax}}} + \frac{1}{k_{\text{b}}} + \frac{1}{k_{\text{e}}} \tag{2.29}$$

Substitute the previously calculated data k_{min} and k_{max} into the formula F_0=$\mu_0 W_1$ to obtain Static friction. In the formula, u_0 is the static friction coefficient; W_1 is the positive pressure.

② Check the stiffness of the transmission system and obtain it by the formula $k'_{min} = 1.6F_0$/reverse difference. It is known that the reverse difference is the same as the value of the repeated positioning accuracy. Compare k_{min} and k'_{min}. Generally, the former is larger than the latter. If not, check whether the previous calculation design is wrong.

③ Positioning error caused by the change of transmission system stiffness

$$\delta_{\text{k}} = F_0\left(\frac{1}{k_{\min}} - \frac{1}{k_{\max}}\right) \tag{2.30}$$

Substituting the previous numerical values into the calculation, we get δ_k.

④ Determine the accuracy.

For the system:

$$V_{300p} \leq 0.8 \times \text{positioning accuracy} - \delta_k \tag{2.31}$$

where V_{300p} is the travel variation within any 300 mm. The positioning accuracy is known. The screw accuracy grade is taken according to the V_{300p} range and the V_{300p} standard value is determined.

⑤ Determine the specification code of the ball screw pair, and determine the relevant parameters of the ball screw pair based on the design calculations above, including model, nominal diameter, lead, thread length, screw length, P-class grade accuracy and selected specification model.

(12) Calculate critical compression load

The maximum axial load F_{max} on the screw is less than the screw pre-tension force F, which means that the critical compression load has met the design requirements and does not need to be verified.

Check critical speed

$$n_{\text{c}} = f\frac{d_2}{L_{\text{e2}}^2} \times 10^7 \tag{2.32}$$

In the formula, n_c is the critical speed, r/min; f is the coefficient related to the support form; d_2 is the bottom diameter of the screw; L_{e2} is the critical speed calculation length, mm. According to the "Modern Machine Tool Design Manual", f, d_z $L_{e2} = L_1 - L_2$. Calculated n_c.

2.5 Design of Intelligent Manufacturing Equipment Support System

The supporting system is a general term for the parts that support and connect in the mechanical system, and can maintain the mutual position relationship between the parts that are supported. Taking machine tools as an example, the supporting system is usually composed of base, column, box, workbench, lifting platform and other basic parts. The static stiffness, dynamic characteristics, thermal characteristics and internal stress should be considered when designing the support system.

The supporting system is composed of supporting parts, and the commonly used supporting parts are usually divided into two categories: cast supporting parts and welded supporting parts. Casting technology can make the supporting parts have a complex shape and inner cavity, with good shock resistance and wear resistance, but the manufacturing process is complex, requires aging treatment, and the production cycle is long. The production of small and medium-sized supporting parts usually uses casting technology, mass production. Using welding technology, the billet can be assembled step by step, which is suitable for manufacturing large and complex supporting parts. The forming process of welded support parts is simple, easy to modify, and usually light in weight. When designing the supporting system, the processing technology and production cost of the supporting parts should be considered under the premise of meeting the working requirements, and the two types of supporting parts should be used reasonably.

It is worth pointing out that the supporting system of modern intelligent manufacturing equipment is increasingly using natural granite, artificial granite and other materials. These materials have better stability and are especially suitable for supporting materials for precision intelligent manufacturing equipment.

2.5.1 *Precautions for Designing Support Systems*

(1) Strength and stiffness

The supporting system is a device that supports and connects all parts of the mechanical system. The deformation of the supporting system will cause the position error of the actuator and affect the normal operation of the equipment. The design should ensure sufficient strength and stiffness.

The static stiffness of supporting parts includes self-stiffness, local stiffness and contact stiffness. The correct design of the section shape of the supporting parts has an important effect on improving the static stiffness of the supporting parts. The moment of inertia of the hollow section is greater than that of the solid section, the square section is more effective against the bending moment, the circular section is more effective against the torque, and the rectangular section has better resistance to the bending moment. The stiffness of the closed section is greater than that of the unclosed section. Reasonable setting of rib and rib can improve the static stiffness of supporting parts.

(2) Dynamic performance

The dynamic performance of the support system mainly refers to natural frequency, vibration shape and damping. In order to make the supporting system have good seismic performance and ensure the smooth operation of the actuator, it is necessary to make the supporting parts have large dynamic stiffness, damping and natural frequency not the same or similar to the excitation frequency, so as to improve the dynamic performance of the supporting system.

(3) Thermal stability

Qualitativeness has great influence on equipment accuracy. When the equipment is working, part of the energy input by the prime mover will be converted into heat, which will heat up the equipment parts, produce uneven thermal deformation, affect the original position relationship of the parts, and make the actuator produce a large error. The supporting system needs to reasonably heat dissipation and heat insulation, or disperse the heat of a certain part of the body to the whole, reduce the impact on a certain point, prevent deformation, and maintain heat soaking.

(4) Processability

When designing the support system, the convenience of processing and assembling the support parts should be considered. Many supporting parts have complex structure, large size, and are not convenient for processing, assembly and transportation, so it is necessary to fully consider whether their manufacturability is reasonable when designing. For large support systems, special design is required.

2.5.2 Design of Support Components

During the design of the supporting parts, the force analysis is carried out first, which is the basis and basis of the structural design of the supporting parts. According to the working force of the actuator and the weight of the machine itself, the stress state of the supporting parts is analyzed to provide a basis for determining its structure and size. According to the force to determine the structure and size, reasonable selection of supporting parts of the cross section shape, to determine its structure and the corresponding size. On the basis of determining the structure and size, the static and dynamic performance of the structure is analyzed. For the determined supporting parts, its three-dimensional model can be drawn, and the static or dynamic performance of the supporting parts can be simulated and optimized by finite element simulation analysis. Finally, the evaluation and modification of the scheme are carried out, and the feasibility of the application of the supporting parts is analyzed, so as to modify and improve the design scheme and determine the final form.

(1) Stiffness calculation of support system

Stiffness is the ability of a material or structure to resist elastic deformation when it is stressed, and it is a representation of the difficulty of elastic deformation of a material or structure. Stiffness is related to material properties, geometry, boundary support and external force. The deformation of supporting parts can be generally divided into three types: self-deformation, local deformation and contact deformation. The deformation of supporting parts can be evaluated by static stiffness, which includes self-stiffness, local stiffness and contact stiffness.

1) Self stiffness. The self stiffness of a support component refers to its ability to resist its own deformation, mainly considering its bending stiffness and torsional stiffness.

Bending stiffness K_w refers to the ratio of bending load to deformation, which can be expressed as

$$K_{\mathrm{w}} = \frac{F}{\delta}\ (\mathrm{N}/\mu\mathrm{m}) \tag{2.33}$$

where F is the bending load, N; δ is the bending deformation, μm.

Torsional stiffness K_n refers to the ratio of torque to the angle of rotation per unit length, which can be expressed as

$$K_{\mathrm{n}} = \frac{M_{\mathrm{n}}}{\theta/l} = \frac{M_{\mathrm{n}}l}{\theta}\ \left(\mathrm{N\,m}^2/\mathrm{rad}\right) \tag{2.34}$$

In the formula, M_n is the torque, N m; l is the length of the torsion section, m; θ is the rotation angle of the torsion section, rad.

2) Local stiffness. Insufficient local stiffness of supporting components can cause significant deformation in their local positions, seriously affecting the working performance of mechanical systems. The local stiffness mainly depends on the structure and size of the load-bearing parts of the supporting components and the arrangement of ribs, etc.
3) Contact stiffness. The contact stiffness of the supporting part refers to the ability of the supporting part to resist contact deformation. Contact stiffness K_j can be expressed as

$$K_{\mathrm{j}} = \frac{P}{\delta}\ (\mathrm{MPa}/\mu\mathrm{m}) \tag{2.35}$$

where P is the average pressure between the two contact surfaces, MPa, and δ is the contact deformation, μm. The contact normal stiffness is

$$k_n = \left[\frac{2\tilde{G}\sqrt{2\tilde{R}}}{3(1-\tilde{\nu})}\right]\sqrt{U_i^n} \tag{2.36}$$

The contact tangential stiffness is

$$k_s = \left[\frac{2\left(\tilde{G}^2 3(1-\tilde{\nu})\tilde{R}\right)^{1/3}}{2-\tilde{\nu}}\right] \times \left|F_i^n\right|^{1/3} \tag{2.37}$$

where, G is the equivalent elastic shear modulus; v is the equivalent Poisson's ratio; R is the equivalent radius; U_i'' is the contact radial deformation; and F_i'' is the normal force.

(2) Structural design of support system

1) Selection of cross-sectional shape of supporting components. The section shapes of the supporting parts are round, rectangular, T-shaped and I-shaped. Generally, the rectangular section has high bending coefficient and the circular section has strong torsional resistance. Using a hollow rectangular or circular section can significantly improve the bending or torsional moment of inertia, and the larger the hollow area in the section, the greater its stiffness. In order to improve the stiffness of the supporting part itself, a closed hollow section should be used as far as possible, but the wall thickness should not be too thin, so as not to cause insufficient local stiffness or thin-wall vibration. The calculated and relative values of moments of inertia of different sections are shown in Table 2.3.

2) The structure of the supporting components. In the structural design of supporting parts, reasonable structure can improve the static stiffness of supporting parts. The structure of the supporting part mainly includes rib plate, rib and window hole. The floor plate usually runs through the entire section of the supporting part, which can transfer the local load borne by the supporting part to other panels, so that the load is evenly distributed. Figure 2.16 shows common floor layout, including longitudinal layout, transverse layout, and oblique layout. The flexural stiffness of longitudinal floor can be increased, the torsional stiffness of transverse floor can be increased, and the flexural and torsional stiffness of inclined floor can be increased simultaneously. Figure 2.17 shows the stiffener.

Rib is a strip structure arranged on the wall panel of the supporting piece, and does not run through the entire section. The main function is to improve the local stiffness of the supporting parts, prevent thin wall vibration and reduce noise. Common ribs are cross ribs, triangular ribs and so on. In order to meet the requirements of installation parts, sand cleaning, etc., sometimes it is necessary to open window holes on the wall panel of the supporting parts, and the window holes should be opened on the thin

Table 2.3 Calculated values and relative values of inertia moment for different cross-sections

Serial	Sectional shape and size	Calculation value of cross-sectional coefficient (mm^4)	
		Bending resistance	Torsion resistance
1	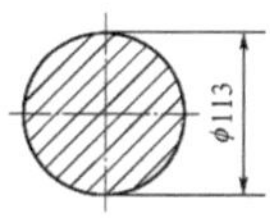	$\frac{800}{1.0}$	$\frac{1600}{1.0}$
2	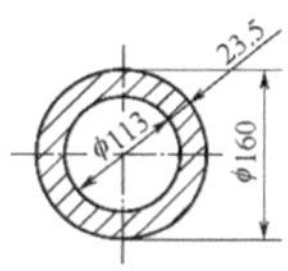	$\frac{2412}{3.02}$	$\frac{4824}{3.02}$
3	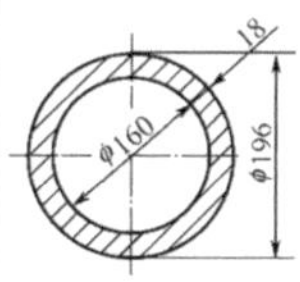	$\frac{4030}{5.04}$	$\frac{8060}{5.04}$
4	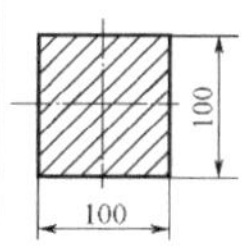	$\frac{833}{1.04}$	$\frac{1400}{0.88}$
5	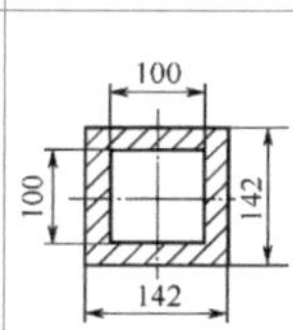	$\frac{2555}{3.19}$	$\frac{2040}{1.27}$
6	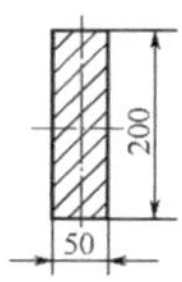	$\frac{3333}{4.17}$	$\frac{680}{0.43}$
7	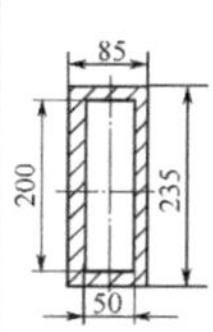	$\frac{5860}{7.325}$	$\frac{1316}{0.82}$

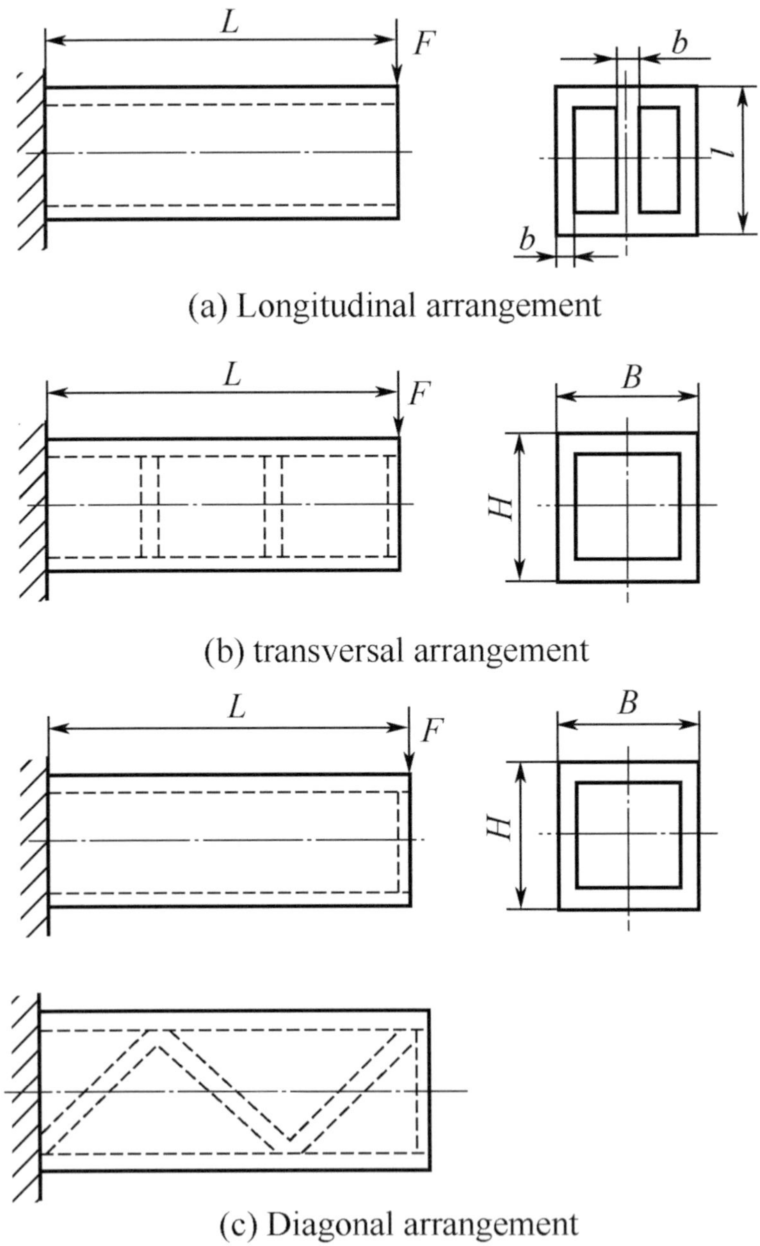

Fig. 2.16 Common layout methods for rib plates

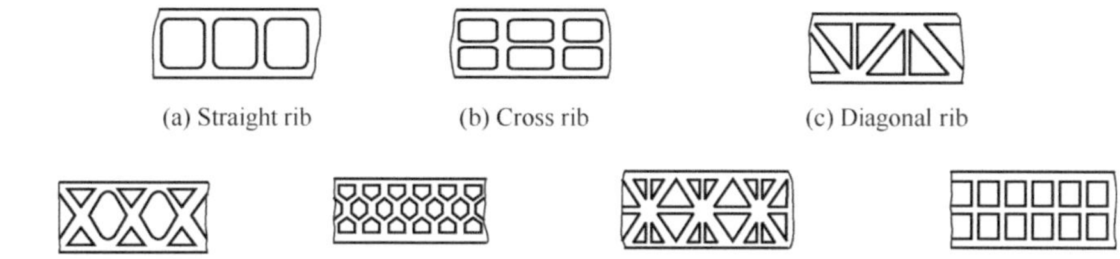
(a) Straight rib (b) Cross rib (c) Diagonal rib
(d) Cross connection rib (e) Honeycombrib (f) '*' shape rib (g) '#' shape rib

Fig. 2.17 Strengthening the form of ribs

wall of the non-main force and the size should be as small as possible. The opening breaks the continuity of the original material and causes the local stiffness of the material to decrease. In the design, the thickness of the wall plate at the opening can be appropriately increased or the cover plate can be connected to compensate for the stiffness loss and meet the requirements of the equipment stiffness.

(3) Material selection of support components

The material of the mechanical system supporting parts should generally be selected according to its function and use requirements, and in the case of meeting the strength and elastic modulus, low-cost materials should be selected as far as possible. The commonly used materials for supporting parts of the equipment support system are cast iron, structural steel, aluminum alloy, etc., which are briefly described respectively.

1) Cast iron. Cast iron is a general term for an alloy of iron, carbon and silicon. The carbon content of cast iron exceeds the carbon content that can be retained in austenitic solid solution at eutectic temperature. Cast iron has low mechanical properties, good wear resistance, good shock absorption performance, good cutting performance, and has good casting performance, mature process, easy to cast a variety of complex structure supporting parts, low material cost, suitable for mass production, in machine tools and other large equipment base and other supporting systems have been widely used. Figure 2.18 shows the supporting parts of an intelligent manufacturing equipment made of cast iron developed by the research team of the author.
2) Structural steel. Structural steel refers to steel that meets specific strength and formability grades. Formability is expressed as the elongation after the break of the tensile test. Structural steel is generally used for load bearing, etc., and the strength of steel is an important design criterion in these uses. Structural steel is generally used for welding support parts, with light weight, easy to form, easy to modify and so on. When structural steel is used for frame or supporting parts, it is usually welded, but it is easy to have defects such as thermal deformation in the welding process, which is suitable for large supporting parts and small batch production.

Whether cast iron or structural steel is used to manufacture supporting parts, residual internal stress will occur in the process of metallurgy or welding, and the

Fig. 2.18 A certain intelligent equipment support component developed by the author's research team

internal stress will be redistributed and gradually disappear, causing the deformation of supporting parts. In addition to the processing after welding, it is also necessary to use annealing, aging treatment and other methods to eliminate the internal stress of the supporting parts.

3) Aluminum alloy. Aluminum alloy is aluminum based on the addition of a certain amount of other alloying elements of the alloy, is one of the light metal materials, low density, has good casting performance and corrosion resistance, is the most used light alloy, has a high strength, its strength close to high alloy steel, stiffness more than steel, has good casting performance and plastic processing properties, good electrical and thermal conductivity, good corrosion resistance andWeldability, can be used as structural materials, in aerospace, aviation, transportation, construction, electrical, light and daily necessities has a wide range of applications. Choosing aluminum alloy material as supporting parts can greatly reduce the weight of supporting parts.
4) Non-metallic materials such as concrete. Reinforced concrete, granite, engineering plastics and composite materials are also used in equipment. Such composite materials usually have the advantages of low material cost, light weight, seismic performance and good wear resistance. The use of reinforced concrete as a supporting part has good dynamic performance, but its surface needs to be painted or sprayed, otherwise it is easy to lead to loose material after oil infiltration; Natural granite has good thermal stability, which is usually used for the base, fuselage and other supporting parts of the measurement or manufacturing system, and is often used in intelligent manufacturing equipment with high precision, such as the base of the coordinate measuring machine (Fig. 2.19); Engineering plastic is a polymer material with resin as the main body, wear-resistant, beautiful, easy to form, the weight is only half of aluminum alloy, but easy to age and harden, thermal deformation is also large, usually only used for the production of supporting parts requiring beautiful appearance and small

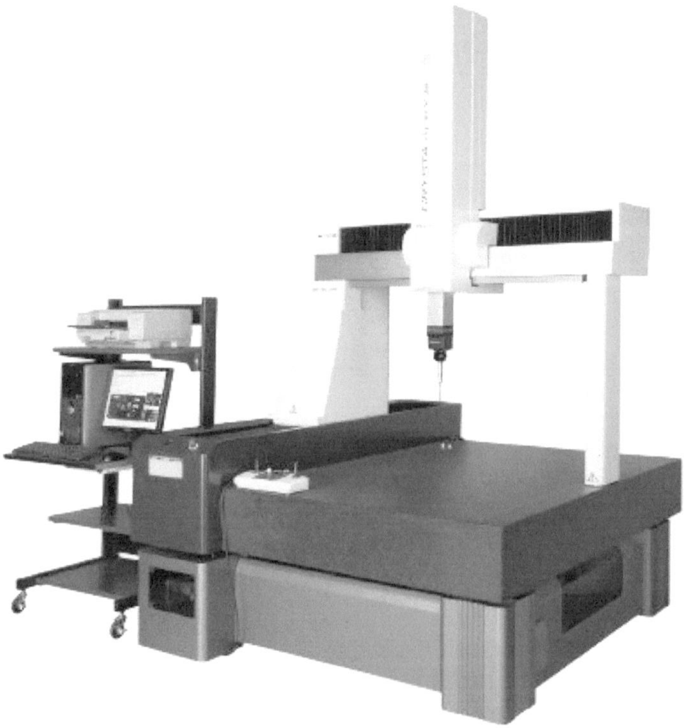

Fig. 2.19 Granite pedestal of coordinate measuring machine

load; Composite materials have the advantages of the above different materials and have been widely used in mechanical systems.

5) Other mineral casting materials. The application of such materials is more and more extensive, mainly including resin concrete, artificial granite, various mineral composite materials and so on. This kind of material can replace cast iron in the machine tool bed (Fig. 2.20), base, beam and other key parts, is widely used in machine tools, electronics, medical, aviation, printing industry and other fields.

2.5.3 Design of Rotating Support Components

The rotating support part consists of a rotating shaft (such as a spindle, a lead screw, etc.), a supporting part (various bearings, bearing seats), and a transmission part and a seal mounted on the rotating shaft. Its main role is to drive other components to carry out accurate rotational motion or indexing, and can withstand a certain load.

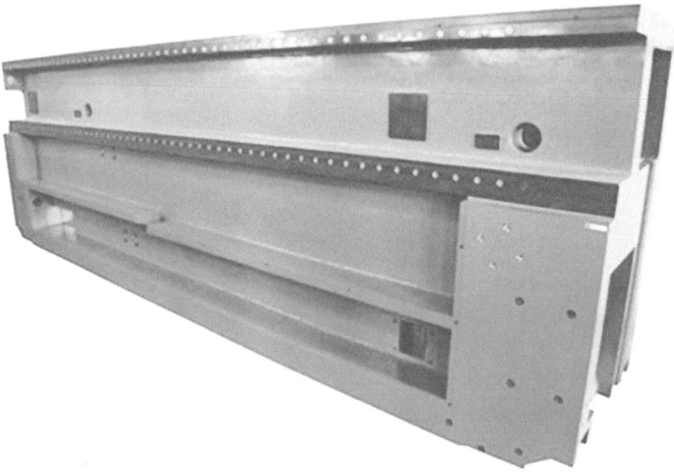

Fig. 2.20 A certain mineral casting material machine tool bed

The rotating support component is an important part of CNC machine tool, its rotating speed affects the efficiency of the system, and the rotating precision determines the accuracy of the system. Therefore, the performance of the rotating support components directly affects the quality and efficiency of CNC machine tools. As a whole system, the working performance and structural connection between the various components of the system need to be coordinated to ensure the rotation accuracy, stiffness, shock resistance, wear resistance, connection of various components, rotation shaft and bearing positioning, bearing clearance adjustment, lubrication, sealing, and easy manufacturing, assembly and maintenance and other common problems.

(1) Basic requirements for the design of rotating support components

The design of rotating support components needs to consider the selection, combination and arrangement of rotating shaft bearings. The number of supports of the rotating support components can be determined according to the needs of the specific situation, and the common is two support structures. The selection of spindle bearing configuration form should first meet the required stiffness and bearing capacity, improving the stiffness of the front support can effectively improve the stiffness of the spindle components, and high stiffness bearings should be configured at the front support. Under the same conditions, the maximum speed of point contact bearing is higher than that of line contact bearing; The maximum speed of cylindrical roller bearings is higher than that of tapered roller bearings. The selected bearing should meet the requirements of stiffness and speed at the same time. Thrust bearings are divided into front thrust, back thrust and two ends thrust. The front thrust is suitable for systems with high axial accuracy requirements for spindle components, the back thrust is suitable for systems with low axial accuracy requirements for spindle components, and the end thrust is suitable for short shafts or intermediate drive shafts.

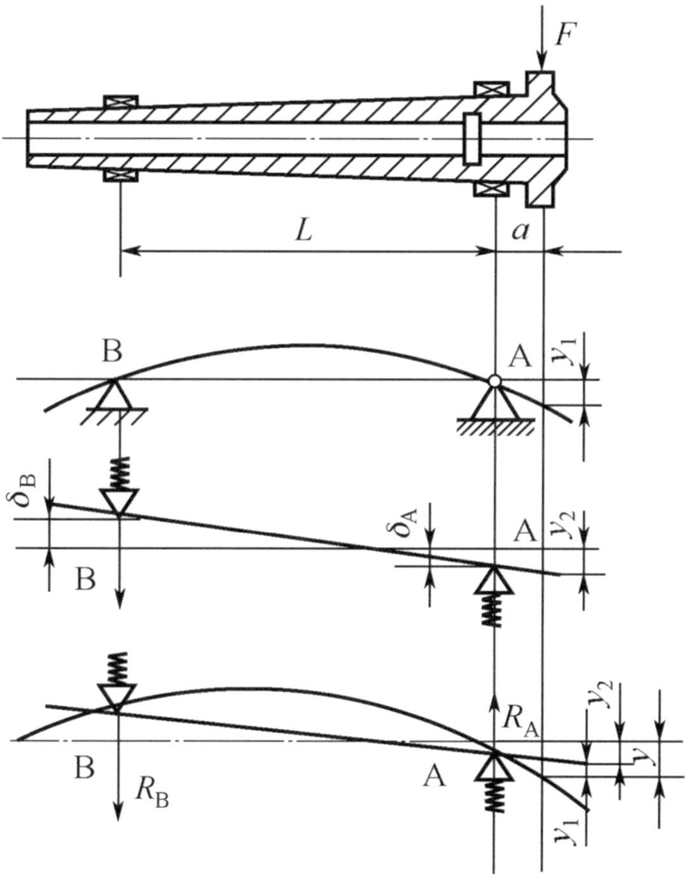

Fig. 2.21 Deformation of the spindle end under stress

(2) The span between the two supports of the rotating support component

The span between the two supports L is one of the important factors determining the stiffness of the rotating support component. After the shaft end is stressed, both the spindle and the support will produce elastic deformation, resulting in displacement of the spindle end, as shown in Fig. 2.21. According to the principle of displacement superposition, the displacement (deflection) y of the spindle end is composed of two parts, namely

$$y = y_1 + y_2 \quad (2.38)$$

y_1 Displacement of the elastic spindle end on the rigid support (assuming the support does not deform), m;

y_2 Displacement of the end of the rigid spindle (assuming the spindle is not deformed) on the elastic support, m.

$$y_1 = \frac{Fa^3}{3EJ_1}\left(\frac{J_1}{J_2} + \frac{L}{a}\right) \tag{2.39}$$

$$y_2 = \frac{F}{k_A}\left(1 + \frac{a}{L}\right)^2 + \frac{F}{k_B}\left(1 + \frac{a}{L}\right)^2 \tag{2.40}$$

In the formula

E Elastic modulus of the spindle material, N/m^2;
J_1 Average moment of inertia of the cross section between the two supports of the main shaft, m^2;
J_2 Average moment of inertia of the cross section of the spindle overhang, m^2;
a Spindle overhang, m;
k_A, k_B Stiffness of the two supporting ends of the main shaft, N/m;
L Span between two supports, m;
F Force on the spindle end, N.

When the overhang of the spindle a is constant and L > L reasonable, the insufficient stiffness of the spindle component is mainly caused by the insufficient stiffness of the spindle, and measures should be taken to improve the stiffness of the spindle. When L < L reasonable, the insufficient stiffness of the spindle component is caused by the insufficient stiffness of the support, and measures should be taken to improve the stiffness of the support. Due to the difficulty of accurately obtaining k_A and k_B, it is generally recommended that L = (1 ~ 5) a in the actual structure, and the upper or lower limit is taken according to the size of the system a value. When L is too small, the radial runout of the bearing has a great impact on the radial runout of the main shaft front end, so it is sometimes recommended that $L/a \geq 2.5$, and the value of L/a can be reduced with the continuous improvement of bearing manufacturing accuracy; If L is too small, the center of gravity of the spindle assembly with accessories will fall on the outside of the two supports, and the spindle is easy to vibrate and the amplitude is large. Ordinary machine tools with good working performance often take L/a = 5 ~ 6.

(3) Bearing selection of rotating support components

The bearing is an important part of the rotating support components, which should have the characteristics of high rotation accuracy, high stiffness, strong bearing capacity, good shock resistance, high speed performance, low friction power consumption, low noisc and long life. Common bearings are rolling bearings, plain bearings and magnetic bearings. The types, codes and characteristics of commonly used rolling bearings are shown in Table 2.4. Lubrication can reduce friction, reduce temperature rise, and together with the sealing device, protect the bearing from foreign intrusion and prevent corrosion. Choosing the appropriate lubrication and

Table 2.4 Types, codes, and characteristics of commonly used rolling bearings

Bearing name	Type code	Structure diagram	Basic rated dynamic load ratio	Extreme speed ratio	Main characteristics
Aligning ball bearing	10,000	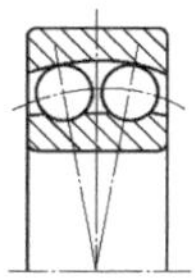	0.6–0.9	Middle	The outer raceway is spherical and can automatically adjust its center. Mainly capable of bearing radial loads, but also able to withstand a small amount of axial loads. The relative deviation of the inner and outer ring axes is allowed to be 0.5–2°. Suitable for shafts with multi-point support and insufficient bending stiffness, as well as shafts that are difficult to align
Self-aligning roller bearing	20,000	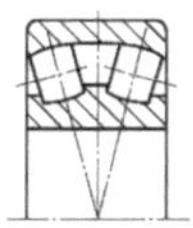	1.8–4	Low	The outer raceway is a surface centered on the center of the bearing, which can automatically adjust its center. Can withstand large radial loads and small axial loads, resistant to vibration and impact. The relative deviation of the inner and outer ring axes is allowed to be 0.5–2°. Suitable for heavy loads that other bearings cannot handle and require centering

(continued)

Table 2.4 (continued)

Bearing name	Type code	Structure diagram	Basic rated dynamic load ratio	Extreme speed ratio	Main characteristics
Tapered roller bearing	30,000	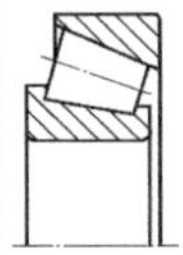	1.1–2.5	Middle	Capable of withstanding both large radial loads and unidirectional axial loads simultaneously. The inner and outer circles can be divided into different heights, and the coolness can be adjusted for easy installation and removal. Usually used in pairs. Suitable for shafts with low rotational speed and high rigidity
Thrust ball bearing	50,000	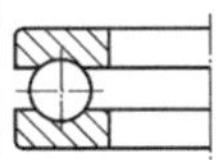	1	Low	Can only withstand unidirectional cat loads. The inner diameters of the two rings are not the same. The smaller inner hole is the tight ring, which fits with the shaft. The larger inner hole is the loose ring, which is fixed to the machine base
Thrust ball bearing	50,000	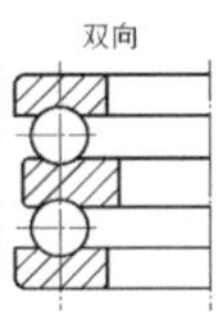	1	Low	Can withstand bidirectional load. The smaller inner hole in the middle ring is called a tight ring, which matches with the transmission, while the other two rings are loosely drawn
Deep groove ball bearing	60,000		1	High	Suitable for situations with large axial loads and low rotational cover

(continued)

Table 2.4 (continued)

Bearing name	Type code	Structure diagram	Basic rated dynamic load ratio	Extreme speed ratio	Main characteristics
Angular contact ball bearing	70,000C $\alpha = 15°$ 70,000AC $\alpha = 25°$ 70,000B $\alpha = 40°$	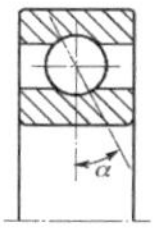	1	high	It mainly bears radial loads and can also withstand certain axial loads. The allowable deviation of the inner and outer axis lines during work is 8 ~ 16′. Low frictional resistance and high maximum speed. The structure is simple, the price is low, and it is widely used. Poor ability to withstand impact. Suitable for high-speed situations It can withstand both radial and unidirectional axial loads simultaneously, and has a high speed. Frequently used in pairs, symmetrically installed. Suitable for situations with high rotational speed and simultaneous radial and axial loads It can only withstand radial loads and cannot withstand axial loads. The inner and outer rings can be separated along the axis. The bearing capacity is larger than that of ball bearings of the same size, with greater impact resistance and higher ultimate speed. Sensitive to shaft deflection, it can only be used for shafts with high rigidity, and requires the bearing holes to be well aligned

(continued)

Table 2.4 (continued)

Bearing name	Type code	Structure diagram	Basic rated dynamic load ratio	Extreme speed ratio	Main characteristics
Cylindrical roller bearing	Outer ring without blocking edgeN0000	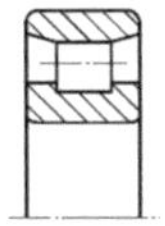	1.5–3	High	The radial structure is compact and has a large radial load-bearing capacity, and the inner and outer rings can be separated. Cannot withstand axial loads, low limit speed, and does not allow deviation of the inner and outer ring axes during operation. Commonly used in situations where the radial size of the surface is limited at lower speeds The outer raceway is spherical and can automatically adjust its center. Mainly capable of bearing radial loads, but also able to withstand a small amount of axial loads. The relative deviation of the inner and outer ring axes is allowed to be 0.5–2°. Suitable for shafts with multi-point support and insufficient bending stiffness, as well as shafts that are difficult to align
	Inner ring without blocking edgeNU0000	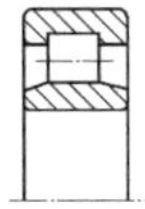			

(continued)

Table 2.4 (continued)

Bearing name	Type code	Structure diagram	Basic rated dynamic load ratio	Extreme speed ratio	Main characteristics
Needle bearing	NA0000	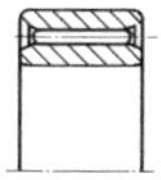	–	Low	The outer raceway is a surface centered on the center of the bearing, which can automatically adjust its center. Can withstand large radial loads and small axial loads, resistant to vibration and impact. The relative deviation of the inner and outer ring axes is allowed to be 0.5–2°. Suitable for heavy loads that other bearings cannot handle and require centering

sealing method can reduce the working temperature of the bearing and extend the service time. Rolling bearings can be lubricated with grease or lubricating oil. Plain bearings should be lubricated with lubricating oil.

1) The size, direction and nature of the load on the bearing. It is the main basis for selecting the bearing type. When selecting the bearing type according to the size of the load, because the main components in the roller bearing are in line contact, it is suitable for bearing a larger load, and the deformation after bearing is small. The ball bearings are mainly point contacts, which should be used to withstand light or medium loads, so in the load is small, ball bearings should be preferred. When the bearing type is selected according to the direction of the load, thrust bearings are generally selected for pure axial loads. Small pure axial load can choose thrust ball bearings; Large pure axial loads can be selected thrust roller bearings. For pure radial loads, deep groove ball bearings, cylindrical roller bearings or needle roller bearings are generally selected. When the bearing bears radial load at the same time, but also bear a small axial load, you can choose deep groove ball bearings or angular contact ball bearings with small contact Angle or tapered roller bearings; When the axial load is large, the angular contact ball bearing or tapered roller bearing with larger contact Angle can be selected, or the structure of centripetal bearing and thrust bearing combined is selected to bear the radial load and axial load respectively.
2) Bearing speed. Under the general speed, the level of the speed has little impact on the selection of the bearing type, but when the speed is high, it has a greater impact. The limit speed nlim values of bearings of various types and sizes are included in the bearing samples.

Compared with child bearings, ball bearings have a higher limit speed, so ball bearings should be preferred at high speed. Under the conditions of the same inner diameter, the smaller the outer diameter, the lighter and smaller the rolling body, the smaller the centrifugal inertia force of the rolling body added to the outer raceway during operation, and the more suitable for working at high speed. At high speed, ultra-light, ultra-light and light series of bearings are selected. Heavy and extra-heavy series of bearings are only used for low-speed and heavy-duty occasions. When the bearing capacity of a light series bearing does not meet the requirements, two can be installed together, or a wide series of bearings can be used. The material and structure of cage have great influence on the bearing speed. Solid cages allow higher rotational speeds than stamped cages. The limit speed of thrust bearings is very low, generally used in low-speed occasions. When the working speed is high, the angular contact ball bearing can be selected to bear the pure axial force when the axial load is not very large; If the working speed slightly exceeds the limit speed specified in the sample, the high-speed performance of the bearing can be improved by improving the tolerance level of the bearing, appropriately increasing the radial clearance of the bearing, choosing circulation lubrication or oil mist lubrication, strengthening the cooling of the circulating oil and other measures, or choosing a special high-speed rolling bearing.

3) Bearing aligning performance. When the center line of the shaft does not coincide with the center line of the bearing seat and there is an Angle error, or when the shaft is bent or tilted due to the force, the inner and outer ring axis of the bearing will be skewed. At this time, aligning ball bearings or aligning roller bearings with certain aligning performance should be used. This kind of bearing can still work normally when there is a small relative deviation of the inner and outer ring axis. Cylindrical roller bearings and needle roller bearings are the most sensitive to the deflection of bearings, and the bearing capacity of such bearings may be lower than that of ball bearings in the deflection state, so when the stiffness of the shaft and the supporting stiffness of the bearing seat hole are low, such bearings should be avoided as far as possible.
4) Installation and disassembly of bearings. Ease of assembly and disassembly is also a factor that should be considered when selecting the bearing type. When the bearing seat has no split surface and the bearing parts must be installed and disassembled along the axis, the bearing with the inner and outer rings can be separated should be preferred; When the shaft is mounted on the long shaft, in order to facilitate assembly and disassembly, the bearing with a 1:12 tapered hole in the inner ring (for mounting on the tight bushing) can be selected.

(4) Bearing combination design

When there is no axial load, deep groove ball bearings should be selected, as shown in Fig. 2.22.

When the radial and axial loads are combined, it is appropriate to use the Angle contact ball bearing or the round cone roller bearing, as shown in Figs. 2.23 and 2.24.

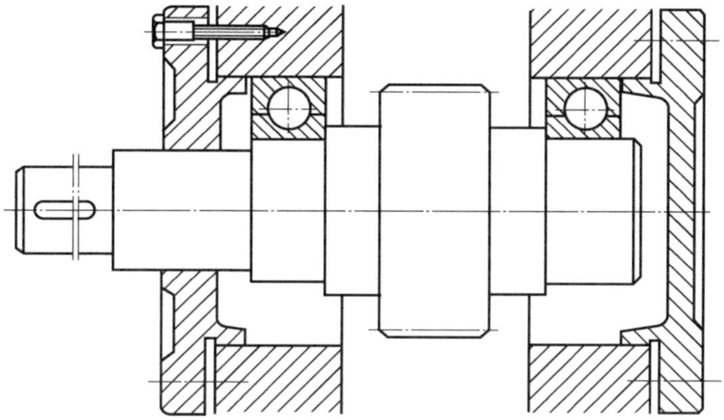

Fig. 2.22 Combination of deep groove bearings

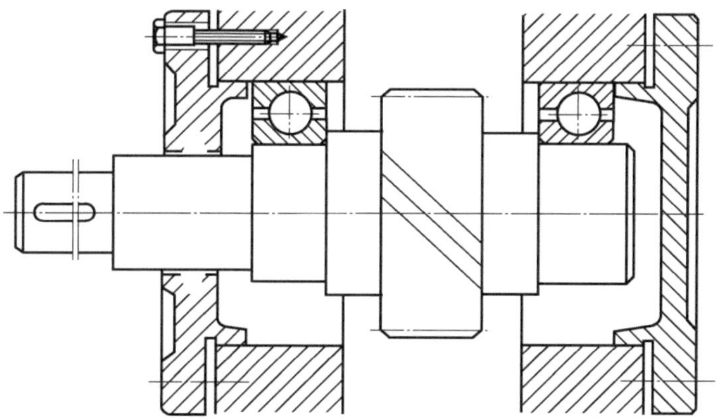

Fig. 2.23 Combination of angular contact ball bearings

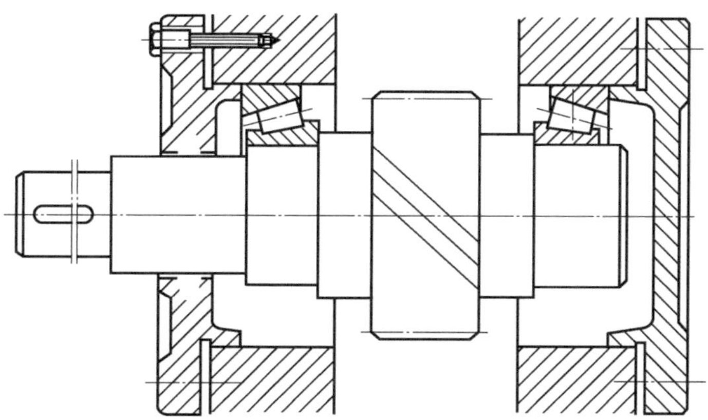

Fig. 2.24 Combination of tapered roller bearings

The bearings are fixed at both ends (Fig. 2.25), one end is fixed and one end is movable (suitable for long shafts with large temperature changes, see Fig. 2.26), etc.

The bearing clearance can be adjusted using gaskets or adjusting nuts, as shown in Fig. 2.27.

For some bearings with adjustable clearance, a certain axial compression force is given during installation, so that the inner and outer rings move relative to eliminate the clearance, and elastic predeformation is generated at the contact between the ring and the rolling body, so as to improve the rotation accuracy and stiffness of the shaft, which is called the pretightening of the bearing. Figure 2.28 shows the bearing preload.

In order to make the parts on the shaft have an accurate working position, the position of the bearing combination needs to be adjusted, as shown in Fig. 2.29.

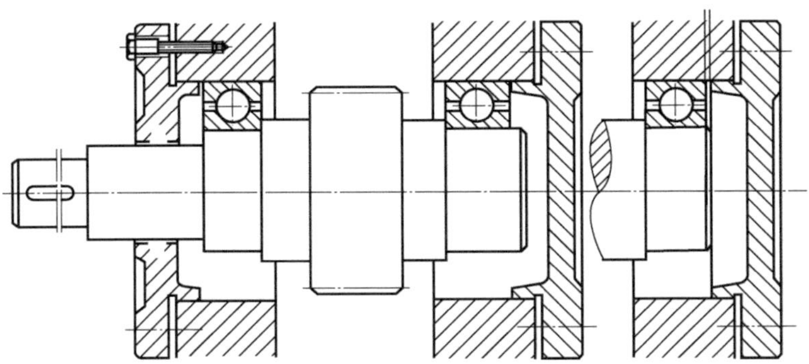

Fig. 2.25 Combination of fixed bearings at both ends

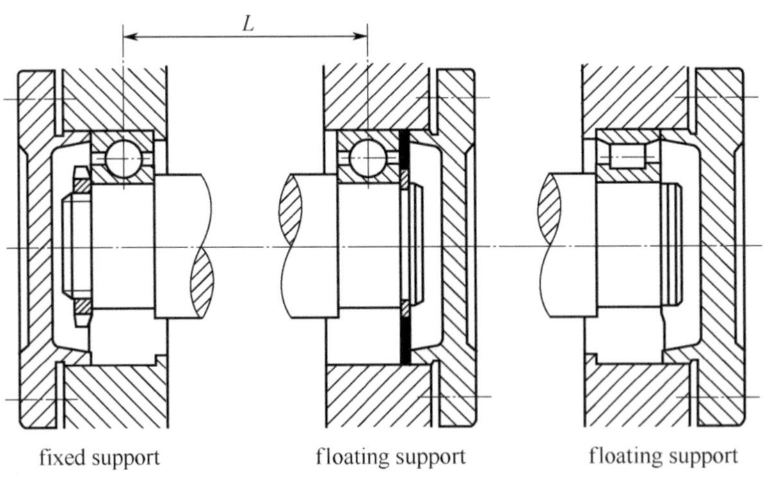

Fig. 2.26 Combination of fixed end and movable end bearings

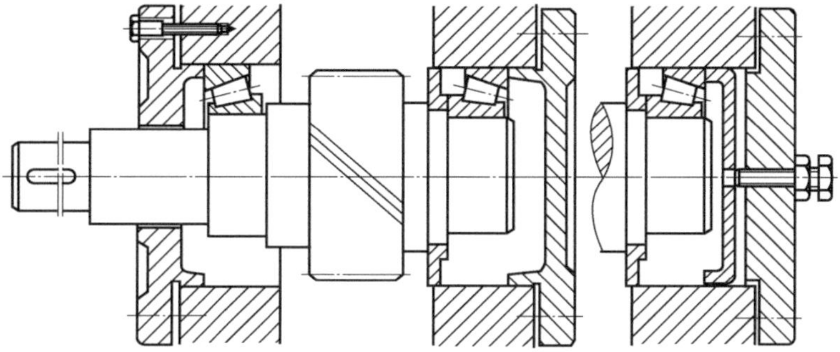

Fig. 2.27 Adjustment of bearing clearance

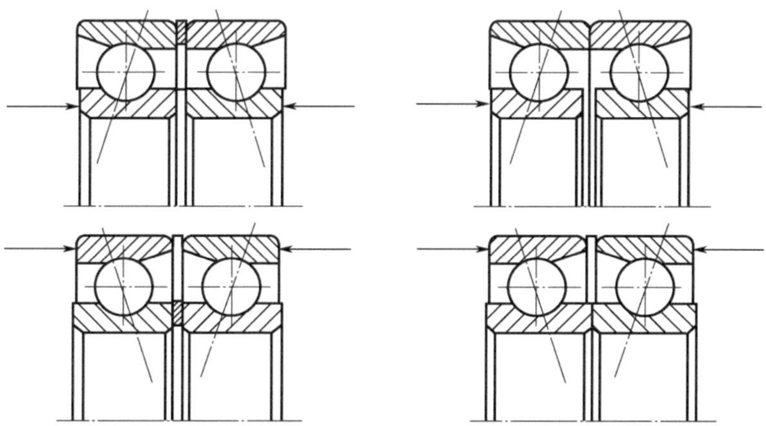

Fig. 2.28 Pre tightening of bearings

For example, bevel gear transmission requires that the vertices of two pitch cones coincide, and one of the methods is to set the cup + adjust the gasket.

The choice of sealing mode of rolling bearing is related to the type of lubrication, working environment, temperature and circular speed of sealing surface. Sealing mode: Contact sealing, such as felt and ring sealing (Fig. 2.30); Non-contact seals, such as labyrinth seals; Combined seal, such as felt labyrinth seal, etc. (Figs. 2.31, 2.32 and 2.33).

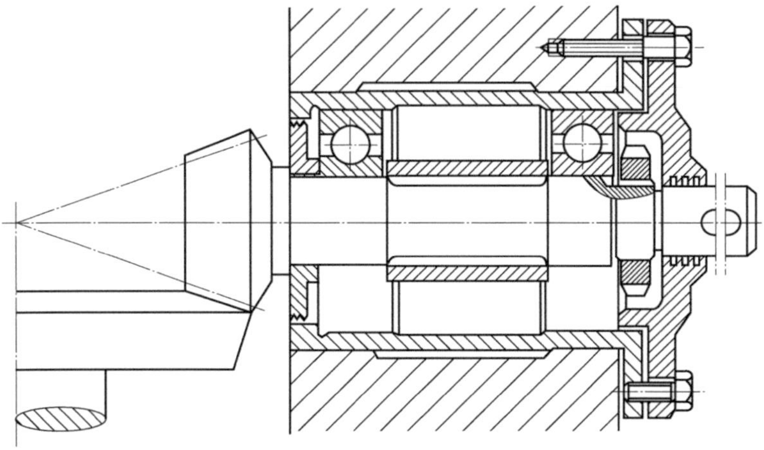

Fig. 2.29 Adjustment of bearing combination position

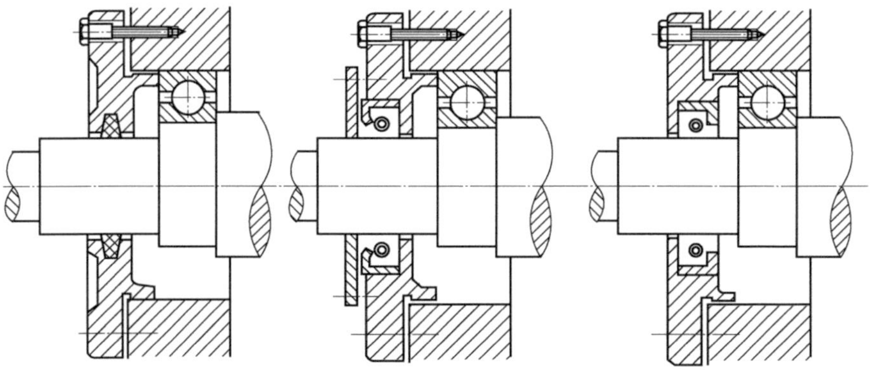

Fig. 2.30 The felt and sealing ring seal of the bearing

2.5.4 Design of Mobile Support Component Structure Scheme

The function of the guide rail is to carry and guide, and it is an important link of the feed system. The moving guide rail is called the moving guide rail, and the stationary guide rail is the supporting guide rail or the static guide rail. The guide surface is the two mating surfaces that have relative motion on the machine tool, so it is also called "guide pair", which is a low pair. The guide rail with relative movement of the mating surface is called the moving guide rail, and the guide rail with relatively fixed mating surface is called the fixed guide rail. The basic requirements of designing mobile support parts are guiding accuracy, wear resistance, stiffness, low speed motion stability and process.

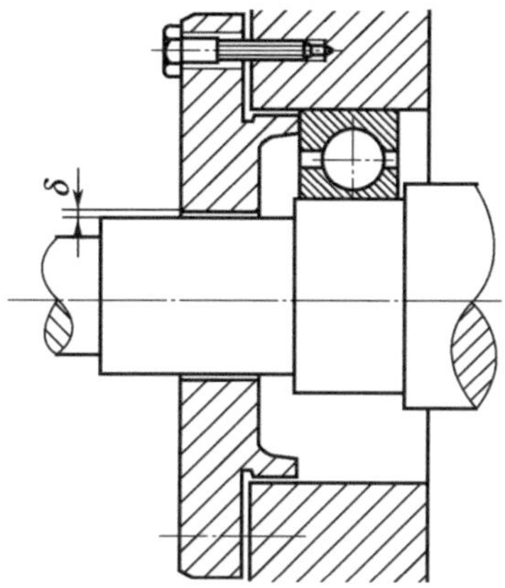

Fig. 2.31 Clearance seal

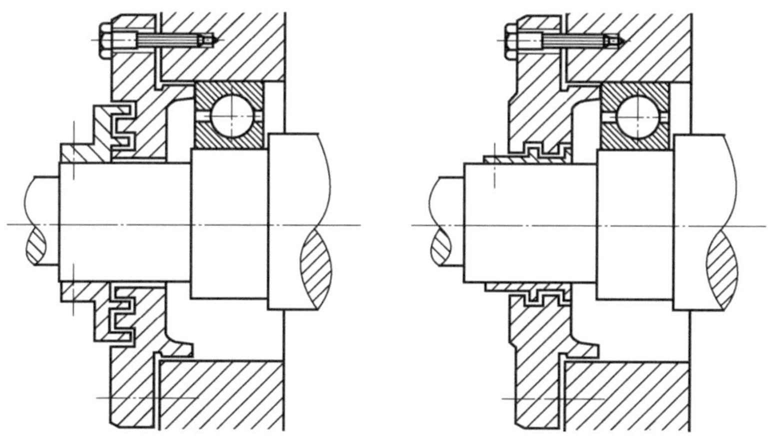

Fig. 2.32 Labyrinth seal

① High guidance accuracy. Guiding accuracy is the straightness of moving parts along the guide rail and the accuracy of their mutual position with the relevant base surface. Only with high guiding accuracy can the working quality of the guide rail be guaranteed, and the guiding accuracy is related to the structure, assembly, material and process of the guide rail.

② The movement is flexible and smooth. When the guide rail is working, it should be lightweight, labor-saving and flexible, and the speed should be uniform. The smoothness of low-speed movement means that when the moving guide rail is moving at low speed or with a small amount of displacement, the guide rail movement has smoothness to prevent crawling.

③ Guide rails should be wear-resistant. The long-term work of the guide rail can still ensure a certain accuracy, only good wear resistance can make the machine

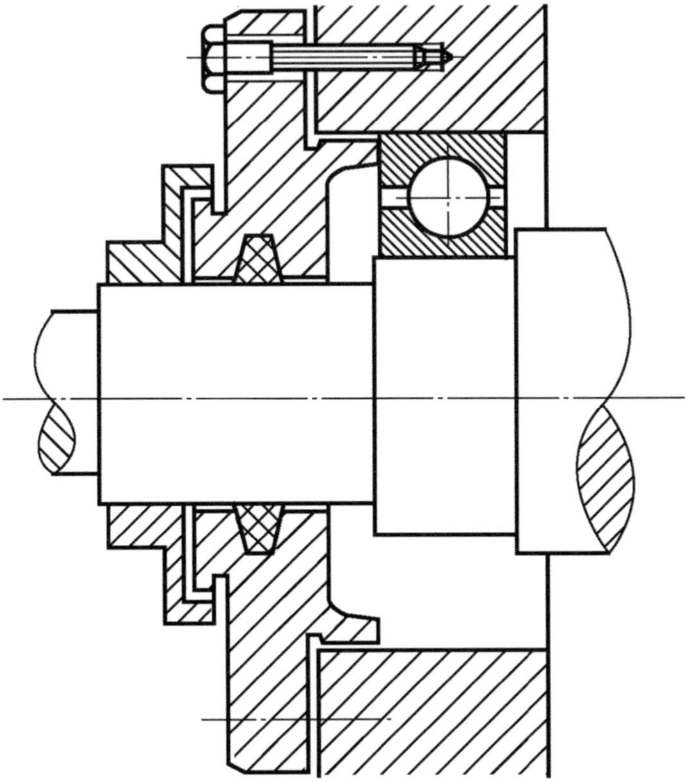

Fig. 2.33 Felt labyrinth combination seal

tool guide rail long-term accuracy, should minimize the wear of the guide rail uneven degree. However, wear is inevitable, and the amount of wear should be as small as possible, which is easy to compensate or adjust after wear and improve the service life.

④ Suitable width, as far as possible to reduce the specific pressure of the guide surface. Auxiliary guide rails can be designed to bear the load.

⑤ Low thermal effect. The main measures to eliminate crawling are to minimize the temperature rise in the working process of equipment, reduce the difference between dynamic and static friction coefficients, change the characteristics of dynamic friction coefficient with speed, and improve the stiffness of transmission mechanism.

The guide rail design tries to make the structure of the guide rail simple, easy to manufacture, assembly and maintenance. According to the working conditions, select the appropriate guide rail type, appropriate section shape, select the appropriate guide rail structure, size, in the rated load and temperature range there is enough stiffness and wear resistance. Choose reasonable lubrication methods to reduce friction and wear. For guides that need to be scraped, the amount of scraping should be minimized.

For the mounted guide rail, it should be easy to replace. The guide rails are classified into open guide rails and closed guide rails, as shown in Fig. 2.34. Under the action of self-weight and load, the working face of the moving-static guide rail is always in contact, and the structure is simple, but it cannot withstand large overturning moment. Closed guide rail can withstand overturning moment under the action of pressure plate.

(1) Design of sliding guide rail

The section shapes of linear motion guides are rectangle, triangle, dovetail and circle, as shown in Tables 2.5 and 2.6.

The cross section shape of rotary motion guide rail includes plane ring guide rail, cone ring guide rail and double cone ring guide rail. Planar ring guide rail is simple in structure, easy to manufacture, and can bear large axial force, but can not bear radial force. It is suitable for machine tools with various rotary motion guide rails centered by spindle, such as high speed vertical lathe with large load.

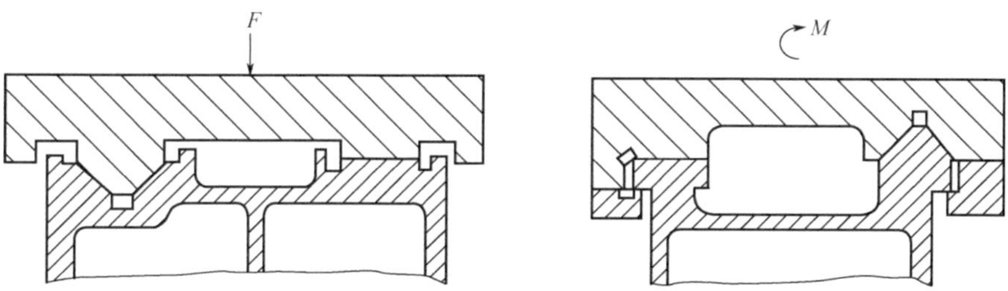

(a) open rail (b) closed guide rail

Fig. 2.34 Open guide rails and closed guide rails

Table 2.5 Sliding rail cross-sectional shape

	Prismatic				Rotundity
	Symmetric triangle	Asymmetric triangle	Rectangle	Swallowtail shape	
Convex shape	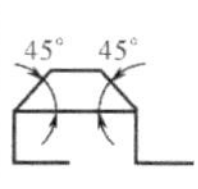	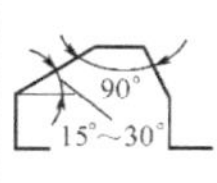	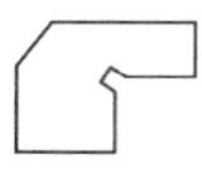	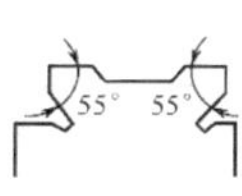	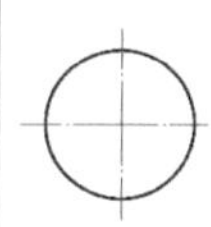
Concave shape	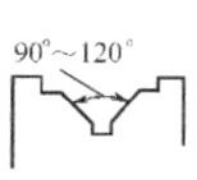	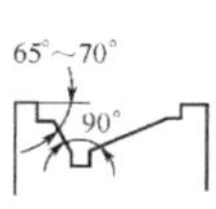	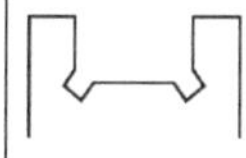	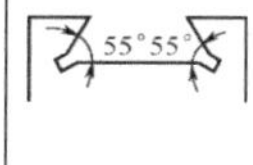	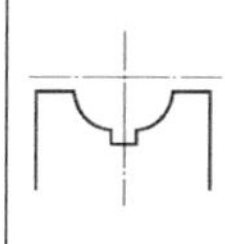

Table 2.6 Characteristics of linear motion guides

Guide rail form	Structural characteristics	Use	Illustration
Rectangular guide rail	Manufacturing is simple, with high rigidity and load-bearing capacity. The displacement in both horizontal and vertical directions does not affect each other, and installation and adjustment are convenient. However, the guide surface cannot automatically compensate for gaps after wear, which affects accuracy	Ordinary precision machine tools or heavy-duty machine tools	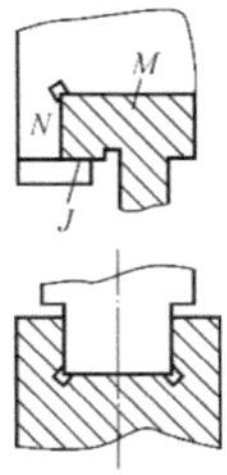
Triangle guide rail	The triangular guide rail can automatically compensate for wear and tear under vertical load, and has good guiding performance; the pressure plate surface needs to have a gap adjustment device; the increase in vertex angle leads to an increase in bearing capacity, but the guidance accuracy is poor	Ordinary precision machine tools or heavy-duty machine tools	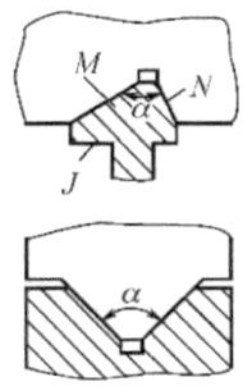
Swallowtail shaped guide rail	After the guide rail wears out, it cannot automatically compensate for the gap and requires a gap adjustment device; the two dovetail surfaces act as pressure plates, and the horizontal and vertical gaps can be adjusted with a single strip; the manufacturing, inspection, and repair of guide rails are complex and have high frictional resistance	Used for multi-layer mobile composite components with low height requirements, widely used in instrument and machine tools	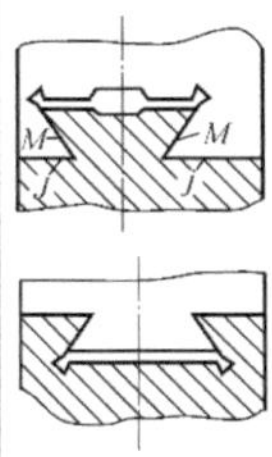
Cylindrical guide rail	Manufacturing is simple, the inner hole can be honed, and the outer circle can achieve precise fit through grinding; it is difficult to adjust the clearance after wear	Commonly used in situations where movement and rotation occur simultaneously. Such as pulling beds, robotic arms, etc.	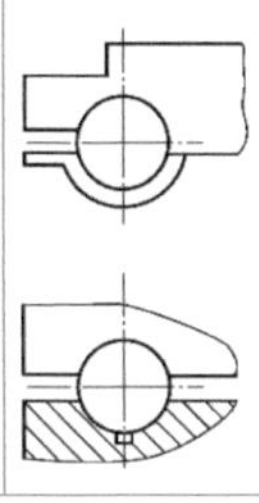

Conical annular guide can bear axial and radial forces at the same time, but can not bear large overturning moment. The guide is better than the plane ring guide, but it is difficult to manufacture. It is suitable for bearing certain radial load and overturning moment. The double conical annular guideway can bear large radial force, axial force and certain overturning moment, but it is difficult to manufacture and grind.

(2) Guide material

Since the guide rail has certain requirements for guiding accuracy, wear resistance, stiffness, low-speed motion smoothness and processability, it needs materials with high wear resistance, good processability and low cost. Common guide materials are cast iron, steel, non-ferrous metals, plastics and so on. Cast iron (such as gray cast iron HT200, inoculation cast iron HT300, etc.) has good shock absorption and wear resistance, low cost, easy to cast and cut, and is often used as a guide material for machine tools. Quenching can improve the surface hardness and wear resistance of the guide rail. The wear resistance of the guide rail can be greatly improved by the steel bearing guide rail, but the process is complicated, the processing is difficult and the cost is high. 20Cr, 45Cr and 40Cr are commonly used steel materials. The non-ferrous metal plate is mounted on the moving guide rail, which is matched with the cast iron supporting guide rail, and the plastic soft belt is mounted on the moving guide rail, and the hardened cast iron supporting guide rail and the inserted steel supporting guide rail constitute the guide rail pair. In order to improve wear resistance and prevent biting and welding, the moving guide rail and the supporting guide rail should be made of different materials. If the same material is used, different heat treatments should also be used to make the two sides have different hardness.

Lubrication and protection are also essential parts of the guide design. The sliding guide rail is generally lubricated with lubricating oil, and the rolling guide rail is commonly lubricated with lubricating grease. Proper lubrication can reduce friction, reduce wear, reduce temperature and prevent rust. The use of scraper or retractable guards can prevent or reduce rail wear.

A plastic coated sliding guide rail (Fig. 2.35) is affixed with a layer of plastic anti-wear tape on the sliding surface of the guide rail, and the other sliding surface of the guide rail is a hardened grinding surface. The anti-wear strip is generally a polymer composite material based on polytetrafluoroethylene and other materials, and then adds alloy powder or oxide to enhance its wear resistance.

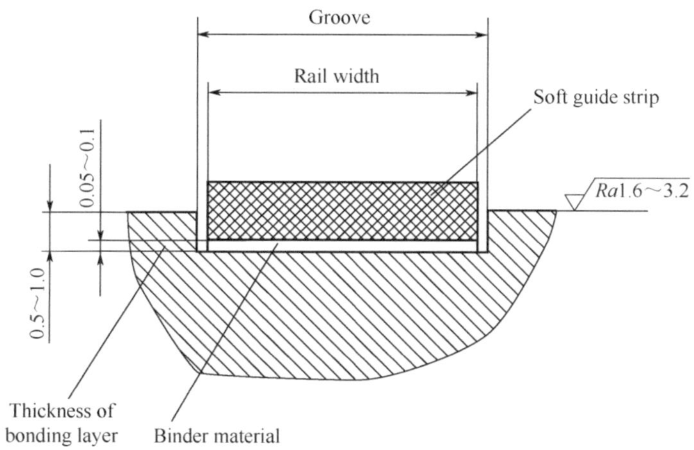

Fig. 2.35 Plastic coated sliding guide rail

(3) Combination of sliding guide rails

Sliding guides are often combined, and the commonly used combination forms are linear guides (usually composed of two guides), double triangle guides, double rectangular guides, rectangular and triangular guides, rectangular and dovetail guides, etc.

1) Double triangular guide rail. Two triangular guide rails act as both support and guide. Due to the symmetrical structure, the drive element can be placed symmetrically in the middle of the two guides, and the two guides are evenly worn, the relative position is unchanged after wear, and the wear in the vertical and horizontal directions can be automatically compensated, so the directivity and accuracy retention are high, and the contact stiffness is good. But the process is poor, processing, inspection, maintenance is more difficult, scraping or grinding of the four surfaces of the guide rail is difficult to fully contact, if the bed and moving parts thermal deformation is different, it is difficult to ensure that the four surfaces contact at the same time. Therefore, it is used for machine tool equipment with higher precision requirements.
2) Double rectangular guide rail. Double rectangular guide rails have high stiffness, strong bearing capacity, small friction coefficient, simple manufacturing and adjustment, convenient inspection and maintenance, but low contact stiffness, low guidance, suitable for heavy duty work. When rectangular and rectangular combinations are used, guide surfaces should be reasonably selected.
3) Combination of triangular and rectangular guide rails. This combination form has the advantages of good guiding ability of triangular guide rail and convenient manufacturing and good rigidity of rectangular guide rail, and avoids the matching changes caused by thermal deformation. However, the guide wear is not uniform, generally the triangular guide wear faster than the rectangular guide, and can not be adjusted to compensate after wear, so it has an impact on the position accuracy. The closed guide rail has a pressing plate surface and can withstand overturning moment. The combination of triangle and rectangle has two forms: V-moment and edge-moment. V-moment combination guide rail is easy to store lubricating oil, low and high speed can be used; Edge-moment cannot store lubricating oil and is only used for low speed movement.
4) Double dovetail guide rail. It can withstand overturning torque, convenient clearance adjustment, but the stiffness is low, the friction is large, the processing, inspection, maintenance is not convenient, suitable for many layers, requiring convenient clearance adjustment in the workplace. In addition, there are rectangular and dovetail, double circular guide combination methods, which are not described here.

(4) Calculation of guide pressure and judgment of the use of the pressure plate

The load on the guide rail can be simplified to a concentrated force F and a concentrated couple M. The pressure caused by F and M on the guide rail is:

$$P_{\mathrm{F}} = \frac{F}{aL} \tag{2.41}$$

And $M = \frac{1}{2}P_{\mathrm{M}} \times \frac{aL}{2} \times \frac{2}{3}L = \frac{P_{\mathrm{M}}aL^2}{6}$ therefore

$$P_{\mathrm{M}} = \frac{6M}{aL^2} \tag{2.42}$$

In the formula

- F The concentrated force on the guide rail;
- M Overturning moment of the guide rail;
- P_M Is the maximum pressure caused by the overturning moment;
- P_F Is the pressure caused by the concentrated force F;
- a Guide rail width;
- L The length of the guide rail.

The maximum, minimum and average pressures on the guide rail are

$$P_{\max} = P_{\mathrm{F}} + P_{\mathrm{M}} = \frac{F}{aL}\left(1 + \frac{6M}{FL}\right) \tag{2.43}$$

$$P_{\min} = P_{\mathrm{F}} - P_{\mathrm{M}} = \frac{F}{aL}\left(1 - \frac{6M}{FL}\right) \tag{2.44}$$

$$P_{平均} = \frac{1}{2}\left(p_{\max} + p_{\min}\right) = \frac{F}{aL} \tag{2.45}$$

When $F_L = 0$, that is, $M = 0, P = P_{max} = P_{min} = P_{average}$, the pressure is distributed in a rectangular shape, and the force condition of the guide rail is the best at this time, but this force condition almost does not exist.

When $6M \neq 0$, that is, $M \neq 0$, due to the overturning moment, the pressure of the guide rail is not distributed in a rectangular shape, and its resultant force point deviates from the center of the guide rail.

When $FL < 1$, that is, M, $min \geq 0$, $P_{max} < 2P_{average}$, the pressure is distributed in a trapezoidal shape. This situation should be ensured as much as possible during design.

When $M = 1$, that is, $M = 1/6, P_{max} = 0, P_{min} < 2P_{average}$, the pressure is distributed in a triangle. Although the pressure difference is large, the guide rail surface can still be in contact over the entire length, which is a critical state.

At that time, open guide rails without pressure plates can be used.

When M > 1, there will be a section on the main guide surface that is not in contact. At this time, a pressure plate must be installed to form an auxiliary guide surface.

The guide rail pressure is shown in Fig. 2.36.

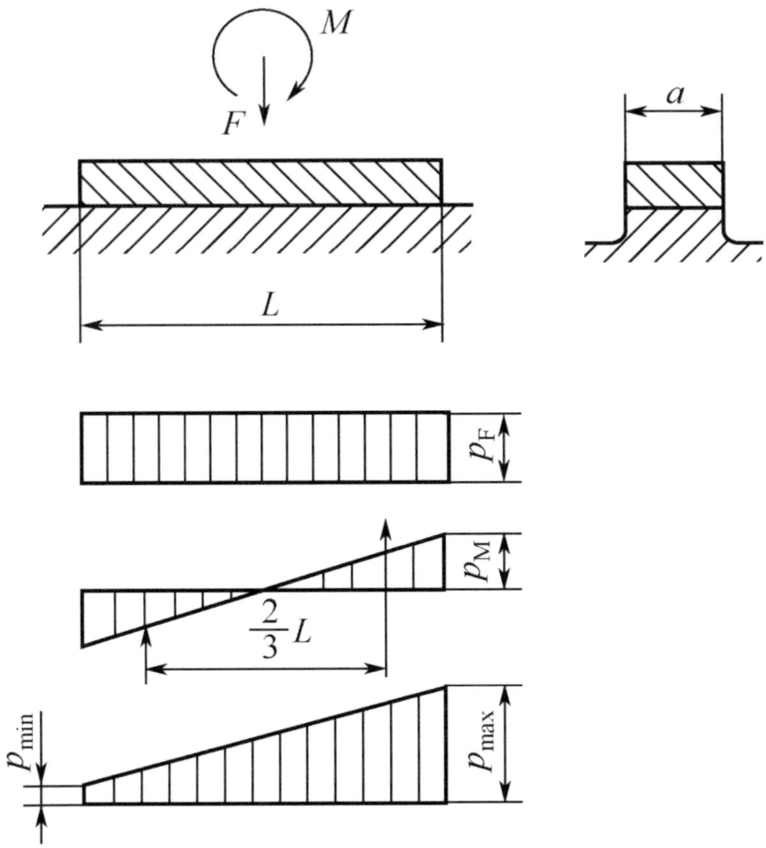

Fig. 2.36 Design of intelligent manufacturing equipment execution system

In addition to sliding guides, rolling parts can also be installed between the working surfaces of the guides to create rolling friction between the two contact surfaces of the guides, reduce the friction coefficient, reduce friction, eliminate "creeping", and have high precision. They are increasingly used in high-end equipment.

2.6 Design of Intelligent Manufacturing Equipment Execution System

The performance of the main components of intelligent manufacturing equipment is directly reflected in the execution system of the equipment. The execution system is a subsystem that directly contacts and interacts with the working object in intelligent manufacturing equipment, and is connected with the transmission system and support system. It is a part of the mechanical system that directly completes the expected function.

The execution system consists of the execution component and the execution mechanism. The execution component is the part that directly completes the function in the execution system, and generally contacts with the working object directly or performs operations on the working object directly. The motion and power of the executive member must meet the functional requirements expected to be realized by the mechanical system, including the form of motion, range, accuracy, load type and size. The actuator is the mechanism that drives the actuator member, which transfers the motion and power transmitted by the transmission system to the actuator member. There are one or more executive mechanisms in the executive system, and the executive mechanisms can drive multiple executive components. The execution system can convert the three motion forms of moving, rotating and swinging into each other, and even the continuous rotation can be changed into intermittent movement. Its functions can be summarized as follows: clamping, handling, conveying, indexing and transposition, detection, transformation of motion forms or motion laws, and completion of complex movements. Mechanical execution system scheme design mainly includes the following contents:

(1) Functional principle design

Any kind of machinery is designed to achieve a certain expected functional requirements, including process requirements and use requirements. The so-called functional principle design is to choose the best working principle according to the expected function of the machine. To achieve the same functional requirements, different working principles can be selected, and the machinery designed according to different working principles will have great differences in working performance, working quality and application occasions.

(2) Motion law design

To achieve a certain working principle, different motion laws can be adopted, and the design of motion laws is usually carried out by decomposing the process action proposed by the working principle. The motion laws obtained by different decomposition methods are also different. To achieve the same working principle, different motion schemes can be selected, and the selected motion schemes are different, and the mechanical products designed are very different.

(3) Mechanism form design

To realize the same movement law, different forms of mechanism can be selected. The so-called mechanism form design refers to the selection of the best mechanism to achieve the above motion laws. A certain motion law can be realized by different mechanisms, but it is necessary to consider the dynamic characteristics of the mechanism, mechanical efficiency, manufacturing cost, shape and other factors, according to the characteristics of the designed machinery for comprehensive consideration, select the appropriate mechanism.

(4) Coordinated design of the system

The coordinated design of the execution system is based on the requirements of the process for each action, analyze how the actuator should coordinate and cooperate, design the motion cycle diagram of the machine, and guide the design, installation and debugging of the actuator.

Complex equipment is usually composed of multiple actuators. Once the modalities of the executive agencies have been chosen, it is also necessary to coordinate their functioning in a certain order so as to unify them as a whole and achieve the desired objectives. If the operation of the various institutions is not coordinated, it will destroy the entire working process of the machinery, fail to meet the work requirements, and even destroy the parts and products, resulting in production and personal accidents.

(5) Mechanism scale design

The scale design of the mechanism refers to the kinematics and dynamics design of each selected actuator, the motion size of each actuator is determined, and the motion diagram of each actuator is drawn.

(6) Motion and dynamic analysis

The motion analysis and dynamic analysis of the whole execution system are carried out to check whether it meets the motion requirements and dynamic performance requirements.

To sum up, different working principles can be adopted to achieve the same functional requirements; To realize the same working principle, different motion rules can be selected; To realize the same movement law, different forms of mechanism can be adopted. Therefore, there can be many different schemes to achieve the same desired functional requirements. The problem to be studied in the scheme design of mechanical execution system is to make reasonable use of the designer's professional knowledge and analytical ability, creatively conceive a variety of possible schemes, and select the best scheme [8].

The following takes the design of shaft (shafting) as an example to illustrate.

The main function of the shaft is to support rotating parts and transfer torque, which is one of the main categories of mechanical design. Figure 2.37 shows the typical shafting structure. The shaft is often equipped with gears, belt wheels or sprockets, etc. These rotating wheels transfer motion between the shafts by engaging gears, belts or chains. The shaft segment that installs the hub of the transmission part is called the shaft head, the shaft segment that is matched with the bearing is called the journal, and the part that connects the shaft head and the journal is called the shaft body. The combination of shaft, bearing and shaft parts constitutes the shafting, which is an important part of the machine and has a significant impact on the normal operation of the machine.

The main tasks of shaft design are material selection, structure design and working capacity calculation. On the one hand, according to the conditions of use, reasonable selection of materials, determine the main size, to ensure that it has enough working

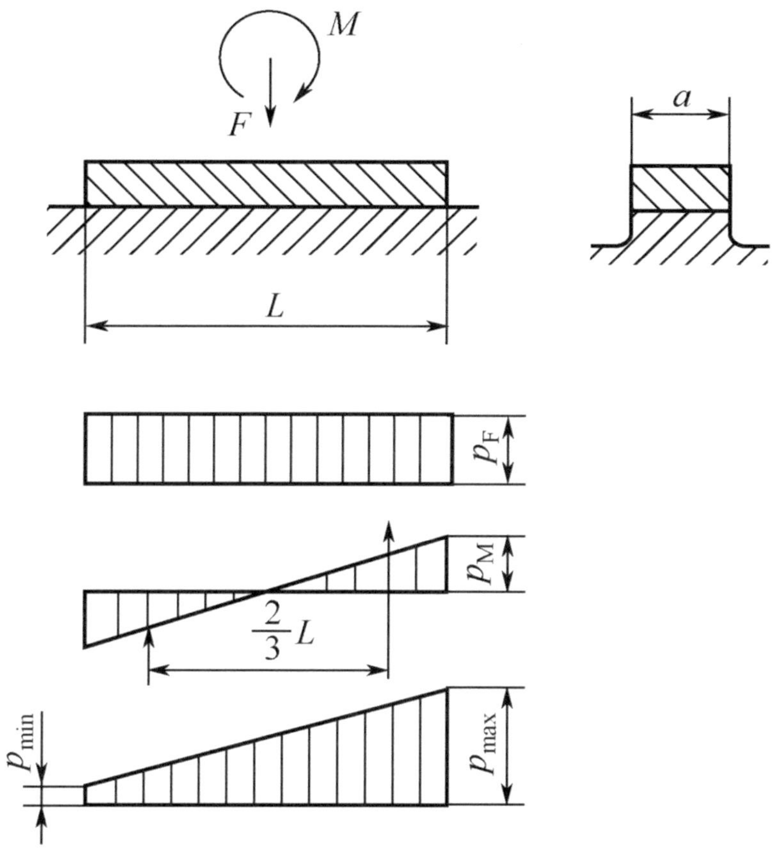

Fig. 2.37 Typical shaft system structure

capacity to meet the requirements of strength, stiffness and vibration stability. On the other hand, the structural design of the shaft should be carried out: according to the requirements of the installation and positioning of the parts on the shaft and the manufacturing process of the shaft, the structural form and size of the shaft should be reasonably determined. The structural design of the shaft also includes the calculation of the working capacity of the shaft. The checking calculation of the bearing capacity of the shaft refers to the checking calculation of the strength, stiffness and vibration stability of the shaft, which should make the force of the shafting reasonable, improve the strength, stiffness and vibration stability of the shaft, save materials and reduce weight. The shaft and the parts on the shaft should be accurately positioned, fixed and reliable, easy to assemble and disassemble and adjust, and should also have good processing and assembly technology, and try to avoid stress concentration. The loading mode of the drive shaft is mainly the torsion caused by the transmission of torque, or the bending caused by the lateral load acting on the gear, belt wheel or sprocket. The most common loading condition of the shaft is the combined action of fluctuating cyclic torque and fluctuating cyclic torque. To design the execution shaft mechanism, it is necessary to determine the structure of the execution shaft, the type of bearings, the combination and arrangement of bearings, the transmission mode and the arrangement of transmission parts, etc. [8, 9].

(1) Selection of bearings

Bearings are used to support rotation of rotating parts such as shafts, reduce bearing friction, and ensure rotation accuracy. Bearing is an important part of shafting and an important part of contemporary mechanical equipment. Bearing selection has been described in the previous section, this section mainly discusses bearing layout. According to the friction form of the relative moving surface, bearings are divided into two categories: rolling bearings and plain bearings. Commonly used rolling bearings have been standardized, mass produced by specialized bearing manufacturers, and have been widely used in mechanical equipment. The design only needs to select the appropriate type according to the working conditions, determine the specification size according to the life calculation, and carry out the combined structure design of the rolling bearing.

When analyzing and designing the combined structure of the rolling bearing, the fixation of the shaft and the parts on the shaft (including axial positioning and circumferential positioning) should be considered; Matching of bearing with shaft and bearing seat; Lubrication and sealing of bearings; Improve the rigidity of shafting and other aspects of the problem. Obviously, the consideration at this time should also be the entire shafting, not just the bearing itself.

Rolling bearings have the characteristics of large speed variation, good bearing capacity, low friction and wear, low lubrication requirements and low cost. According to the load, speed and other requirements, choose the appropriate form. When the precision of the execution shaft is very high or the load is very large, because the shock resistance of the execution shaft mechanism mainly depends on the front support bearing, the front support bearing of the execution shaft is generally selected by plain bearings, and the rear support and thrust bearings are still selected by rolling bearings.

(2) Bearing arrangement

In order to ensure that the shafting can withstand the axial force without axial movement, it is necessary to design the axial support and fixed structure of the shafting reasonably. The commonly used shafting support and fixed forms are:

1) Fixed at both ends (double fulcrum one-way fixed). The two ends of the shafting are supported by two bearings, each bearing bears an axial force in one direction. This kind of structure is simple and suitable for shafting with low working temperature and small supporting span (span ≤ 400 mm). In order to compensate for the thermal extension of the shaft, the bearing should be left with an axial clearance of 0.25 ~ 0.4 mm during assembly. The size of the clearance is usually adjusted by the adjusting gasket under the bearing cover or the adjusting screw screwed on the bearing cover, which is very convenient to adjust. Usually, two

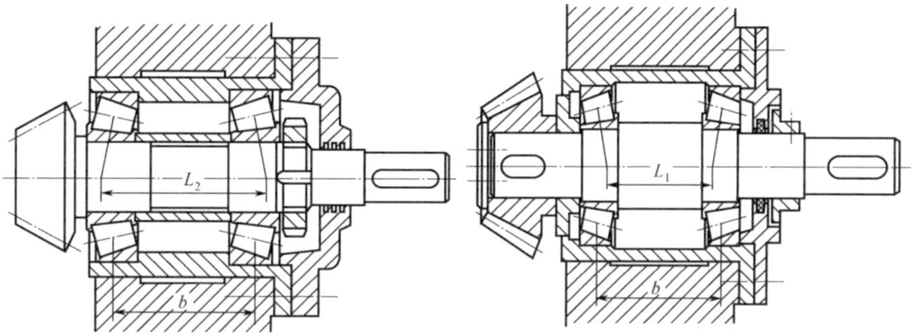

Fig. 2.38 Double pivot unidirectional fixation

tapered roller bearings or angular contact ball bearings are reversed at both ends of the execution shaft, as shown in Fig. 2.38. This configuration is simple in structure and easy to adjust, but after being heated, axial elongation will cause axial loosening of the bearing, affecting the accuracy and stiffness. It is suitable for high precision, small load and short execution shaft length.

2) One end fixed and one end moving (single fulcrum fixed bidirectional). With one end fixed and one end moving shafting, the bearings with two directions fixed end bear the axial force and control the clearance, and the floating floating bearings ensure that the support can move freely when the shaft expands. In order to avoid loosening, the inner ring of the running end bearing should be fixed with the shaft. This kind of structure is suitable for shafting with higher working temperature and larger supporting span.

When the shaft is long or the operating temperature is high, the thermal expansion elongation of the shaft is large, and the fulcrum structure with one end fixed and one end moving should be used. The bearings are fixed on both sides of the rear support for large loads and low accuracy in the workplace; The bearing is fixed on both sides of the front support, which is mostly used in work occasions with high precision and small load; When the bearing is fixed on the inside of the front support, the structure is more complex and the adjustment is not convenient, and it is generally only used in the workplace with large loads and high precision. Figure 2.39 shows the bidirectional fixed layout of a single fulcrum.

3) Floating support at both ends. The supporting bearings at both ends of the shafting (cylindrical roller bearings) can move in the axial direction to meet the requirements of the main and slave wheels when the herringbone gear is driven. Of course, this structural form is rarely used, and is only used for similar special occasions. This support method is used for the shaft that requires the right and left bidirectional swimming, and the shafting structure with both ends swimming can be used. This configuration is usually used for high-speed shafts of gear drives to prevent gear sticking or uneven forces on both sides of the herringbone teeth.

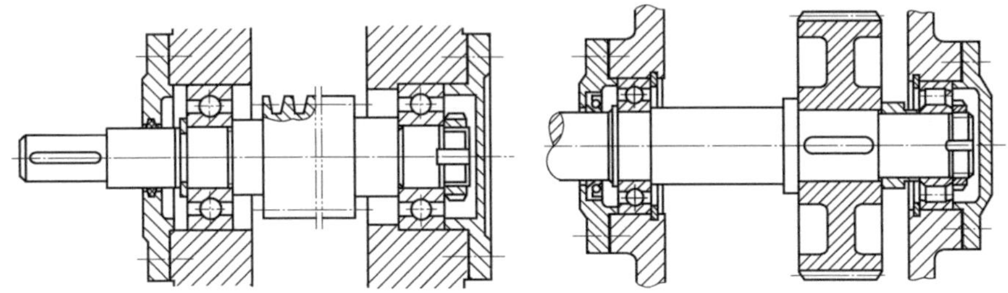

Fig. 2.39 Single pivot bidirectional fixation

As shown in Fig. 2.40, both ends of the herringbone gear shafting are cylindrical roller bearings.

It is worth pointing out that the axial position of the shaft parts and bearings on the shaft is mostly located by the shaft shoulder or sleeve, and the positioning end face should maintain a good perpendicularity with the axis; The radius of the shaft shoulder must be less than the radius or width of the fillet of the corresponding shaft part or bearing. For the positioning of rolling bearings, the height of the shaft shoulder should be less than 3/4 of the height of the bearing inner ring to facilitate the removal of the bearing, as shown in Fig. 2.41.

In addition to conventional bearings, ceramic bearings, maglev bearings and other high-performance advanced bearings can also be used in some occasions. Ceramic bearing has good thermal stability, large elastic modulus, large stiffness, light weight, small centrifugal force acting on the rolling body at high speed, and small friction. The rolling body is ceramic, the coefficient of thermal expansion is small, the temperature

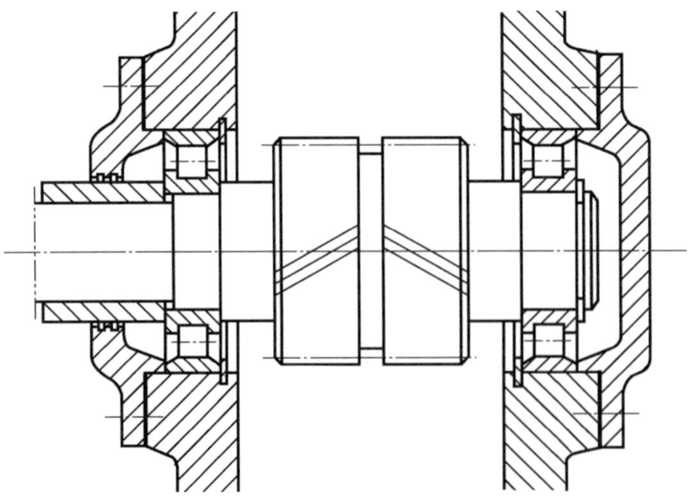

Fig. 2.40 Double end swimming support

Fig. 2.41 A structure that facilitates the disassembly of bearings

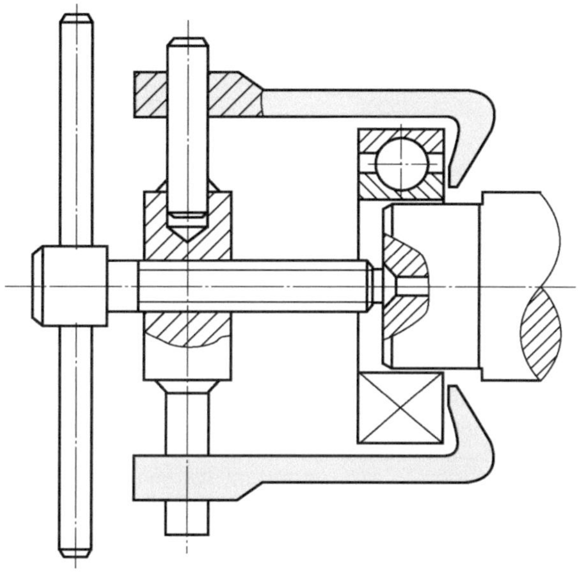

rise is slow, and the movement is stable. Ceramic bearings are especially suitable for rotating parts of high-speed, ultra-high-speed and ultra-precision equipment.

Maglev bearings use magnetic force to support moving parts so that they are non-contact with fixed parts to achieve bearing function (Fig. 2.42). Maglev bearing is composed of stator and rotor (Fig. 2.43), no mechanical wear, no noise, small temperature rise, no need for lubrication. Maglev bearings are especially suitable for high-speed and ultra-high-speed machining equipment.

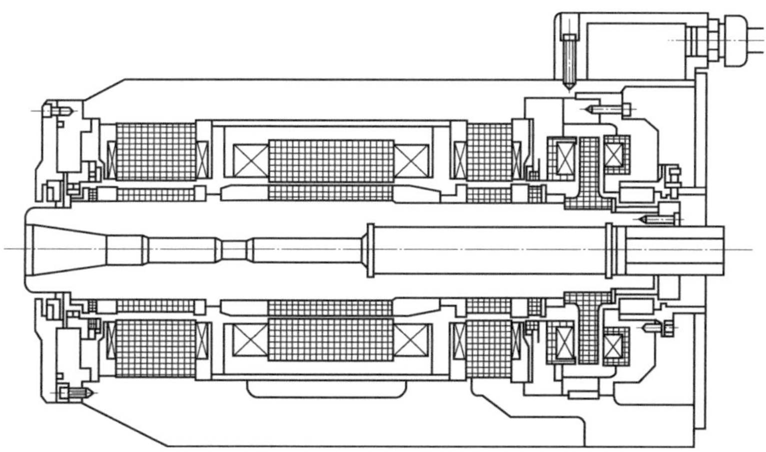

Fig. 2.42 Magnetic levitation bearing spindle

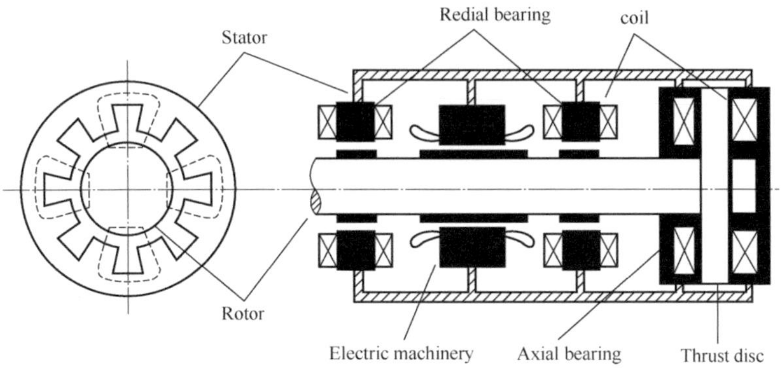

Fig. 2.43 Schematic diagram of magnetic levitation bearing

Lubrication is very important for bearings, lubricants and lubrication methods should be selected according to the operating conditions of the equipment, and the operating conditions considered mainly include operating temperature and operating speed. For equipment with general speed, lubricating grease can be used for bearing lubrication; For high-speed equipment, in order to improve the life of lubrication, lubricating oil should be used, and other auxiliary devices should be added if necessary, such as high pressure oil injection.

In addition, it is also necessary to protect the bearings, such as the equipment does not work for a long time, and the bearings need to be treated with rust prevention; Reasonable design of sealing device, reasonable selection of lubricating grease, bearing insulation measures to prevent current through.

This section only makes a brief description of the selection of rolling bearings, and the design of bearings can be consulted in detail in the mechanical design manual.

(3) Selection of transmission mode

1) Gear drive. Gear transmission refers to the device that transmits motion and power by gear pairs, which is the most widely used mechanical transmission mode in modern equipment. Its transmission ratio is accurate, high efficiency, compact structure, reliable operation, long life, can transmit large torque, suitable for variable speed, variable load work occasions. But the tooth shape has the inevitable error, resulting in the transmission is not smooth and the transmission precision is limited.

Gear transmission has high transmission accuracy, wide application range, can achieve parallel shaft, intersecting shaft, staggered shaft and other space between any two shaft transmission, reliable work, long service life, high transmission efficiency (generally 0.94 ~ 0.99), manufacturing and installation requirements are higher (high cost), the environmental conditions are more stringent, not suitable for far apart between the two shaft transmission, vibration reductionAnd impact resistance is not as good as flexible transmission such as belt drive.

Gear transmission is carried out by the meshing of teeth and teeth, and the teeth are the direct working part of the gear, so the failure of the gear mainly occurs on the teeth. The main failure forms are tooth break, tooth pitting, tooth surface wear, tooth surface bonding and plastic deformation.

① The tooth is broken. Tooth fracture usually has two cases: one is due to repeated bending stress and stress concentration caused by fatigue fracture; The other is the overload fracture caused by sudden severe overload or impact load. In particular, gears made of brittle materials (cast iron, hardened steel, etc.) are more prone to tooth breakage. Both kinds of fracture start from the side of the tooth under tensile stress. Increasing the radius of tooth transition, improving the mechanical properties of the material, reducing the surface roughness to reduce the stress concentration, and strengthening the tooth root (such as shot peening, rolling extrusion) can improve the fracture resistance of the tooth.

② pitting of tooth surface. When the gear teeth work, the front meshing is under the repeated action of alternating contact stress, and several small cracks will be produced on the tooth surface near the pitch line. With the expansion of the cracks, small pieces of metal will be spalling off, which is called pitting of the tooth surface. The further expansion of pitting on the tooth surface will affect the smoothness of the transmission, and produce vibration and noise, resulting in the gear can not work normally. Pitting is a common failure form of well-lubricated closed gear drive. The pitting resistance of the tooth surface can be improved by increasing the hardness of the tooth surface and decreasing the surface roughness. Open gear drive, because the tooth surface wear faster, no pitting.

③ Tooth surface wear. When the gear teeth engage, due to the relative sliding, especially when the external hard particles enter the working face between the gear teeth, it will cause the gear tooth surface wear. After the tooth surface is gradually worn, the tooth surface will lose the correct tooth shape, and in serious cases, the tooth is too thin and broken, and the tooth surface wear is the main failure form of open gear drive. In order to reduce wear, important gear drives should be closed drive, and pay attention to lubrication.

④ The tooth surface is glued. In high-speed and heavy-duty gear transmission, the pressure between the tooth surfaces is large, the temperature rise is high, and the lubrication effect is poor. When the instantaneous temperature is too high, the two tooth surfaces are partially melted and the metals are glued to each other. When the two tooth surfaces are relative to each other, the adhesive part is torn, thus forming a band or a large area of scars along the sliding direction on the tooth surface. The transmission of low speed and heavy load is not easy to form oil film, and the friction heat is not large, but it may also appear cold gluing due to heavy load. The use of large viscosity or good adhesion resistance lubricating oil, reduce the surface roughness to form good lubrication conditions; The anti-bonding ability of tooth surface can be enhanced by increasing the hardness of tooth surface.

⑤ Plastic deformation of tooth surface. The soft tooth surface gear with low hardness loses its original tooth shape under the action of friction due to the excessive pressure on the tooth surface at low speed and heavy load. Increasing the hardness of the tooth surface and using lubricating oil with higher viscosity can help to prevent or reduce the plastic deformation of the tooth surface.

The different failure modes of gear transmission do not necessarily occur at the same time on a pair of gears, but they affect each other. Pitting of the tooth surface will aggravate the wear of the tooth surface, and serious wear will lead to tooth fracture. The statistics show that the main failure forms of tooth break and tooth pitting are under certain conditions, so the main failure forms that may occur should be analyzed according to the actual working conditions to determine the corresponding design criteria.

According to the failure form of gear transmission, the following design criteria are proposed:

① Closed soft tooth surface (hardness $\leq$ 350HBW) gear drive. When the lubrication condition is good, the pitting of the tooth surface will be the main failure form, which is usually designed according to the contact fatigue strength of the tooth surface, and then checked according to the bending fatigue strength of the tooth root.

② Closed hard tooth surface (hardness > 350HBW) gear drive. The pitting resistance is strong, the possibility of tooth fracture is large, in the design calculation usually according to the tooth root bending fatigue strength design, and then according to the tooth surface contact fatigue strength check.

③ Open gear drive. The main failure form is the wear of the tooth surface, but because the wear mechanism is more complex, there is no mature design calculation method, so it can only be calculated according to the bending fatigue strength of the tooth root, with an increase of 10 ~ 20% modulus to increase the tooth thickness, so that it has a longer service life, so as to consider the impact of wear.

2) Belt drive. Belt drive has the characteristics of simple structure, smooth transmission, cushioning vibration absorption, large shaft spacing and multi-axis power transmission, low cost, no lubrication, easy maintenance, etc., which is widely used in modern mechanical transmission. The friction belt drive can be overloaded and slip, the running noise is low, but the transmission ratio is not accurate; Synchronous belt drive can ensure transmission synchronization, but the absorption capacity of load changes is slightly poor, high-speed operation has noise, occupy a large space.

According to different uses, belt transmission can be divided into general industrial belts, automotive belts, agricultural machinery belts and household appliances belts. Friction belt is divided into flat belt, V-belt and special belt (multi-wedge belt, circular belt) according to its different cross-section shape.

The type of transmission belt is usually selected according to the type of the working machine, the use of the environment and the characteristics of various belts. When there are a variety of transmission belts to meet the transmission needs, the optimal scheme can be selected according to the compactness of the transmission structure, production costs and operating expenses, as well as market supply and other factors.

① Flat belt drive. When the flat belt drive works, the belt is set on the smooth wheel face and driven by the friction between the belt and the wheel face. The transmission forms are open transmission, cross transmission and semi-cross transmission, respectively to meet the needs of different relative positions and different rotation directions of the drive shaft and the driven shaft. Flat belt transmission structure is simple, but easy to slip, usually used for transmission ratio of about 3.

Flat with rubber belt, braided belt, strong nylon belt and high-speed ring belt. Rubber belt is the most commonly used type of flat belt. It has high strength and wide transmission power range. The braided belt is flexible, but easy to relax. The strong nylon belt has high strength and is not easy to relax. The section size of the flat belt has standard specifications, and can be selected to any length and connected into a ring with gluing, stitching or metal joints. High-speed ring belt thin and soft, good flexibility, good wear resistance, and can be made into endless ring, smooth transmission, dedicated to high-speed transmission.

② V-belt drive. When the V-belt drive works, the belt is installed in the corresponding groove on the belt wheel, and the transmission is realized by the friction between the two walls of the belt and the groove. The V-belt is usually used in several, with a corresponding number of grooves on the pulley. When V-belt drive, the belt and pulley contact is good, the slip is small, the transmission ratio is relatively stable, and the operation is stable. V-belt drive is suitable for short center distance and large transmission ratio (but generally not more than 7) occasions, in the vertical and inclined transmission can also work better. In addition, because several V-belts are used, one of them will not cause an accident.

V-belt (also known as triangle belt) is an endless ring tape made of strong layer, stretch layer, compression layer and wrapping layer (Fig. 2.44). The strong layer is mainly used to withstand the tension, the stretching layer and the compression layer play the role of stretching and compression when bending, and the role of the coating layer is mainly to enhance the strength of the belt. Figure 2.45 shows the standard dimensions and lengths of a triangular belt. In addition, there is a flexible triangle belt, its section size standard is the same as the triangle belt, but the length specification is not limited, easy to install and tighten, local damage can be replaced locally, but the strength and smoothness are not as good as the triangle belt. The triangular belt is often used in parallel, and the type, number of roots and structure size of the belt wheel can be determined according to the transmitted power and the speed of the wheel.

a. Standard V-belt. Mostly used in household facilities, agricultural machinery, heavy machinery. The ratio of top width to height is 1.6:1. Belts using cord and fiber bundle as pulling elements transmit much less power than equal-width and narrow triangular belts. Due to their high tensile strength and lateral stiffness,

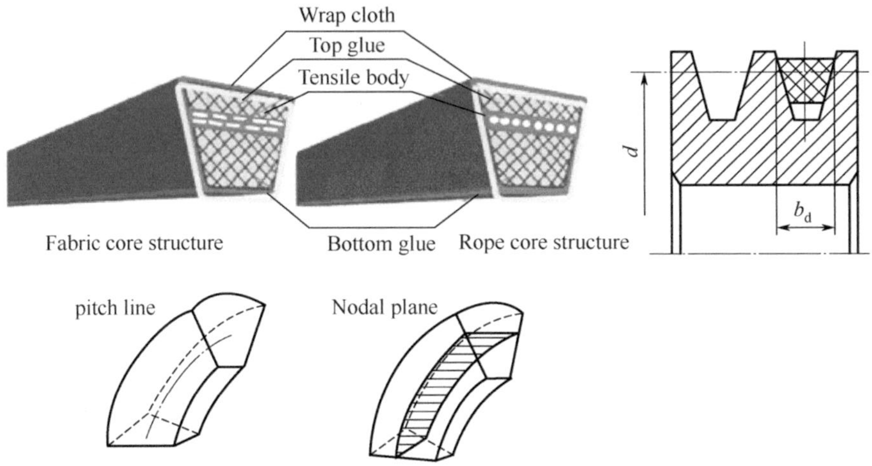

Fig. 2.44 Structure of V-belt

type	Y	Z	A	B	C	D	E	F
Top width b/mm	6	10	13	17	22	32	38	50
Section width b_d/mm	5.3	8.5	11	14	19	27	32	42
height h/mm	4	6	8	10.5	13.5	19	23.5	30
Wedge angle φ	40°							
Cross-sectional area A/mm^2	18	47	81	138	230	476	692	1173

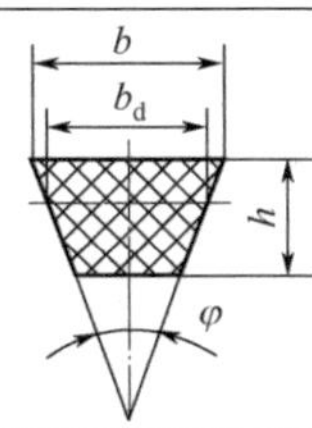

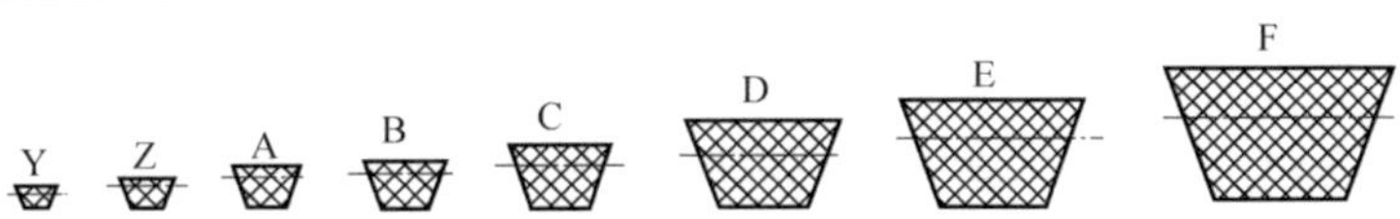

Fig. 2.45 Part of V-belt cross-sectional dimensions

these belts are suitable for harsh operating conditions with sudden changes in load. Belt speed up to 30 m/s, bending frequency up to 40 Hz.

b. Narrow V-belt. The ratio of the top width to the height of the V-belt is 1.2:1. The narrow V-belt is a variant of the standard V-belt, which eliminates the central part that has little effect on power transmission, and the power transmitted is higher than that of the standard V-belt of the same width. Band speed up to 42 m/s, bending frequency up to 100 Hz.

c. V-belt with thick edge. Narrow V-belt with thick edge for automobile, the fiber under the surface is perpendicular to the direction of movement of the belt, so that the belt has high flexibility, but also has excellent lateral stiffness and high wear resistance. These fibers also provide good support for specially treated tension elements. When used on the small diameter belt wheel than the narrow V-belt wrapped edge transmission ability is stronger, longer life.

d. Other. The latest development of the V-belt is the V-belt made of Kevlar fiber bearing elements. Kevlar has high tensile strength, low elongation and can withstand high temperatures.

③ Multi-wedge belt (multi-slot belt) drive. The flexibility is good, and the back of the band can also be used to transfer power. If the containment Angle around each driven pulley is large enough, it is possible to use one such belt to drive several accessories of the vehicle (alternator, fan, water pump, air conditioning compressor, power steering pump, etc.) at the same time. It is available in 5 sections, allowing the use of a narrower pulley than the narrow V-belt (diameter dmin $\approx$ 45 mm). In order to be able to transmit the same power, the preload force of this belt is preferably about 20% greater than that of the narrow triangle belt.

④ synchronous belt drive. Synchronous belt drive is a special belt drive. The working face of the belt is made into a tooth shape, and the rim surface of the belt wheel is also made into a corresponding tooth shape. The belt and the belt wheel are mainly driven by meshing. Synchronous toothed belt is generally made of thin steel wire rope as a strong layer, coated with polyurethane or neoprene rubber. The center line of the strong layer is the pitch line of the belt, and the perimeter of the belt line is the nominal length. The basic parameters of the tape are pitch p and module m. The circumference pitch p is equal to the dimension measured along the pitch line between the corresponding points of the adjacent two teeth, and the modulus $m = p/\pi$. China's synchronous gear belt adopts the modular system, and its specification is expressed by "module X bandwidth X number of teeth".

Synchronous gear belt drive is mainly used in situations requiring accurate transmission ratios, such as external equipment in computers, movie projectors, video recorders and textile machinery. Compared with the ordinary belt drive, the synchronous gear belt drive has the advantages of compact structure and good wear resistance. The deformation of the strong layer made of steel wire rope is small after loading, the circumference pitch of the toothed belt is basically unchanged, there is no relative sliding between the belt and the pulley, and the transmission ratio is constant and accurate. Toothed belt thin and light, can be used for high speed occasions, transmission line speed up to 40 m/s, transmission ratio up to 10, transmission efficiency up to 98%.

The disadvantage is that due to the small pretension, the carrying capacity is also small; High manufacturing and installation accuracy requirements, strict center distance requirements, high cost.

In addition to gear transmission and belt transmission, common transmission forms are chain transmission and direct prime mover transmission (electric spindle). Chain drive is a meshing drive with chain as the middle flexible part. Compared with belt drive, chain drive has the advantages of stable average transmission ratio, small axial force and high efficiency. Compared with gear transmission, it has the advantages of good environmental adaptability, low cost and suitable for long-distance transmission. Figure 2.46 shows the structure of a sleeve roller chain. Direct prime

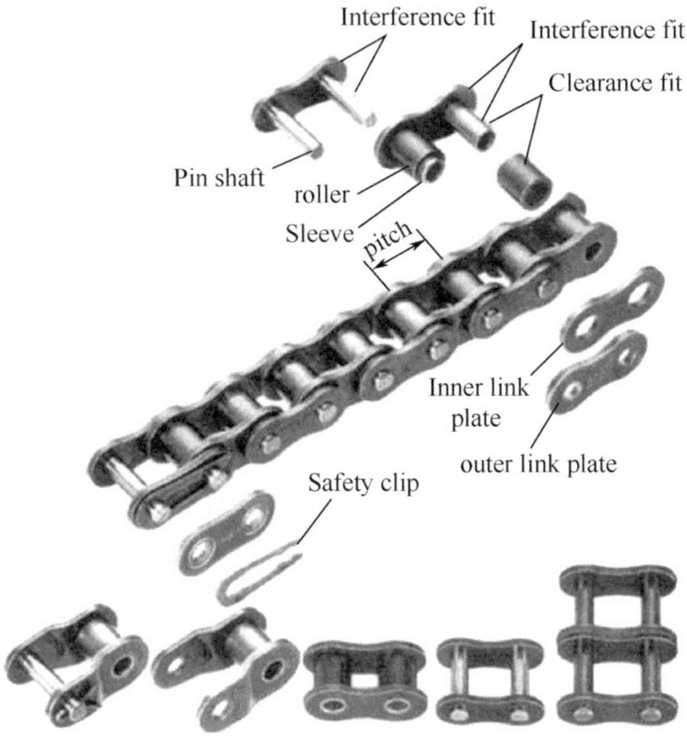

Fig. 2.46 Structure of sleeve roller chain

mover transmission is the use of prime mover directly drive the execution shaft rotation transmission mode. The rotor shaft of the prime mover can be directly connected with the execution shaft, which is suitable for high-speed execution shaft mechanism. Figure 2.47 shows the structure of the electric spindle of a CNC machine tool. The relationship of space, here will not be repeated.

(4) Layout of shaft transmission parts

To reduce the bending deformation of the execution shaft caused by power transmission, arrange the transmission member between the front and rear supports as close as possible to the front supports to reduce the length of the torsion section of the execution shaft, as shown in Fig. 2.48. When there are several transmission parts on the execution shaft, the larger transmission part should be arranged near the front support to reduce the deformation of the execution shaft. If the stiffness of the shaft is required to be high but the stiffness of the bearing is low, the direction of the power transmission and the direction of the external load should be reversed to reduce the support reaction of the front support. Otherwise, arrange them in the same direction to reduce the deformation of the execution shaft, as shown in Fig. 2.49.

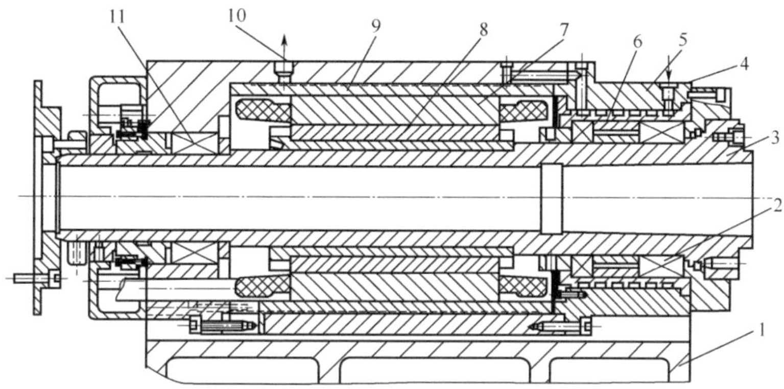

Fig. 2.47 Structure of CNC machine tool electric spindle, 1 main axle box, 2 front bearing of main shaft, 3 main shaft, 4 coolant inlet; 5 main spindle front bearing seat, 6 front bearing cooling sleeve, and 7 fixing element; 8 rotors; 9 fixed cooling jacket; 10 coolant outlet; 11 rear bearing of spindle

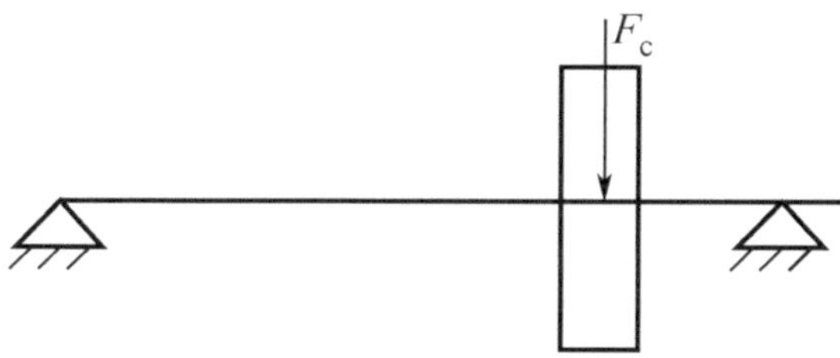

Fig. 2.48 Arrangement of actuator transmission

Fig. 2.49 Execution axis force diagram

2.7 Dynamic Design of Intelligent Manufacturing Equipment Ontology

The improvement of the level of automation has greatly reduced the labor intensity and production difficulty of human beings, and reduced the consumption of human and material resources. At present, industrial production has put forward more requirements for the performance, accuracy, automation and intelligence of mechanical equipment. The vibration problem of mechanical system is becoming increasingly prominent, and the design of mechanical structure is facing new challenges. The traditional design mode and method can not meet the development needs of the current equipment industry, so it is imperative to integrate the dynamic design theory. The dynamic performance of mechanical system has become one of the important indexes in product development and design.

Dynamic design refers to the dynamic performance of mechanical structure and mechanical system in the drawing design stage has been fully considered, the whole design process is essentially using dynamic analysis technology, with the help of computer analysis, computer-aided design and simulation to achieve the purpose of improving design efficiency and design quality.

The traditional mechanical design belongs to static design, with insufficient consideration of various parameters. The design process is based on theory and experience, and lacks pertinence and applicability, which is difficult to meet the needs of market competition and social development. The dynamic design of mechanical structure is based on the action of dynamic load, design the parameters of the structure, determine the structure form, design a safe and economic structure to meet the actual use requirements. This design mode can fully understand the dynamic characteristics of the structure, quickly find out the problems that may occur in the running state of the structure, provide a basis for modifying the structure, solve the running problems of the structure, effectively improve the stability, reliability and safety of the structure, ensure the working life of mechanical equipment, and optimize the performance of the structure.

The dynamic design of intelligent manufacturing equipment structure involves computer technology, design technology, dynamic analysis technology, mechanical modeling and so on. In the specific design, it is necessary to build a dynamic model and carry out dynamic characteristics analysis according to the functional requirements and design standards, and the structure meets the dynamic characteristics design requirements under the dynamic load state, so that the mechanical structure has excellent dynamic performance.

2.7.1 Principles of Dynamic Design

Intelligent manufacturing equipment itself is produced in order to facilitate user operation, so the design of the body should be from the user's point of view to meet the needs of users for different product functions. At the same time, with the development of science and technology, designers need to collect a large amount of technical information, master the current development trend and trend of intelligent manufacturing equipment, and pay attention to the collection of information on new technologies, new processes and new materials when designing the body, and maintain the advanced nature of the product. In the further design, it is necessary to consider the system of the body, start from the whole, and ensure the coordination between each subsystem. Under the condition of meeting the functional requirements, the body with simple structure and small number of parts should be made as far as possible, so that it is conducive to processing and assembly and easy to operate.

2.7.2 Steps for Dynamic Design of the Machine Body

The steps of dynamic design of intelligent manufacturing equipment body mainly include overall scheme design, calculation of main parameters, overall structure design, analysis and evaluation, modification and improvement.

(1) Overall scheme design

In the program design stage, the designer must clarify the functional requirements to be realized by the body, do a lot of research and analysis, comprehensively consider the limitations and influences of the external environment, put forward as many design schemes as possible, and determine an optimal scheme by comparing the advantages and disadvantages of different schemes.

(2) Main parameter calculation

The main parameters of the body refer to the size parameters, motion parameters and dynamic parameters, which reflect the working characteristics and technical performance of mechanical products, when calculating the main parameters, it is necessary to reasonably determine the size of the parameters, to avoid unreasonable design, the size is too large not only takes up large space is bulky, but also causes material waste and increases the cost; The size is too small, the mechanical properties may not meet the requirements, and the product life and reliability are difficult to guarantee.

(3) Overall structural design

After determining the scheme and calculating the main parameters, the designer needs to consider the overall layout of the equipment and the selection of parts, determine the size of the main mechanism, and draw the overall structure. The selection of parts should be generalized, standardized and serialized as far as possible to reduce design and manufacturing costs. The structure drawing should reflect the basic structure, relative position relationship and transmission mode of the main parts.

(4) Analysis and evaluation

After the overall design of the body is completed, it also needs technical personnel to analyze, discuss and evaluate it. This paper mainly evaluates the principle scheme, technical design and structural scheme, and points out the shortcomings and improvement directions in the design.

(5) Modification and improvement

After finding out the problems or deficiencies in the design, the designer also needs to modify and improve the overall design. The design of the body structure is only a preliminary design, there are inevitably a variety of problems, so in the whole life cycle of the body, the product is constantly modified and improved, only in this way can the product gradually mature.

(6) Provide the final design plan

After continuous discussion and modification, the detailed design scheme is given, including assembly drawing, parts drawing, design specification, software and so on.

2.7.3 Dynamic Performance Analysis of 3 Intelligent Manufacturing Equipment Ontology

The dynamic characteristics of the system are the natural frequency and damping characteristics of the mechanical system itself, the vibration modes corresponding to each order of natural frequency and the mechanical response under dynamic load. Modes are the natural vibration characteristics of mechanical structures, and each mode has a specific natural frequency, damping ratio and mode shape. The process of analyzing these modal parameters is called modal analysis, which is a method to study the dynamic characteristics of structures and an application of system identification methods in the field of engineering vibration. These modal parameters can be obtained by calculation or experimental analysis, such a calculation or experimental analysis process is called modal analysis. Modal analysis can evaluate the dynamic characteristics of existing structural systems, diagnose and predict the faults of structural systems. In the new product design, the dynamic characteristics of the structure are estimated and optimized, the load of the structural system is identified, and the radiation noise of the structure is controlled.

To design the mechanical system of intelligent manufacturing equipment, not only to meet the static requirements, but also to ensure good dynamic performance. Therefore, after the design of the airframe is completed, it is also necessary to conduct dynamic performance analysis. The theoretical basis of dynamic performance analysis is modal analysis and modal synthesis theory. The main methods used include finite element analysis, model test and transfer function analysis.

(1) Finite element analysis method

Finite element analysis method (FEA) is used and the mathematical method of approximate to the real physical system (geometry and loading conditions were simulated). By using simple and interacting elements (i.e. units), a real system of infinite unknowns can be approximated with a finite number of unknowns. With the development of computer technology and calculation methods, finite element analysis has been paid more and more attention in the field of engineering design and scientific research and has been widely used, and has become an effective way to solve complex engineering analysis and calculation problems. From automobiles to space shuttles, almost all design and manufacturing have been inseparable from finite element analysis and calculation, which is used in mechanical manufacturing, material processing, aerospace and automobile. The wide use in the fields of civil engineering, electrical and electronic engineering, national defense, shipbuilding, railway, petrochemical, energy and scientific research has made a qualitative leap in the design level.

(2) Model testing method

Model test method refers to a method of using various technologies and analog devices to carry out realistic test on the operating system and get the required data in line with the actual situation. The model test method is to reduce the mechanical phenomenon acting on the body according to a certain similarity relationship, and then reproduce it on the model, perform vibration excitation input on the mechanical system, obtain parameters expressing the dynamic characteristics of the mechanical system through measurement and calculation (natural frequency, damping ratio, mode shape), and then convert the similarity relationship to the prototype to judge the dynamic performance of the body and verify the rationality of the design.

The process of parameter identification to obtain modal parameters of the collected system input and output signals through experiments is called experimental modal analysis. In general, modal analysis refers to experimental modal analysis. Vibration mode is an inherent and integral characteristic of elastic structure. If the modal analysis method is used to understand the characteristics of the main modes of the structure in a certain susceptible frequency range, the actual vibration response of the structure can be predicted under the action of various external or internal vibration sources in this frequency band. Therefore, modal analysis is an important method for structural dynamic design and equipment fault diagnosis. The ultimate goal of modal analysis is to identify the modal parameters of the system, and provide the basis for the analysis of the vibration characteristics of the structural system, the diagnosis and prediction of the vibration fault, and the optimal design of the structural dynamic characteristics.

Hammer modal testing, also known as hammer modal testing, is a simple and intuitive way to measure and process input force and response data, and display the results. Two hammering methods are provided: fixed knock point moving response point and fixed response point moving response point. The structure is stimulated by the hammer, and the acceleration and force signals are collected and processed at the same time, and the transfer function matrix of the structure is obtained in real time. At the same time, the measurement parameters can be set easily.

The mode test of the exciter method is mainly to control the exciter through the output signal source of the analyzer, and stimulate the tested part. The output signal includes the swept frequency sine, random noise, sine, frequency modulation pulse and other signals.

The first step of the test modal analysis is to establish the test system, determine the test object, select the excitation mode, select the force sensor and response sensor, and calibrate the entire test system. The second step is to measure the response data of the tested system, which is the key step of the test mode. The accuracy and reliability of the measured data directly affect the results of the modal test. Under the action of a certain exciting force, once the system under test is excited, the time domain signal of the exciting force or response can be measured by the test instrument, and the average estimate of the frequency response function of the system can be obtained. The third step is to estimate modal parameters, including natural frequency, modal shape, modal damping, modal stiffness and modal mass, by using the measured frequency response function or time history. Finally, the modal model verification and the correctness of the results obtained from the third step of modal parameter estimation are necessary for the evaluation of modal test results and further dynamic analysis of the measured system.

The excitation modes include natural excitation and artificial excitation. Natural vibration sources include earthquake, ground pulsation, wind vibration, sea wave and so on. Ground pulsation is often used as the excitation of large structures, which is characterized by a wide frequency band, including various frequency components, but the randomness is large, and the sampling time is long. Artificial vibration sources include vibration starter, vibration exciter, seismic simulator, vehicle vibration, blasting, tension release, mechanical vibration, human body shaking and pile driving. The two methods of blasting and tension release are widely used. In engineering practice, the appropriate excitation mode should be selected according to the characteristics of the tested object.

(3) Transfer function analysis method

The transfer function is the ratio of the Laplace transform of the linear system response (i.e., the output) to the Laplace transform of the excitation (i.e., the input) under zero initial conditions. It is denoted as $G(s) = Y(s)/U(s)$, where $Y(s)$ and $U(s)$ are the Laplace transforms of the output and input, respectively. Transfer function is one of the basic mathematical tools to describe the dynamic characteristics of linear systems. The main research methods of classical control theory, frequency response method and root locus method, are based on transfer function. Transfer function is one of the main tools to study classical control theory. The transfer function of the system corresponds to the differential equation describing its motion. The transfer function of the whole system can be derived according to the transfer function of each unit of the system and the relationship between them, and it can be used to analyze the dynamic characteristics and stability of the system, or according to the given requirements of the comprehensive control system, and design a satisfactory controller. The method of analyzing and synthesizing control system with transfer function is called frequency domain method. It is not only the basis of classical control theory, but also in the development process of modern control theory based on time domain method, it also develops and forms the multi-variable frequency domain control theory, which becomes a powerful tool for the study of multi-variable control system.

Dynamic design of mechanical structure is an advanced design method which combines theory with practice and involves multi-disciplinary technology. At present, dynamic design technology of mechanical structure has been widely popularized in many developed countries. Through the dynamic design of mechanical structure, the accurate dynamic model is established to ensure the design and calculation accuracy, and the dynamic characteristics of the structure are analyzed, which has obvious advantages for the design of complex mechanical structures. In dynamic design, real-time dynamic test data can be obtained by modeling, which provides a basis for structural modification and has great popularization and application value.

2.8 Intelligent Manufacturing Equipment Ontology Optimization Design

2.8.1 The Main Content of Ontology Design

For precise, complex and large structural parts, it is difficult to calculate their static and dynamic properties by general mechanical methods to meet the engineering requirements. Structural optimization design is to find the best design scheme according to a certain goal (such as the lightest weight, the lowest cost, the maximum stiffness, etc.) under the given constraints, which was once called the optimal design of the structure or the optimal design of the structure, relative to the "structural analysis", also known as "structural synthesis". If the minimum weight of the structure is the goal, it is called the minimum weight design.

Optimization design mainly studies the theory and method of structural design, with a wide range of contents, including structural size optimization, structural form optimization, topology optimization, layout optimization, etc., and can also include reliability index optimization, material performance optimization, dynamic performance optimization, control structure optimization, etc. Mechanical structure optimization design is a comprehensive method and technology integrating finite element analysis technology, numerical optimization method and computer graphics technology, which is a multi-disciplinary mechanical structure design theory and technology.

Structural shape optimization design is a mechanical structure optimization design that determines the two-dimensional and three-dimensional structure shapes, aiming at improving the structural characteristics, improving the distribution of stress concentration, stress field and temperature field, and increasing the fatigue strength of members. The contents include determining the boundary shape and internal shape of the continuous structure, and the structural layout of the structural parts.

The structural optimization methods mainly include numerical method, variational method, sensitivity analysis and finite element analysis.

The axial end deformation y and natural frequency ω of the example shown in Fig. 2.50 were calculated by the finite element analysis method using the equation of state, that is, the values of Di, li and a were obtained to minimize the mass and meet the condition $y \leq [y]$.

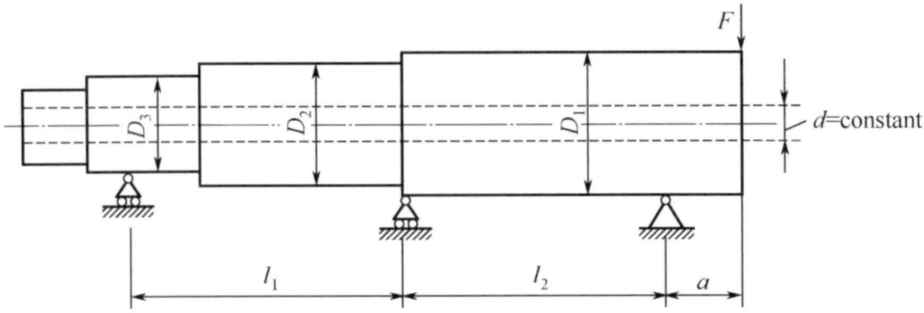

Fig. 2.50 Example of optimization design

2.8.2 *Optimization Design of Structural Modules*

This optimization method combines different aspects of product planning to propose different optimization methods. In other words, structural module optimization is from the perspective of product planning, decomposing the planning into separate tasks, solving the optimization problem from the source, and achieving higher work efficiency on the basis of reducing product problems. In this way, the quality of optimal design can be guaranteed and the efficiency of optimal design can be improved. Mechanical mechanism design should respect the original design and planning of the product, and optimize in terms of details. Combined with Feldman's concept, when optimizing product structure, it is necessary to involve four stages of product task functions, including functional component stage, functional component stage, functional component stage, and product stage. Excellent high-quality module structure can achieve coordination and connection parallel, with standardized interfaces, can produce compatibility, interchangeability and relevance in the process of convenience, generalization, economy, hierarchy, serialization and integration. On this basis, the design and optimization of mechanical structure should be combined with CAD drawing technology and software design to realize deformation design and combination design in the optimization process. According to the classification principle, the mechanical mechanism module is classified according to its size into four grades: components, components, components and products. It is a common method to distinguish functional areas by functional modules in the design of mechanical mechanism optimization. Functional decomposition can granulate the foundation so that the mechanical mechanism corresponds to the function one by one. In this way, the mechanical mechanism module can realize the mapping effect, which is an important prerequisite for improving the smooth optimization of the mechanical mechanism [10].

2.8.3 Optimization Design of System Modules

System optimization is a hierarchical design, the designer first regards the mechanical mechanism as a whole, the optimization process as a complete structure, each optimization element can be regarded as a separate component, and the close relationship between these separate components is clear. Hierarchical design considers both the whole and the part, which can realize the design element is not only a single individual, but also a common whole, and achieve the purpose of overall optimization.

At present, the mechanical mechanism optimization system model is generally in accordance with the German VDI2221 design method (Figs. 2.51 and 2.52), and some mechanism optimization systems use the system idea of independent research and development in China. Design is regarded as a system composed of several design elements, each design element is independent, and there are hierarchical connections between the elements. According to user needs for mechanical structure design, quality function segment first according to user needs to clarify the basic functions and characteristics of the mechanical structure, and then according to the functional characteristics of the mechanical structure to determine the characteristics of each component. The characteristics of parts are taken as the core, and the process characteristics of parts are defined in the process of discussion and experiment. Finally, according to the analysis of the process characteristics, the characteristics of the system optimization operation are obtained, and the final system optimization method is determined. System optimization and design is based on the product as a whole system, through the planning and definition of different areas and different characteristics, to achieve the overall design of the product, and finally get the design result of the finished product. Representative design methods include bonding graph method, proof design method, ideation design method, graphic modeling method, design element method, etc. Each method needs to be combined with the system to complete the formulation of optimal design methods and schemes.

2.8.4 Optimization Design of Product Features

In the optimization design of mechanical system structure, the corresponding optimization design should be carried out according to the characteristics of product development. Computer aided design can improve the optimization design effect and automation level, complete the management, coordination and expression of mechanical mechanism optimization, and play a decisive role in this process, throughout the optimization design of mechanical products.

Nowadays, product feature optimization methods mainly include instance method, coding method and hybrid method. The case method completes the description of conceptual entity and engineering example through the frame structure, obtains candidate resources in the process of reasoning, and matches the candidate resources

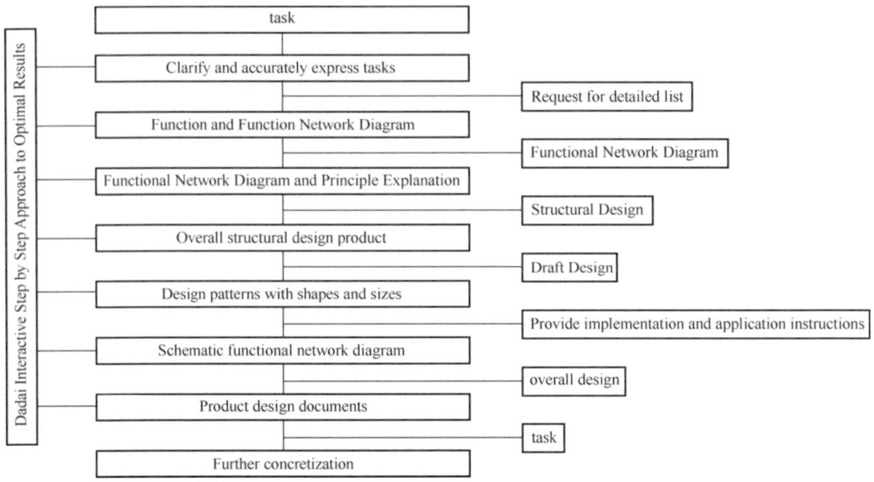

Fig. 2.51 The general process of product development and design

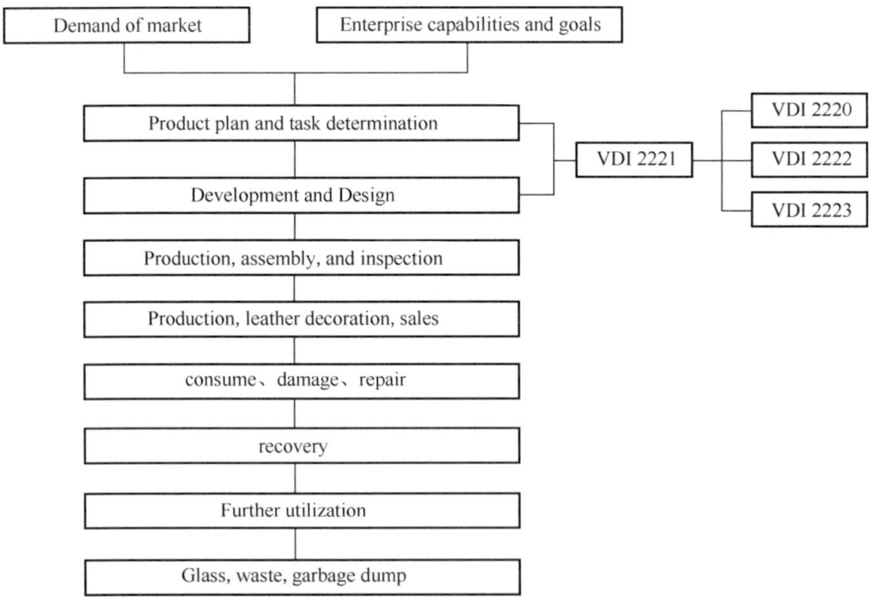

Fig. 2.52 Product life cycle

into the optimization prevention and matching design. Coding method is combined with motion conversion to realize mechanism sorting and separation. Through the collection and arrangement of knowledge base, the optimization design of mechanical structure is completed, and the optimization design scheme is determined. Finally, there is the hybrid expression, which integrates the network system, framework, process and rules to achieve high-quality product feature design.

2.8.5 Mechanical Structure Optimization Methods

Mechanical structure optimization mainly includes design parameters and design rules, which is the key to the success or failure of the design.

(1) Design parameter optimization

Mechanical structure optimization involves thousands of design variables and hundreds of design functions. Due to the large number of functions and variables involved, it is very difficult to design programs. To solve this problem, it is necessary to establish parameter information model in design, so that parameters can be used for reference in optimization and improve the effect of design optimization.

(2) Design rule extraction

By extracting optimization rules and changing model design parameters, an effective optimization example is obtained. In the calculation, the results are included in the database, the data is discrete through the management system, and the finite element of the data is collected. Rough set theory is used to analyze finite element, and data mining is used to extract data optimization rules.

2.9 Digital Design of Intelligent Manufacturing Equipment

Intelligent manufacturing equipment belongs to high-end equipment, which has an important role in promoting social production and economic development. Digital design is to transform traditional product design methods through digital means, aiming at establishing a set of design methods based on computer technology and network information technology to support the whole process of product development and production. The connotation of digital design is to support the whole process of product development, product innovation design, relevant data management, development process control and optimization. In the digital design of intelligent manufacturing equipment, product modeling is the foundation, optimization design is the main body, and data management is the core.

With the development of computer technology and information technology, digital design has entered every link of industrial design, inspection and production in our country. In 2020, the penetration rate of digital R&D and design tools in key areas of China's manufacturing industry exceeds 70%, the CNC rate of key processes exceeds 50%, the penetration rate of digital workshops/smart factories exceeds 20%, and the operating cost, product development cycle and product defect rate are greatly reduced. Figure 2.53 shows the architecture of the manufacturing digital factory.

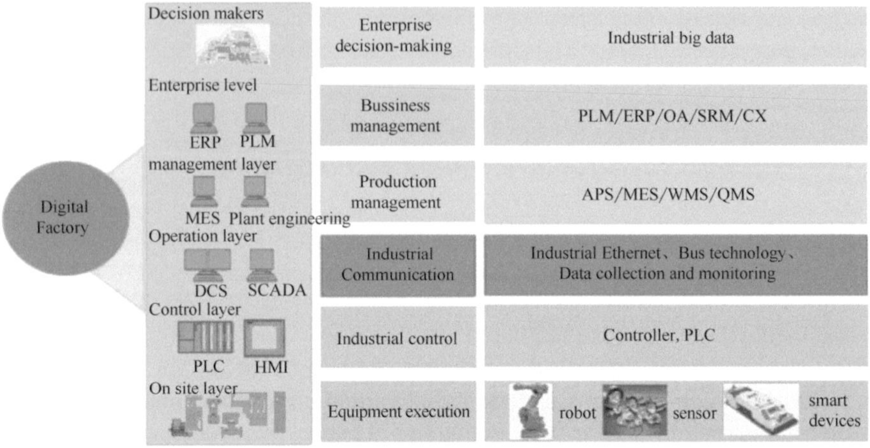

Fig. 2.53 Architecture of digital factory in manufacturing industry

Although digital design has developed rapidly, there is still a certain distance to realize the digitalization of the entire design production. The analysis of digital design of intelligent manufacturing equipment can help us rationally understand the current development level of digital design and intelligent manufacturing equipment design and production, and provide suggestions for its development direction.

2.9.1 Current Situation of Digital Design

The working mode of digital design and manufacturing is mainly divided into serial design and parallel design. The organization mode of serial design is hierarchical structure, the activities of each stage are carried out in time order, each stage is arranged in turn, and each has its own input and output; The working mode of parallel design is to consider the related work in the subsequent stages, including processing technology, assembly, inspection, etc., while designing the product. The work in each stage of the product development process of parallel design is carried out cross-over, as shown in Fig. 2.54. Compared with the traditional design and manufacturing process, digital design and manufacturing has the characteristics of process extension, high intelligence level and high integration, and its performance requires seven main aspects: stability, integration, agility, active sharing ability of manufacturing engineering information, digital simulation ability, ability to support heterogeneous distributed environment and expansion ability.

Digital manufacturing refers to the digitalization of the manufacturing field, which is the result of the intersection, integration, development and application of manufacturing technology, computer technology, network technology and management

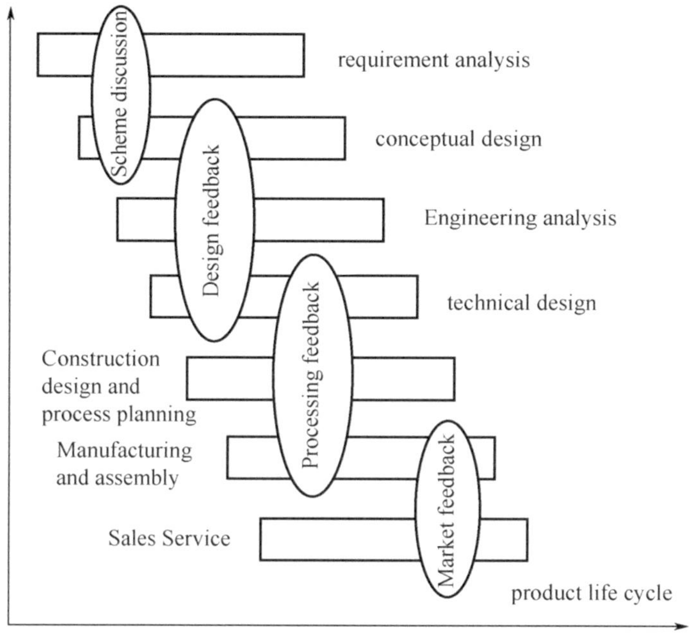

Fig. 2.54 Concurrent design

science, and is also the inevitable trend of manufacturing enterprises, manufacturing systems and production processes, and production systems to continuously realize digitalization. Its connotation includes three levels: digital manufacturing technology with design as the center and control as the centerDigital manufacturing technology, management-centered digital manufacturing technology. Digital manufacturing uses digital equipment such as CNC machine tools, machining centers, measuring equipment, transport cars, three-dimensional warehouses, and multi-level distributed control computers to complete manufacturing activities such as parts processing, assembly, material storage and transportation, automatic detection and monitoring through computer scheduling and control according to the product's engineering and technical information and workshop processing instructions. It can realize flexible automated manufacturing of multi-variety, small and medium-sized batch products, improve production efficiency and product quality, shorten production cycle, and reduce costs to meet the rapid response needs of the market. Key technologies include rapid process preparation, high-speed cutting of complex structural parts, rapid forming, flexible and reconfigurable production lines, and manufacturing execution systems.

As the foundation and main link of digital manufacturing technology, digital design uses digital product modeling, simulation, multidisciplinary comprehensive optimization, virtual prototype, information integration and process integration and other technologies and methods to complete the conceptual design of products, engineering and structural analysis, structural performance optimization, process design and numerical control programming. Digital design can realize the optimal design of mechanical equipment, improve the development decision-making ability, accelerate the product development process, shorten the development cycle, and reduce the development cost.

Digital design can reduce the production of mock-ups in the design process. The traditional design needs to go through the repeated process of "prototype trial–production–prototype test–modification design" in product development, which prolongates the product development cycle, consumes time and energy, and costs high. In digital design, before the physical prototype is manufactured, the digital model is simulated, analyzed and tested, and the irrational part of the design is found and optimized in time, and the parallel design is easy to realize. Digital design can make a design work by multiple design teams in different regions, parallel design, joint assembly, which is of great significance to improve the quality and speed of product design.

Design is a modern product development mode which uses computer technology, communication technology and management technology to assist product design. It breaks the traditional division of departments and closed organizational model, emphasizes the collaborative work of multi-functional teams, attaches importance to the reorganization and optimization of the product development process, and requires product developers to consider various factors in the product life cycle from the beginning of design. By setting up a product development team composed of multidisciplinary personnel, improving the product development process, using various computer-aided tools, etc., various factors in the product life cycle can be considered at the early stage of product development to ensure the success of product design and manufacturing.

The key technologies of digital design include full life cycle digital modeling, knowledge-based innovative design, multidisciplinary comprehensive optimization, parallel engineering, virtual prototype, remote collaborative design and so on.

2.9.2 Development of Digital Design

With the development of computer technology, CAD technology was born in the 1950s, which is a comprehensive high-tech technology integrating computer graphics, database, network communication and other computer and other fields of knowledge. It assists in the design and analysis of engineering and products in order to achieve ideal goals or achieve innovative results [11]. Product design realized the transition from manual drawing to computer digital drawing, and designers got rid of the drawing board that had been used for more than 200 years. In the 1970s, the rapid development of the aircraft and automobile industry requires the use of computers to deal with a variety of curves and space curves, digital design has been developed by leaps and bounds, the late addition of mass, center of gravity, moment of inertia and other parameters, the birth of CAE, CAM model expression, the development

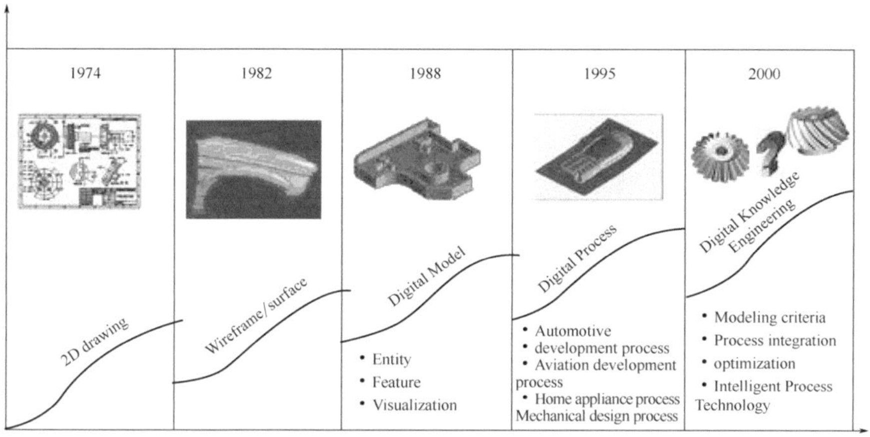

Fig. 2.55 The development history of digital design

of digital design and manufacturing technology. Figure 2.55 shows the development of digital design.

At present, most enterprises only use CAD drawings for design, production, etc., some enterprises use three-dimensional as an auxiliary design, and engineering drawings still use two-dimensional drawings, which is not high design efficiency. Some enterprises based on MBD digital design and manufacturing technology, that is, model-based digital definition technology, 3D manufacturing information PMI and 3D design information are jointly defined into the product 3D digital model, so that CAD and CAM can achieve a real high degree of integration, so that the production and manufacturing process can no longer use two-dimensional drawings, this part of the enterprise is still a small number, mainly concentrated in aviationAir as well as automotive fields. Figure 2.56 shows the MBD application.

MBD technology, also known as 3D labeling technology, is the only way for the development of 3D design, and is the core technology in which 3D models replace 2D engineering drawings as the only data source for processing and manufacturing. The concept of MBD technology and the establishment of corresponding specifications have been available for many years, originated from Boeing, and have been applied in many foreign enterprises. In the development process of Boeing 787 aircraft, MBD technology is fully adopted, and three-dimensional product manufacturing information and three-dimensional design information are jointly defined into the three-dimensional digital model of the product, abandoning the two-dimensional pattern and directly using the three-dimensional annotation model as the manufacturing basis, realizing the height set of product design, process design, tooling design, parts processing, parts assembly, parts inspection and inspectionIntegration, synergy and integration [12].

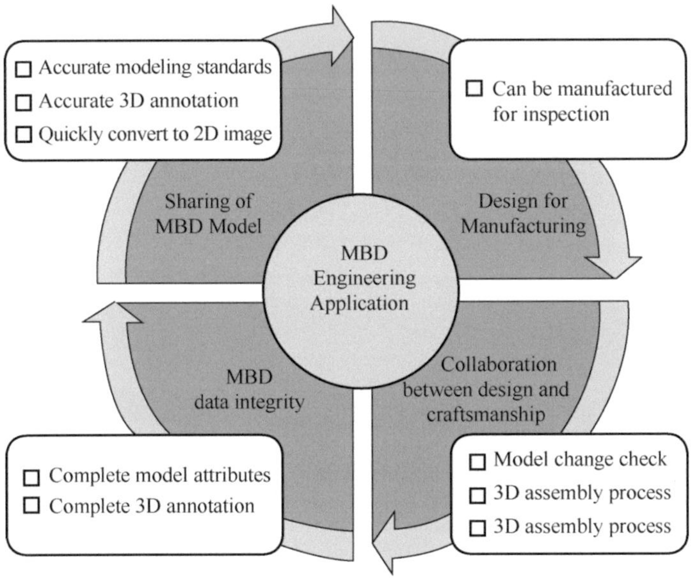

Fig. 2.56 Application of MBD

2.9.3 Main Methods and Commonly Used File Exchange Types for Digital Design and Manufacturing

(1) Feature modeling

A feature is a set of geometric entities with definite constraint relations, which also contains some specific functional semantic information. Features can be expressed as: product features = shape features + semantic information. Figure 2.57 shows the structure of the feature modeling framework.

Among them, product feature is a geometric entity with certain attributes, including feature attribute data, feature function and the relationship between features. Feature design is based on the solid model, according to the feature classification, to define a feature, describe the operational feature, specify the representation of the feature, and use the solid modeling to achieve specific, is the carrier of a variety of product information, including geometric information and non-geometric information. There are four types: shape feature, material feature, precision feature and assembly feature.

With feature technology, design intent can be incorporated into the product model and can be adjusted at any time. In addition, due to the use of engineering unit characteristics for modeling, the designer's randomness in the design is reduced, which helps to eliminate the conflict between the design result and the manufacturing realization. The essence of feature modeling is still entity modeling, but the engineering

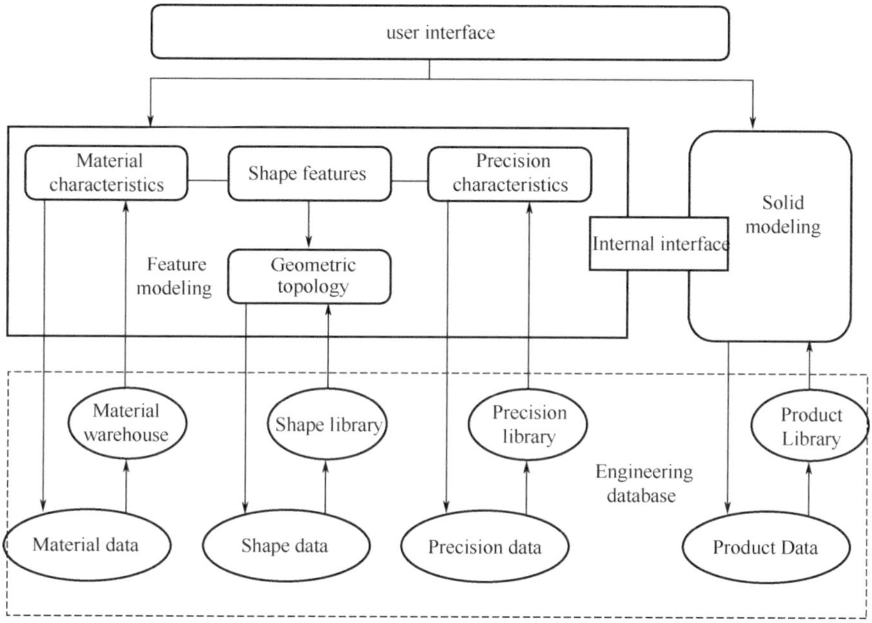

Fig. 2.57 Feature modeling framework structure

semantic abstraction is carried out, that is, semantic + shape features. At present, the best and most mature application is shape feature design.

The modeling system requires that the product part model established should include geometric data, topological data, shape feature data, accuracy data and technical data. The method is flexible and changeable, and it can easily realize the establishment, modification, deletion and update of the feature and part model. It can separately define and reference the data of each level in the product model and associate it to form a new featureCharacteristics and parts model to meet the needs of various application fields. Figure 2.58 shows the shape and feature classification.

(2) Parametric design and variable design (Table 2.7)

Parametric design refers to the method that the structure shape of the design object is basically unchanged, and the size relationship is defined by a set of parameters. Parametric design is based on features, full size constraints, size driven design modification, full data correlation, control size has a display correspondence, the modification of the result is driven by size.

Variable design provides users with a three-dimensional environment for interactive operation models. Designers can directly operate when defining relationships on components, and no longer care about how two-dimensional design information becomes three-dimensional, thus simplifying the process of design modeling. Designers can directly edit and modify any features on the parts graphically, which can achieve dynamic capture of design, analysis and manufacturing intentions.

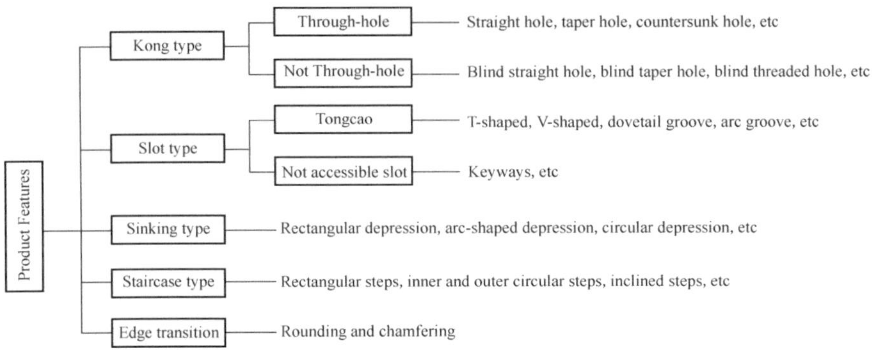

Fig. 2.58 Shape feature

Table 2.7 Comparison between variable technology and parameterization technology

Variable technology	Parameterization technology
The process involves separating shape constraints and size constraints for processing	The design process considers both shape and size together, and achieves control over geometric shape through size constraints
You can first decide on the shape of interest, and then provide some necessary dimensions. Whether all dimensions are specified does not affect subsequent operations	When not fully constrained, the modeling system does not allow subsequent operations to be performed
Engineering relationships can be directly coupled with geometric equations as constraints, and finally solved uniformly through constraint solvers	The engineering relationship does not directly participate in constraint management, but is processed externally by a separate processor
The mathematical method of simultaneous solving is adopted, and the order of solving equations does not matter	Demanding full constraints, each equation must be an explicit function, that is, the variables used must have been defined and assigned to a certain size parameter in the previous equation, and the solution of its geometric equation can only be solved sequentially
Variable technology solves product design problems under arbitrary constraints, which can not only achieve size driven but also constraint driven	Solves geometric shape problems under specific conditions (full constraints), manifested in size driven geometric shape modification

(3) Common file exchange types for digital design and manufacturing

There are three types of file exchange commonly used in digital design and manufacturing: IGES, STEP and DXF.IGES initial graphics exchange specification is the earliest and most widely used graphics data exchange standard in the world. The basic unit of IGES file information is the entity. STEP product model data exchange standard is a product data expression and exchange standard formulated by the International Organization for Standardization (ISO), product model data covers the whole life cycle of products, shape feature information model is the core of STEP product model, and geometric information exchange is a widely used part of STEP standard. DXF (Data Exchange File) is an open vector data format, DXF is widely used in CAD systems, most CAD systems can read or output DXF files.

(4) The main process of digital design and manufacturing

Product digital model is the concentrated embodiment of digital design and manufacturing, and is the carrier of product function information, performance information, structure information, part geometry information, assembly information, technology and processing information. The design process first sets one or a group of part models as the main model, and other models are based on the main model, on which the new model is constructed.

The model of product design stage includes four stages: conceptual design, part geometry model, product model simulation and product model assembly.

Design includes product scheme composition, innovative design, etc. The designer follows the design specification, proposes the product design scheme from the analysis of functional requirements, and completes the design in the form of scheme report, sketch, etc. At this stage, it is not necessary to consider the precise shape and geometric parameter design of the product.

Part geometric model stage is the core of product detailed design, is the key content of conceptual design refinement, is the basis of all subsequent work, and is the most suitable stage for computer representation of product models. The geometric information of a geometric model is represented by a two-dimensional or three-dimensional model, the non-geometric information is represented by an attribute, and the definition of the attribute information is explained in text form. Part geometry model is the information model produced in the detailed design stage, and is the information carrier of other stages of design, usually as the main model.

The geometric model can be expressed by wire-frame model, surface model and solid model. Wire frame model refers to describing a 3D wire frame model in a computer and giving two kinds of information: vertex table and edge table. Its disadvantage is that the information is too simple, there is no surface information, can not be blanking, there is ambiguity in the understanding of the model when displaying, it is not easy to describe objects containing curved surfaces, and can not be applied to engineering analysis and automatic calculation of CNC machining tool trajectory. The surface model is an exponential data structure represented by three layers of information: surface, edge and point. The surface model avoids the ambiguity of the wire-frame model and represents the shell of the geometric shape of the part, which does not have the solid characteristics of the part and cannot calculate the physical characteristics, such as moment of inertia, volume, etc. Solid model generally refers to the model represented by five layers of structural information: "body–surface–ring–edge–point". The most common solid modeling methods are boundary description method and constructive solid geometry method. The solid

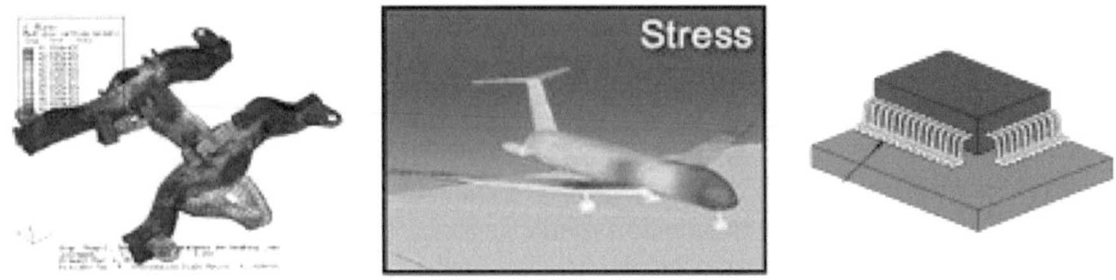

(a) Automobile chassis (b) Aircraft whole machine (c) Chip

Fig. 2.59 Product CAE simulation

modeling method is completely effective in representing the shape and geometric properties of objects.

The product model simulation stage is generally not carried out directly on the geometric model of the part generated in the detailed design stage. The product simulation model expresses the information of the simulation analysis stage, and the simulation model is based on the geometric model of the design stage, which needs to be optimized and the details deleted to reduce the calculation amount. Therefore, it is necessary to constantly feed back to the relevant design engineers to adjust or modify the product model. Figure 2.59 shows the CAE simulation.

In the assembly stage of the product model, the product assembly model represents the structural relationship between the components of the product, the assembly bill of materials, the assembly constraint relationship, the assembly sequence and the path planning for the reality, and so on. The assembly structure tree reflects the overall structure of the product. Attribute information table is used to represent the non-geometric information of the product. The assembly constraint model includes assembly feature description, assembly relation description, assembly operation description and assembly constraint parameter. Assembly planning model is used for assembly sequence planning and path planning.

The model of product manufacturing stage includes three stages: process information model design stage, tooling model stage and CNC machining model stage.

The process information model provides basic information for CAPP in the design stage, and establishes the process information model according to the machining requirements of parts and the information such as size, roughness, datum and machining method. Process design data is derived from geometric models and assembly models generated during the detailed tolerance design phase. The tooling model stage is an intermediate state model stage produced by continuous evolution. Tooling model includes two parts: tooling design model and product process model. NC machining model stage is the stage of NC machining design model and corresponding NC program generation.

Compared with the traditional two-dimensional blueprint design and manufacturing process (Fig. 2.60), the digital design and manufacturing process (Fig. 2.61) is different from the traditional two-dimensional blueprint design and manufacturing

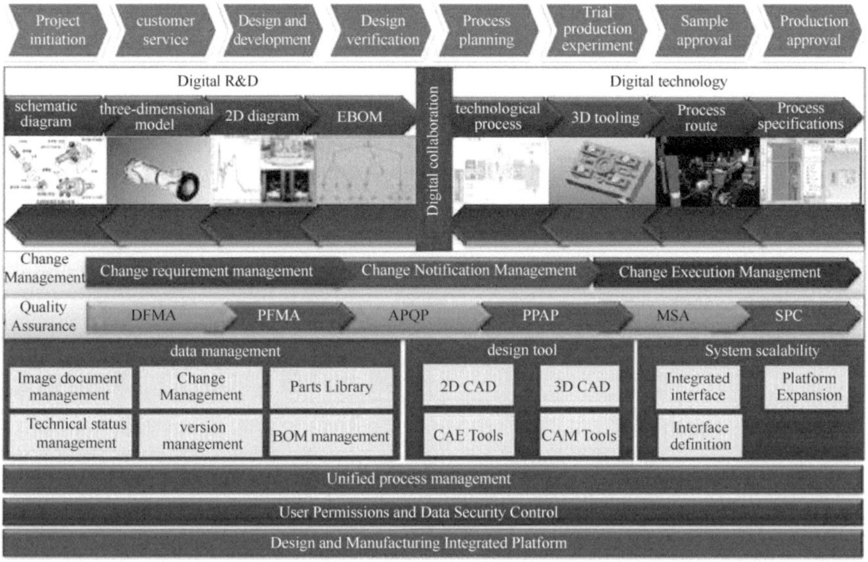

Fig. 2.60 Digital design and manufacturing process

process (Fig. 2.61). In addition to the digitization of the information conversion process, the digital design and manufacturing mostly uses digital prototypes for analysis and inspection, while the traditional design and manufacturing process requires the assistance of physical prototypes.

The physical prototype is generally a product model made of material materials. The digital prototype refers to the full digital model of the machine or subsystem of the mechanical product expressed by the computer as the carrier. It accurately expresses the real physical product with the size ratio of 1:1, and its role is to verify the function and performance of the physical prototype with the digital prototype. A digital prototype is a digital model of a product expressed on a computer compared to a physical prototype. In the field of CAD, the concept of virtual prototype is actually the meaning of digital prototype. The virtual reality technology used in the establishment of virtual prototype greatly enhances the practicability of digital design and

Fig. 2.61 2D blueprint design and manufacturing process

manufacturing because of its characteristics of autonomy, interactivity and immersion. The characteristics of digital prototype are: (1) authenticity. The purpose of the digital prototype is to replace or simplify the physical prototype, equal to the physical prototype in important aspects such as simulation, and maintain the same geometric appearance, physical characteristics and behavioral characteristics as the physical prototype. (2) For the whole life cycle of the product. The traditional engineering simulation only analyzes a certain aspect of the product, while the digital prototype is a comprehensive simulation of the product. Digital prototyping is a consortium of distributed, different tool-developed and even heterogeneous sub-models, including CAD models, appearance models, functional and performance simulation models, various analysis models, use and maintenance models, and environmental models. (3) Interdisciplinarity. Complex product design usually involves many different fields such as machinery, control, electronics, fluid power, etc. It is necessary to conduct complete and accurate simulation analysis of subsystems in many different disciplines as a whole, so that the digital prototype can meet the designer's requirements for functional verification and performance analysis.

Different classification standards, different categories of digital prototypes, the following is the classification of common digital prototypes.

According to the different degree of digital prototype reflecting the integrity of mechanical products, it can be divided into full machine prototype and subsystem prototype. The full machine prototype is a digital prototype that completely describes all the structural parts, system equipment, functional components and accessories of the system. The subsystem prototype is a digital description of all the information contained in the subsystem divided according to the different functions of the mechanical product.

According to the different development process of digital prototype, it can be divided into scheme prototype, detailed prototype and production prototype. Scheme prototype refers to the prototype of product scheme design stage, including the digital description of all the information of product scheme design; Detailed prototype refers to the prototype of product detailed design stage, including the digital description of all information of product detailed design; Production prototype refers to the prototype of the product production stage, including the digital description of all the information of product manufacturing and assembly.

According to the special use or purpose of the digital prototype, it can be divided into geometric prototype, functional prototype, performance prototype and special prototype. The geometric prototype focuses on the geometric description of products. Functional prototype focuses on product function description; Performance prototype focuses on product performance description; Special prototype can support simulation, training, marketing and other special purposes.

2.9.4 Future Trends of Digital Design and Manufacturing

With the continuous development of network technology, various types of terminals with environmental awareness and computing modes based on network technology have promoted the application of the Internet of Things in the industrial field more and more widely, and continue to integrate into all aspects of industrial production, upgrading traditional industry to intelligent industry. Production process detection, parameter collection, equipment monitoring, material consumption and other production links can be monitored in real time to achieve intelligent monitoring, intelligent control, intelligent diagnosis, intelligent decision-making and intelligent maintenance of the production process, so as to achieve the purpose of establishing a digital factory (Fig. 2.62). Enterprises can also use the industrial cloud platform (Fig. 2.63) to achieve digital integration such as collaborative R&D, manufacturing, and supply.

At present, the penetration rate of intelligent equipment in automobile production line is very high, and the process of body assembly and other processes are automatically completed by intelligent robots. Because the intelligent manufacturing equipment used comes from different suppliers, the management, monitoring and control of all intelligent manufacturing equipment are relatively isolated, and if the unified system platform of intelligent manufacturing equipment status, failure and control can be centralized management, it can maximize the intelligence and efficiency of automobile production.

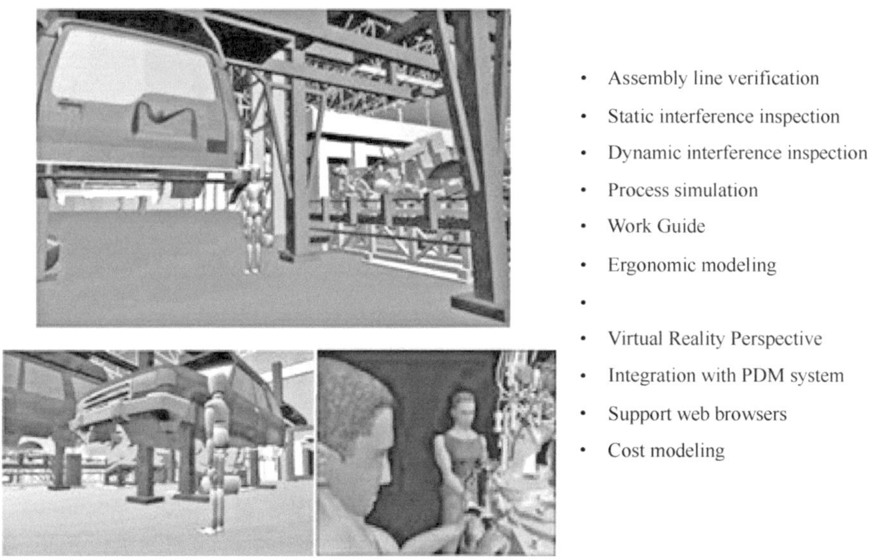

Fig. 2.62 Digital factory

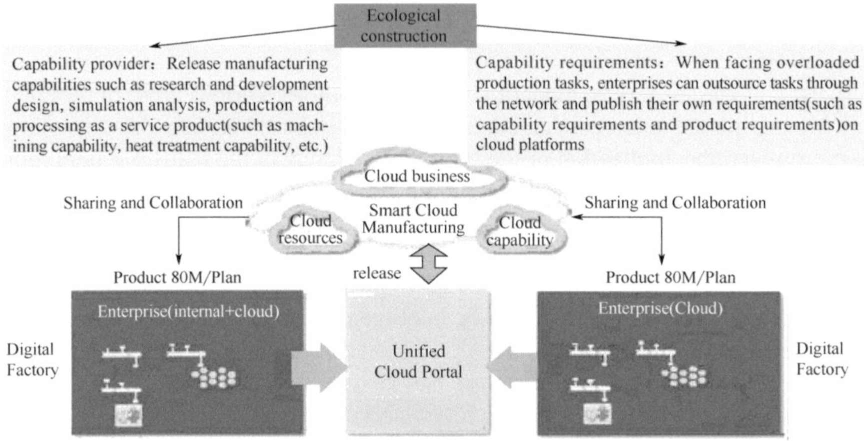

Fig. 2.63 Digital cloud platform

The automatic control scheme of intelligent manufacturing equipment of automobile enterprises adopts collaborative middleware system as the basic system, which is composed of data service layer, IoT perception layer, platform service layer, etc. Through the IoT sensing layer, various intelligent devices can be connected, including robots, I/O devices, sensor devices, etc. The sensing layer can parse the data of intelligent manufacturing equipment into platform data through protocol converter and send it to the platform service layer. The platform service layer processes the intelligent equipment data and sends it to the data service layer. The data services layer is responsible for big data analysis and data storage. Platform services are the "brain" of the entire platform, responsible for all equipment management, data management, communication management, rights management, etc., and can release platform services in standard communication protocols to support the collaborative call of third-party systems. The IoT awareness layer supports the access and control of all devices.

2.9.5 Digital Design of Intelligent Manufacturing Equipment

Digital design refers to the application of computer-aided design technology in the field of product design, through the digital platform based on product description, the establishment of digital product models, and the application in the product development process, to reduce or avoid the physical model of product development technology. Digital design avoids the "prototype production-prototype test-modified design" link of traditional design, and simulates and tests digital models by computer means before the prototype is born, eliminating unreasonable design factors in the bud, which can reduce the manufacture of physical models in the design process,

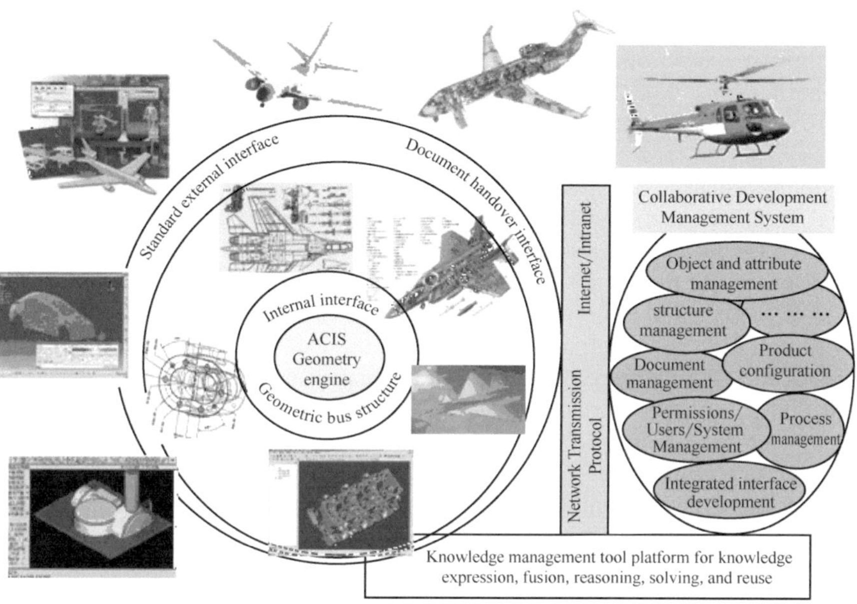

Fig. 2.64 Digital design of intelligent manufacturing equipment

speed up the development cycle, and reduce design costs. Figure 2.64 shows the digital design of intelligent manufacturing equipment.

The digital design method of intelligent manufacturing equipment can adopt the "1 + 3 + X" comprehensive design method, that is, the use of functional optimization, dynamic optimization, intelligent optimization and visual optimization and the design of a certain product has special requirements to complete the design work. The twenty-first century is a relatively mature industrial era of automation, with the "Industry 4.0" and "Made in China 2025" proposed, automation has made standardized mass production reached a very high level, but when the production of personalized, small batch demand becomes more and more, there are new challenges, from the lean point of view, quality, cost and delivery have become difficultHard.

At present, the interactive interface of intelligent manufacturing equipment is mostly designed by engineers and technicians based on functional requirements, ignoring the impact of human, machine and environment on the interactive interface, and simple technical information is difficult to achieve correct and efficient guidance. This requires designers to conduct in-depth and systematic research on the interface interaction design of intelligent manufacturing equipment terminal. The information transmission of intelligent manufacturing equipment is output or fed back to the user by the interactive interface of the mobile terminal, and then the user inputs the instructions. The human–machine interface is friendly, the logic operation is accurate, the database is reasonable and other performance indicators directly affect the overall use effect and intelligence of the intelligent manufacturing equipment system. The application and popularization of digital design and manufacturing and intelligent manufacturing equipment is an inevitable trend of historical development.

The result of digital design is digital prototype and virtual prototype, which combines virtual technology and simulation method on the basis of CAD model, and provides a new design method for product development. Virtual prototype is a prototype system or subsystem model based on a computer. Before establishing a physical prototype, designers use computer technology to establish a mathematical model of the mechanical system, conduct simulation analysis, and display various characteristics of the system under real engineering conditions in a graphical way, so as to modify and obtain the optimal design scheme. Virtual prototype design environment is a collection of models, simulations and simulators. Using the advantages of virtual environment in visualization and interactive exploration of virtual object functions, interactive modeling and analysis of products in geometry, function, manufacturing and other aspects are carried out. To a certain extent, the functional authenticity of virtual prototype is comparable to that of physical prototype, and the development is rapid. Combined with rapid prototyping technology (additive manufacturing, 3D printing), compared with traditional prototype manufacturing, it can save a lot of manpower and material resources, and the product development cycle is greatly shortened. Figure 2.65 shows an example of the device digital design.

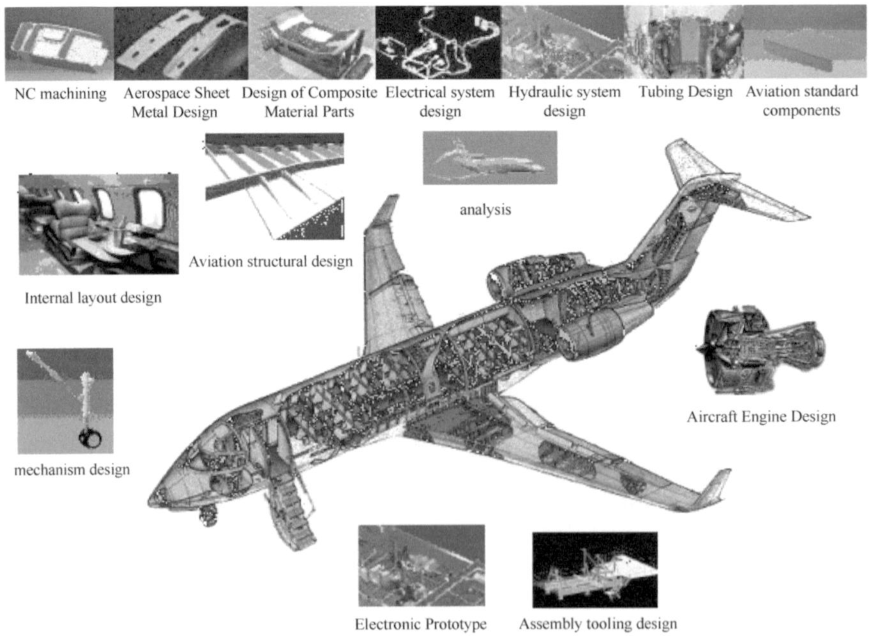

Fig. 2.65 Example of equipment digital design

2.9.6 *Green Design*

Green design is a design technology extended by green products, also known as ecological design and environmental design, which refers to the design of the whole process of the product and its life cycle, to fully consider the impact on resources and the environment, while considering the function, quality, development cycle and cost of the product, but also to optimize various related factors, so that the product and its manufacturing process on the overall negative impact on the environment. The minimum noise, so that the indicators of the product meet the requirements of green environmental protection, in the design stage will be environmental factors and pollution prevention measures into the product design, the environmental performance as the design goal and starting point of the product, and strive to make the impact of the product on the environment is minimal. Its core can be summarized as 3R1D (Reduce, Recycle and Reuse, Degradable) [13].

2.10 Summary of This Chapter

This chapter briefly introduces the relevant contents of the intelligent manufacturing equipment mechanical ontology design, and expounds the task, basic requirements, design criteria and main contents of the mechanical ontology design. On the basis of clear design tasks and contents, the design of feed transmission system, support system, execution system and dynamic design are carried out. In addition, this chapter also briefly introduces the mechanical optimization design and digital design. With the development of artificial intelligence and computer network, new design methods, design tools and design means continue to emerge. Due to the limited space, a lot of content did not expand in detail, readers can codnsult the relevant information if they are interested in further learning.

References

1. Yao Qihuai. Precision Mechanical Engineering [M]. Beijing: China Machine Press, 2015.
2. Li Huilai. Tutorial on Product Structural Design Examples: Introduction, Improvement, Proficiency, Job Application [M]. Beijing: Electronic Industry Press, 2013.
3. Lu Yaoshun, Zhang Yuchuan. Design Decision Analysis and Evaluation Model [J]. Journal of Wuhan University of Technology (Transportation Science & Engineering), 2000, 24(002):185–188.
4. Guan Huizhen. Design of Mechanical Manufacturing Equipment [M]. Beijing: China Machine Press, 2019.
5. Mechanical Design Branch of Chinese Mechanical Engineering Society, Xie Liyang. Modern Mechanical Design Methods [M]. Beijing: China Machine Press, 2007.
6. Wang Xiankui, Yi Hong, Tang Xiaoqi. Handbook of Machine Tool Numerical Control Technology [M]. Beijing: National Defense Industry Press, 2013.

7. Cheng Hu. Engineering Design of Linear Motion Rolling Guides [J]. Hebei Journal of Industrial Science and Technology, 2003, 20(3):38–42.
8. Geng Shuguang, Zhang Lixin. Scheme Design of Mechanical Execution System [J]. Journal of Jilin Engineering Normal University, 2004(06):29–31.
9. Robert L. Norton. Mechanical Design [M]. Translated by Huang Ping, Li Jingrong, Zhai Jingmei, et al. Beijing: China Machine Press, 2017.
10. Wu Kangping. Discussion on the Concept and Method of Mechanical Mechanism Optimization Design [J]. Internal Combustion Engines and Parts, 2018(12):34–35.
11. Wang Dakang. Computer-Aided Design and Manufacturing Technology [M]. Beijing: China Machine Press, 2005.
12. Liu Jianhua, Sun Liansheng, Zhang Xu, et al. Connotation and Key Issues of 3D Digital Design and Manufacturing Technology [J]. Computer Integrated Manufacturing Systems, 2014, 20(03):494–504.
13. Wang Longtai. Advanced Manufacturing Technology [M]. Beijing: China Machine Press, 2019.

Chapter 3
Design of Intelligent Manufacturing Equipment Drive System

Equipment actuator is a driving device that can provide linear or rotational motion, which uses a certain drive energy and works under a certain control signal. The driving mechanism of intelligent manufacturing equipment is mainly composed of two parts: the executive mechanism and the control system. The actuator is the terminal of information, so it is also called the actuator as the terminal element. After the information of the production process is processed by the controller, the operation instructions are output to the actuator to complete the production process, or the manual operation instructions issued by the operator station are given to the actuator to control the production process. Great attention must be paid to the design, installation, commissioning and maintenance of the actuator. The rational design of the drive mechanism, improper selection or use will affect the automation of the production process, lead to the decline of the control quality of the automatic control system, control failure, and even cause serious production accidents.

The transmission of energy provided by the drive system of traditional mechanical equipment to the working parts often requires a series of complex transmission structures, and these structures will bring problems such as wear, noise and energy loss, reduce processing accuracy and drive efficiency, resulting in increased production costs. In the future, the intelligent equipment drive system will develop in the direction of high-speed and efficient transmission.

M. Xu et al., *Design and Case of Intelligent Electromechanical Equipment System*, Advanced and Intelligent Manufacturing in China,
https://doi.org/10.1007/978-981-95-3833-1_3

3.1 Classification and Characteristics of Driving Mechanisms

3.1.1 Classification of Driving Mechanisms

(1) Classification by energy form used

According to the form of energy used, the drive mechanism can be divided into three categories: pneumatic actuator, electric actuator and hydraulic actuator. The pneumatic actuator uses compressed air as an energy source, the electric actuator uses electric energy as an energy source, and the hydraulic actuator uses high-pressure liquid as an energy source.

Electric actuators are mainly motors, electric cylinders, etc., with small size, fast signal transmission speed, high sensitivity and accuracy, simple installation and wiring, easy to transmit signals, etc., often used with DCS (distributed control system) and PLC (programmable logic controller). The pneumatic actuator has the advantages of simple structure, safe and reliable, large output torque, cheap price, intrinsically safe and explosion-proof, etc. Compared with electric actuator, pneumatic actuator output torque is large, can be continuously controlled, there is no frequent action and damage to the actuator shortcomings. The hydraulic actuator has the largest output torque and can also withstand the frequent action of the actuator, which is often used for the control of the main valve and steam control door, but its structure is complex, its volume is huge and its cost is high.

(2) Classification by output displacement

According to the different output displacement, the drive mechanism is divided into angular displacement actuator and linear displacement actuator. The angular displacement actuator is divided into partial rotation actuator and multi-rotation actuator. The output Angle of the partial rotation actuator is up to 90°, and the multi-rotation actuator can continuously output 360° full rotation.

(3) Classification by dynamic characteristics

According to the different dynamic characteristics, the driving mechanism can be divided into proportional actuator and integral actuator. The integrated actuator is an actuator whose output linear displacement or angular displacement is proportional to the input signal. This type of actuator has no preamplifier and directly controls the servo motor by the action of the switch. The output Angle is the integration of the speed to the time. The integrated actuator is an actuator whose linear or angular displacement of the output is integrated with the input signal, mainly used in remote control, which belongs to open-loop control, such as using it to remotely open and close the globe valve or gate valve. Corresponding to this, the general actuator with a preamplifier and valve position feedback is a proportional actuator.

(4) Classification by presence or absence of microprocessors

According to whether there is a microprocessor in the actuator, it can be divided into analog actuator and intelligent actuator. The circuit of the analog actuator is mainly composed of electronic devices such as transistors or operational amplifiers, and the circuit of the intelligent actuator is equipped with chips such as microprocessors. Intelligent actuator, that is, fieldbus actuator, is based on DCS control, fieldbus control, flow characteristic compensation, self-diagnosis and variable speed and other aspects of the requirements developed, to achieve multi-parameter detection control, mechatronics structure and perfect configuration function, can be the controller, servo amplifier, motor, reducer, position transmitter and control valve and other links set. The signal is controlled by fieldbus.

(5) Classified by polarity

According to polarity, it can be divided into positive acting actuator and negative acting actuator. When the input signal (or operating variable) of the actuator increases, the adjustable quantity increases, that is, with the increase of the operating pressure, the output rod extends out, the pressure decreases and the actuator itself returns to the positive acting actuator; Otherwise, it is a counter-acting actuator.

(6) Classification by speed

According to whether the output shaft speed of the actuator is variable, it is divided into constant speed actuator and variable speed actuator. The speed of the output shaft of all analog (traditional) electric actuators is immutable, while the speed of the output shaft of an electric actuator with a frequency converter can be changed.

Taking the classification of robot drive systems as an example, common robot drive systems can be divided into electrical drive, hydraulic drive, and pneumatic drive according to the motion principle, as shown in Table 3.1.

Hydraulically driven manipulator is usually composed of hydraulic motor (various cylinders, oil motors), servo valve, oil pump, oil tank and other driving system, which is carried out by the driving manipulator actuator. Usually, it has a great snatch ability, compact structure, smooth action, impact resistance, shock resistance, and good explosion-proof, but the hydraulic components require high manufacturing accuracy and sealing performance, otherwise the oil leakage will pollute the environment. The pneumatic drive system is usually composed of cylinders, air valves, air tanks and air compressors, which is characterized by convenient air source, rapid action, simple structure, low cost and easy maintenance. However, it is difficult to control the speed, and the air pressure cannot be too high, so the snatch ability is low.

Electric drive is one of the most commonly used drive methods for manipulators. It is characterized by convenient power supply, fast response, large driving force, convenient signal detection, transmission and processing, and can adopt a variety of flexible control schemes. The driving motor generally adopts stepper motor and DC servo motor (AC) as the main driving mode. The deceleration mechanism has

Table 3.1 Classification of robot drive systems

Drive form	Advantages	Disadvantages	Application area
Motor drive	The energy used is simple, the speed range of the mechanism is large, the volume is small, the efficiency is high, the control and accuracy are high, and the use is convenient, the noise is low, and the control is flexible. The installation and maintenance are easy, and there is no leakage	Low thrust, low stiffness, requiring deceleration device, braking device in case of power shortage, high cost in case of high thrust	Widely used in almost all fields of robotics, such as electrical servo drive, information processing, transportation, etc.
Hydraulic drive	Easy to obtain significant thrust or torque; stable and reliable work, high positioning accuracy; easy to achieve automatic control; the structural dimensions are smaller than those of pneumatic transmission, with a longer service life and the ability to achieve precise control of speed and position. The transmission is smooth and does not require a deceleration device	The viscosity change of oil affects its working performance; there is a risk of leakage, combustion, and explosion; strict requirements for oil filtration equipment and hydraulic components, high cost, and easy leakage	Suitable for large robots and heavy loads, mostly used for high-power robot systems, heavy-duty, low-speed drive
Pneumatic drive	No need to add power equipment; even if there is a leak, it does not pollute the environment and is safe to use; the manufacturing requirements are also lower than hydraulic components	The air has high compressibility, poor working stability, and difficult speed control; steel parts are prone to rusting; noise pollution	Used for high-precision point control systems, small and medium-sized fast drives

harmonic drive, RV cycloidal pin wheel drive, gear drive, spiral drive and multi-bar mechanism. Mechanical drive is only used when the action is fixed. Generally, CAM linkage mechanism is used to realize the specified action. It is characterized by reliable action, high working speed and low cost, but it is not easy to adjust. There is also a hybrid drive, that is, a liquid–gas or electric-hydraulic hybrid drive.

3.1.2 Technical Characteristics of the Driving Mechanism

(1) Characteristics of electric actuators

The electric actuator is divided into electromagnetic and electric two categories, the former is mainly driven by electromagnetic valves and electromagnets, the latter is powered by the motor, the output Angle or linear displacement, used to drive valves or other devices. The characteristics of the electric actuator are as follows:

① There must be sufficient torque (force) moment. There should be enough torque for the actuator whose output is angular displacement, and enough force for the actuator whose output is linear displacement in order to overcome the resistance of the load. In order to increase the output torque or force, the output shaft of many motors has a reducer. The function of the reducer is to convert the high speed and small torque power output by the motor into the low speed and large torque power output by the actuator.
② It must have self-locking characteristics. There should be self-locking characteristics in the drive system of the reducer or motor, when the motor does not rotate, the unbalanced force of the load can not cause changes in the rotation Angle or displacement of the actuator. The electric actuator is often equipped with an electromagnetic brake, or the execution end is a worm gear and worm mechanism, which has self-locking.
③ Can be manually operated. In the event of a power outage or controller failure, manual operation on the actuator should be possible to take emergency measures. For this, a clutch and handwheel are necessary.
④ There should be a valve position signal. When the actuator is manually operated, in order to provide the controller with the basis for automatic tracking, there should be a valve position output signal on the actuator, which can not only meet the needs of the actuator's own position feedback, but also meet the needs of the valve position indication.
⑤ Product series combination. Modern electric actuator mostly uses the modular design idea, that is, the reducer and some functional units are designed into standard modules, and according to different needs, they are combined into three series of electric actuator products with various Angle stroke, straight stroke and multi-rotation. This combined actuator series, variety is complete, common parts, high degree of standardization, can meet a variety of industrial supporting needs.
⑥ Function is perfect and intelligent. It can receive both analog signal and data communication signal; It can be used in both open loop and closed loop.
⑦ With valve position and force (torque) limit. In order to protect the valve and transmission mechanism from damage due to excessive operating force, there should be mechanical limit, electrical limit and force or torque limit device on the actuator. It can effectively protect the safe operation of equipment, motors and valves.
⑧ Strong adaptability and high reliability.

(2) Characteristics of pneumatic actuators

The characteristics of the pneumatic actuator are:

① Working medium. With compressed air as the working medium, the working medium is easy to obtain and friendly to the environment.
② Working pressure. The working pressure of the medium is low, and the material requirements of the pneumatic components are low.

③ Speed of action. The operation speed is fast, but the speed will slow down when the load is increased.
④ Reliability. The pneumatic actuator has high reliability and can adapt to frequent start-stop action. The load change has no effect on the actuator, but the valve cannot be maintained after the air supply is interrupted (it can be maintained after the holding valve is added).
⑤ Safety valve position without external power. When the power source or control signal is lost, the safety valve position action can be realized, and the position can be fully opened, fully closed or kept unchanged, with positive and negative function.
⑥ Adjustment control. Equipped with intelligent positioner, it can realize intelligent closed-loop control, high control precision, set output characteristic curve and other advanced diagnostic functions, and support digital bus communication.
⑦ Environmental adaptability. With cylinder as the main body, with explosion-proof function, and can withstand high temperature, dust, air pollution and other harsh environmental conditions. When compressed air is used as a power source, the pneumatic actuator is suitable for explosion-proof dangerous areas, suitable for petrochemical, petroleum, oil processing and other industries.
⑧ Technology maturity. The equipment technology is mature, standardized, convenient installation and construction, and the project investment is less.
⑨ Maintenance. The pneumatic actuator has the advantages of simple structure, easy operation, low failure rate, less maintenance and long service life.
⑩ Working speed stability. Because the air has compressibility, the stability of the working speed is slightly poor, but the gas–liquid linkage device will get more satisfactory results.
⑪ Total output. Because the working pressure is low (generally 0.3 ~ 1.0 MPa), and because the structure size should not be too large, the total output force of the pneumatic transmission device should not be greater than 10 ~ 40 kN.
⑫ Noise. The noise is large, and the muffler should be added when the high-speed exhaust.
⑬ Transmission efficiency. Pneumatic transmission efficiency is low.

In modern industry, the application of electric equipment is far more common than pneumatic equipment, because pneumatic equipment needs to spend a large investment in the air source, and the laying of pipelines is also more troublesome than laying wires, and the transmission speed of pneumatic signals is far less than that of electrical signals. However, in some specific occasions, the superiority of pneumatic equipment is highlighted. For example, in explosion-proof safety, pneumatic equipment will not have sparks and heating problems, the air medium also helps to disperse flammable and explosive and toxic and harmful gases; Pneumatic equipment in the occurrence of pipeline blockage, airflow short circuit, mechanical parts stuck when the fault will not heat damage; Adaptability in harsh environments such as humidity is also better than electric actuators.

3.2 Motor Drive System

3.2.1 Overview of Motor Drive System

Electricity in modern industrial and agricultural production, transportation, science and technology, information transmission, national defense construction, and daily life. It has been widely used in daily life, and electric motors are the main equipment for producing, transmitting, distributing, and applying electrical energy prepare [1]. The electrical servo system using electric motors as the power source is flexible and convenient, easy to obtain driving energy, and pollution-free. Dyeing, with a large power range, has become the main form of servo systems. And the motor is the most important component among them. One of the most widely used drive systems in intelligent manufacturing equipment is the motor drive system, mainly consisting of servo motors drive system, variable frequency motor drive system, stepper motor drive system, linear motor drive system, etc. Form [2].

(1) Servo motor drive system

The servo motor is the prime mover that controls the mechanical motion in the servo system and converts the voltage signal into torque and speed to drive the control object. The servo motor has AC servo and DC servo, among which the AC servo drive system is closed-loop control, which has better control performance and good acceleration performance, and is widely used in all walks of life. Servo motors can be selected for equipment requiring production accuracy, such as machine tools, printing equipment, packaging equipment, textile equipment, high-precision processing equipment, robots, automated production lines and other equipment with relatively high requirements for process accuracy, processing efficiency and work reliability.

The torque and speed of the servo motor are controlled by the signal voltage, and when the size and direction of the signal voltage change, the speed and rotation direction of the motor will sensitively follow the change. AC servo motor is a two-phase asynchronous motor, the stator has two windings: excitation winding and control winding. The rotor is divided into cage type rotor and cup type rotor. The cage type rotor is the same as the three-phase cage type rotor mentioned below. The cup type rotor is a hollow thin-walled cylinder made of aluminum alloy or copper alloy, as shown in Fig. 3.1. Figure 3.2 shows the internal structure of the servo motor.

The AC servo motor should not only have the servomotor controlled by the control signal, but also have the controllability of the speed change. There are three main control methods of AC servo motor: amplitude control, phase control and amplitude phase control. Amplitude control means that the phase difference between the control voltage and the excitation voltage is nearly unchanged at 90°, and the speed of the motor is changed by changing the size of the control voltage. The control voltage is large, the motor speed is fast, the control voltage is slow, and the motor speed is slow. If the control voltage is zero, the motor will stop immediately. Phase control is to keep the rated value of the control voltage and the excitation voltage unchanged,

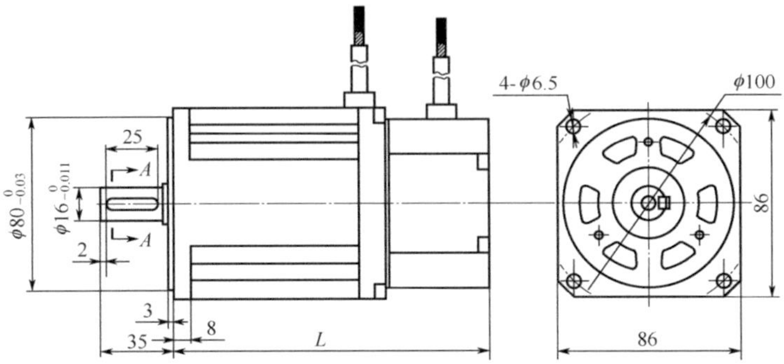

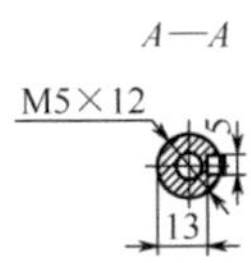

	90CB075C-700000	90CB075C-705000
L	148	148
L(with brake)	190	190

Fig. 3.1 Hollow thin-walled circular tube structure

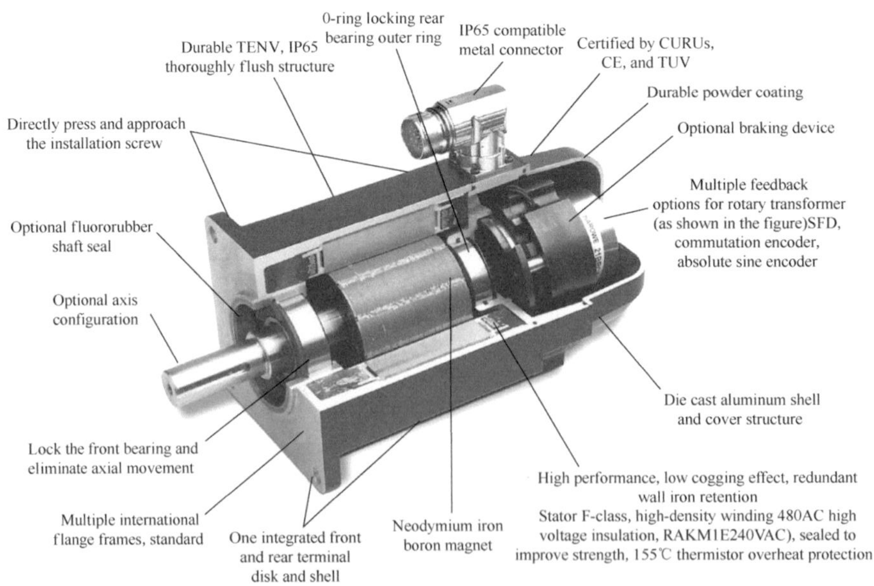

Fig. 3.2 Internal structure of servo motor

and change the speed of the motor by changing the phase difference; Amplitude-phase control adopts capacitive phase separation, which not only changes the size of the control voltage, but also changes the phase difference between the control voltage and the excitation voltage, so as to realize amplitude-phase control. The

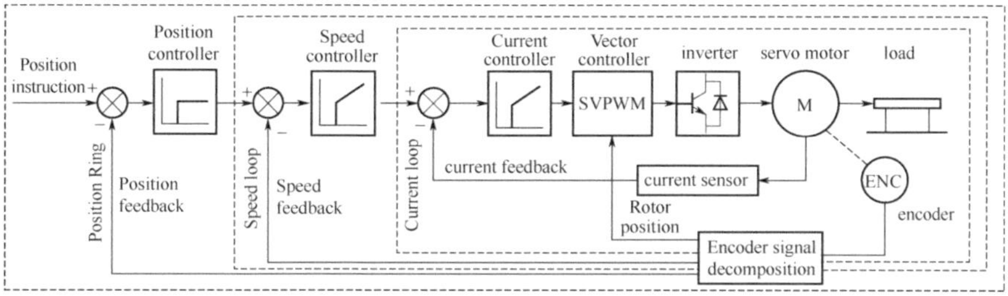

Fig. 3.3 Control system of serve motor

method has simple equipment, large output power and wide application. AC servo motor is a multi-variable, strong coupling, nonlinear, variable parameter complex object, the traditional control method is difficult to accurately control it, in order to improve the dynamic response characteristics of the motor, has developed a variety of professional control algorithms, the modern extension, variable structure and other nonlinear control methods into the motor control system, effectively solve the servo motor drive system control problems. Figure 3.3 shows the servo motor control system.

The design of servo motor drive system must first determine the performance requirements achieved by the design object. The main parameter of servo motor design is power, and the maximum load torque of the system shall not exceed the rated torque of the motor; The rotor inertia of the motor should match the moment of inertia of the load. In the servo system design and selection, the load and motor inertia matching is very important, the moment of inertia has an impact on the accuracy, stability, dynamic response of the system. The moment of inertia of the mechanical system should be minimized in the design to achieve the purpose of precise control of the motor. After selecting the motor, it is necessary to check its carrying capacity.

The structure of the DC servo motor is similar to that of the general DC motor, but in order to reduce the moment of inertia, its excitation winding and armature are respectively powered by two independent power sources and are controlled by armature. Figure 3.4 shows the section view of the DC servo motor.

The permanent magnet of permanent magnet DC servo motor is very thin and can provide enough magnetic induction intensity, the motor is small in size, light in weight, the permanent magnet material has strong anti-demagnetization ability, the motor will not be demagnetized due to vibration, and the magnetic stability is high, so it has been widely used. DC servo motors are often used in systems with slightly larger power, with an output power of 1 ~ 600 W.

(2) Variable frequency motor drive system

Variable frequency speed regulating motor is referred to as variable frequency motor, which is a general term for the motor driven by the frequency converter. The motor can be driven by the inverter to achieve different speed and torque to adapt to the load demand change. The variable frequency motor is developed from the traditional cage motor, which changes the traditional motor fan into an independent fan, and improves

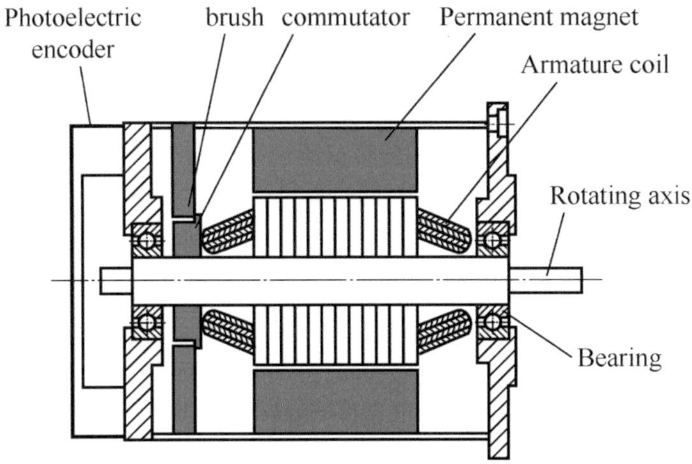

Fig. 3.4 Sectional view of DC servo motor

the insulation performance of the motor winding. In situations where requirements are not high, such as low power and at rated operating frequency, ordinary cage motors can be used instead.

Frequency converter is a power control device that converts the power frequency power supply to another frequency by using the on–off action of power semiconductor devices. The inverter mainly adopts AC-DC-AC mode (VVVF frequency conversion or vector control frequency conversion), first converts the power frequency AC power into DC power through the rectifier, and then converts the DC power into the frequency and voltage can be controlled AC power to supply the motor. The circuit of the inverter is generally composed of four parts: rectification, intermediate DC link, inverter and control. The rectifier part is a three-phase bridge uncontrollable rectifier, the inverter part is an IGBT three-phase bridge inverter, and the output is PWM waveform, and the intermediate DC link is filtering, DC energy storage and buffer reactive power.

After years of rapid development, AC speed regulation has become the mainstream of motor speed regulation, and can be widely used in all walks of life stepless transmission. Because the frequency conversion motor has advantages over the ordinary motor in frequency conversion control, where the frequency converter is used, there are frequency conversion motor applications, and its application field has been extended from the high-performance field to the general drive and special drive occasions, and even frequency conversion air conditioning, refrigerators, washing machines and other household appliances. AC drives have been widely used in industrial robots, automated publishing equipment, processing tools, transmission equipment, elevators, compressors, steel rolling, fan pumps, electric vehicles, lifting equipment and other fields.

Variable frequency speed regulation technology has been widely used in various fields for its excellent speed regulation, starting and braking performance, high efficiency, high power factor and power saving effect, and has become the mainstream of modern speed regulation technology. Variable frequency motor refers to the motor that runs continuously with 100% rated load in the range of 10–100% rated speed under standard environmental conditions, and the temperature rise will not exceed the allowable value of the motor calibration. AC variable frequency motor has simple structure, small size, small inertia, low cost, easy maintenance, easy speed regulation, energy saving, and can realize soft start and fast braking, and strong environmental adaptability. Frequency conversion speed control has developed rapidly and has gradually replaced most of the DC speed control transmission devices.

There are many components and control methods of variable frequency speed regulation system of synchronous motor. In principle, all kinds of power electronic inverters can be used for variable frequency speed regulation of synchronous motor. Frequency converters can be divided into AC-AC and AC-DC-AC two categories. From the frequency control mode of inverter, the speed control system is often divided into two kinds: independent control and automatic control [3].

To design a variable frequency motor drive system, we must first design the main circuit to convert the three-phase AC voltage into the AC voltage with adjustable frequency and amplitude. AC-DC-AC circuit is the most widely used frequency conversion circuit, the front stage is equipped with a rectifier filter circuit, the rear stage has an inverter. The rectifier filter circuit is divided into controllable rectifier and uncontrollable rectifier. The controllable rectifier circuit can control the output DC voltage, but it interferes with the power grid, the DC voltage harmonic is large, and the input power factor is low. The voltage obtained by uncontrollable rectification is stable, the DC harmonics are small, but the size is uncontrollable, and the size is only controlled by the inverter.

Frequency conversion speed regulation usually has constant torque speed regulation and constant power speed regulation two kinds.

① Constant torque speed regulation. When the speed is lower than the rated speed of the motor, the ratio of voltage and frequency should be kept unchanged, and the two should be proportionally adjusted for constant torque speed regulation.
② Constant power speed regulation. When the speed is higher than the rated speed of the motor, the voltage should be kept unchanged, the magnetic flux and torque are reduced, the power is unchanged, and the speed is adjusted for constant power.

When designing variable frequency motor drive system, control circuit is the core part of AC variable frequency motor speed regulation system. There are two control strategies: vector control and direct torque control.

① Vector control. Also known as magnetic field oriented control, it is a technology that uses the inverter to control the three-phase AC motor, and controls the output of the motor by adjusting the output frequency of the inverter, the size and Angle of the output voltage. Its characteristic is that the magnetic field and torque of the motor can be controlled separately, similar to the characteristics of the separately excited DC motor. When processing, the three-phase output current and voltage are represented by a vector, so it is called vector control. Suitable for AC induction motor and DC brushless motor, can be used with AC motor, motor size is small, cost and energy consumption are low. The universal inverter with vector control mode can not only match the DC motor in speed range, but also control the torque generated by the asynchronous motor. Since the vector control mode is based on the accurate parameters of the controlled asynchronous motor, some general frequency converters need to accurately input the parameters of the asynchronous motor when in use, and some general frequency converters need to use speed sensors and encoders. Since the motor parameters may change, which will affect the control performance of the inverter, the relevant parameters in the control algorithm are adjusted according to the identification results, so as to carry out effective vector control of the ordinary asynchronous motor. In addition to being used in high-performance motor applications, vector control has also been used in some home appliances.

② Direct Torque Control (DTC). Torque as the center to carry out comprehensive control, not only control the torque, but also for the flux control and flux self-control. The difference between direct torque control and vector control is that it does not indirectly control torque by controlling current and flux equivalent, but the torque is directly controlled as the controlled quantity. Its essence is to directly control the stator flux and electromagnetic torque by using the analysis method of space vector and the stator magnetic field orientation. This method does not need complex coordinate transformation, but directly calculates the magnitude of flux mode and torque on the stator coordinates of the motor, and realizes PWM pulse width modulation and high dynamic performance of the system through direct tracking of flux and torque. The control method reduces the problem that the control performance is easily affected by parameter change in vector control technology. The main control circuits are:

   ① Operation circuit. The external speed, torque and other instructions are compared with the current and voltage signals of the detection circuit to determine the output voltage and frequency of the inverter.
   ② Voltage and current detection circuit.
   ③ Drive circuit.
   ④ Speed detection circuit. Speed detector mounted on asynchronous motor shaft.
   ⑤ Protection circuit.

(3) Ordinary AC and DC motor drive system

AC motor is divided into asynchronous motor and synchronous motor, and three-phase asynchronous motor is mainly used in production. Synchronous motors are mainly used in equipment with large power, no speed regulation and long working time. Single-phase induction motors are used in household appliances with low power. AC induction motor is to convert electrical energy into mechanical energy, which is mainly composed of stator, rotor and air gap between them. According to the rotor structure of the asynchronous motor is divided into cage type asynchronous motor and wound rotor asynchronous motor. A three-phase alternating current is applied to the stator to generate a rotating magnetic field and cut the rotor to obtain torque. It has the advantages of simple structure, reliable operation, convenient maintenance and installation, and is widely used in various fields.

Three-phase asynchronous motor (Fig. 3.5) has three-phase symmetrical winding in the stator core and connected into a star, connected to the three-phase power supply, and the magnetic field produced by them together with the current alternating in space constantly rotating, which is the rotating magnetic field. As long as one end of any two of the three wires connected to the three-phase power supply is reversed, the rotating magnetic field will be reversed. The number of poles of the three-phase asynchronous motor is the number of poles of the rotating magnetic field, and the number of poles of the rotating magnetic field is related to the arrangement of the three-phase winding. The speed of the induction motor is related to the speed of the rotating magnetic field, and the speed of the rotating magnetic field is related to the number of poles of the rotating magnetic field, when the rotating magnetic field has p opposers, the magnetic field speed is:

$$n_0 = \frac{60f}{p}$$

The speed of the rotor of the asynchronous motor is always less than the speed of the rotating magnetic field, which is the origin of the name of the asynchronous motor, the speed of the rotating magnetic field n_0 is called the synchronous speed, and the slip rate s is used to indicate the difference between the rotor speed n and the rotating magnetic field speed n_0.

$$s = \frac{n_0 - n}{n_0} \tag{3.1}$$

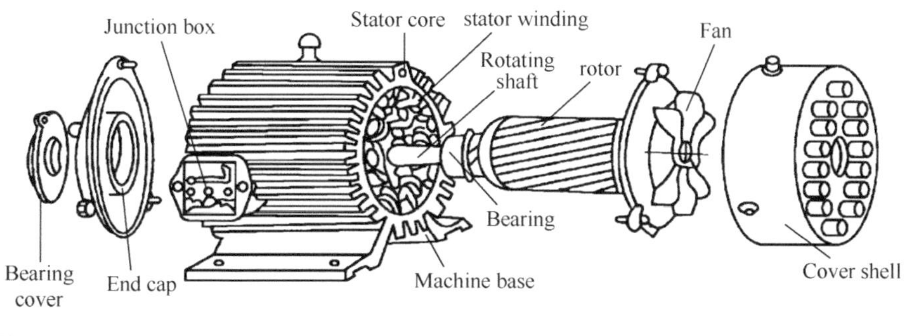

Fig. 3.5 Structure diagram of three-phase asynchronous motor

The slip rate is a very important physical quantity. Various physical quantities of the rotor circuit, such as electromotive force, current, frequency, inductive reactance and power factor, are related to it.

The circuit of the asynchronous motor is divided into the stator circuit and the rotor circuit. The stator current frequency is

$$f_1 = \frac{n_0 p}{60} \tag{3.2}$$

where, p is the number of magnetic poles of the rotating magnetic field.

Rotor current frequency:

$$f_2 = s f_1 \tag{3.3}$$

Rotor electromotive force (effective value):

$$E_2 = s E_{20} \tag{3.4}$$

where E_{20} is the induced electromotive force.

Rotor inductance:

$$X_2 = s X_{20} \tag{3.5}$$

where X_{20} is the inductive reactance.

Rotor current:

$$I_2 = \frac{s E_{20}}{\sqrt{R_2^2 + (s X_{20})^2}} \tag{3.6}$$

where R_2 is the resistance of each phase of the rotor winding,

Power factor of the rotor circuit:

$$\cos \varphi_2 = \frac{R_2}{\sqrt{R_2^2 + (s X_{20})^2}} \tag{3.7}$$

The mechanical characteristics of electromagnetic torque T are the most important physical quantity for analyzing motor characteristics.

$$T = K_{\mathrm{T}} \phi I_2 \cos \varphi_2 \tag{3.8}$$

where K_T is a constant, I_2 is the rotor current, Ø is the magnetic flux, $\cos \varphi_2$ is the power factor, and torque T is proportional to the square of the stator phase voltage U_1 and is also affected by the rotor resistance. The mechanical characteristics of the motor are used to characterize the relationship between the torque M generated

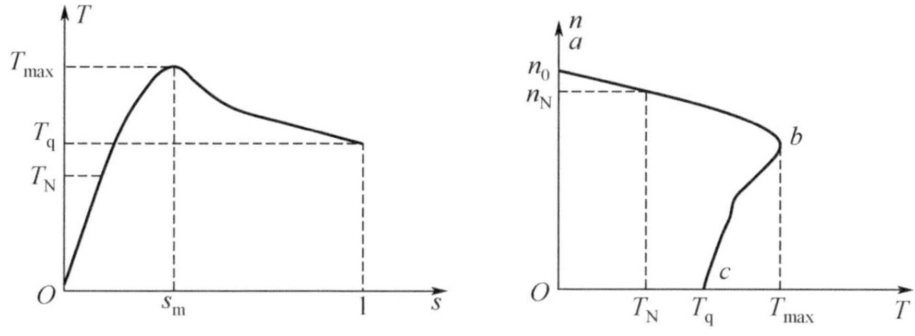

(a) T=f(s)curve (b) N=f(T)curve

Fig. 3.6 Mechanical characteristic curve of electric motor

on the motor shaft and the corresponding operating speed n. It is expressed by the function $n = f(M)$. It is an important characteristic that characterizes the operation of the motor. Under a certain voltage and rotor resistance, the relationship curve between torque and slip is called the mechanical characteristic curve of the motor (Fig. 3.6).

Mechanical characteristic curves need to analyze rated torque T_N, maximum torque $T_{\max}$, starting torque T_{at}

$$T_{\mathrm{N}} = 9550\frac{P_2}{n} \tag{3.9}$$

where, P_z is the mechanical power output on the motor shaft, and the rated torque can be obtained by applying the formula from the rated power and rated speed on the motor nameplate.

$$T_{\max} = K\frac{U_1^2}{2X_{20}} \tag{3.10}$$

In the formula, X_{20} is the rotor inductive reactance; K is a constant. The rated torque of the motor is smaller than the maximum torque, and the ratio of the two is the overload coefficient.

$$\lambda = \frac{T_{\max}}{T_{\mathrm{N}}} \tag{3.11}$$

When selecting a motor, the maximum torque of the motor must be calculated based on the overload factor of the selected motor, and it must be greater than the maximum load torque.

$$T_{\mathrm{st}} = K\frac{U_1^2 R_2}{R_2^2 + X_{20}^2} \tag{3.12}$$

The starting characteristics of the motor need to analyze the starting current and starting torque. Too much starting current will affect the normal operation of the adjacent load, and too much starting torque will damage the transmission mechanism. When the motor starts, directly start the motor with a power of less than 10 kW, and connect the motor directly to the power supply with a rated voltage with a knife switch or contactor. Step-down starting is to reduce the voltage on the stator winding of the motor when starting, so as to reduce the starting current, commonly used star-delta switching starting, auto step-down starting and other ways.

The cage motor can adjust speed by changing the frequency of power supply and the number of magnetic poles. If the rotor motor is wound, the speed can be adjusted by changing the slip rate. Variable pole speed regulation is widely used in machine tools, and variable slip speed regulation is the most widely used in lifting equipment. When the motor stops and reverses rapidly, it is necessary to overcome inertia to brake the motor and the opposite direction of torque and rotor rotation, and the torque is called brake torque. The braking methods include energy consumption braking, reverse braking, power generation feedback braking, etc.

① Energy consumption braking. Energy consumption Small braking energy consumption, smooth braking. During braking, the DC power supply is switched on while the three-phase power supply is cut off, so that the DC current passes into the stator winding, and the braking torque is related to the current of the DC power supply, and the current of the general DC power supply is 0.5–1 times of the rated current.

② Reverse braking. The reverse braking operation is simple, the braking effect is good, but the energy consumption is large. When the motor is stopped, one end of any two of the three wires connected to the power supply is adjusted to reverse the rotating magnetic field, and the rotor is still rotating in the original direction due to inertia, and the torque direction is opposite to the motor rotation direction to achieve the purpose of braking. The method must have a resistor inserted into the stator or rotor circuit.

③ Power generation feedback braking. When the rotor speed exceeds the rotating magnetic field speed, the torque is also braking, and the motor has been transferred to the generator state, and the potential energy is converted into electrical energy and fed back to the power grid. This braking method is in line with the concept of green environmental protection because the braking energy is returned to the power grid in the form of electrical energy.

Motor nameplate indicates the motor model, connection, voltage, current, power and efficiency, power factor, speed, insulation level and working mode and other parameters, to be carefully read before design and use.

To select the motor, we must first choose the appropriate power according to the working condition requirements. Too much power will cause a waste of resources, not in line with the mainstream of today's environmental protection and energy saving; The power is too small, resulting in a "small horse cart", the motor is easy to overload and heat or even damage.

The selection of the motor should also be considered according to the actual working conditions. The motor that needs continuous operation in the production should first calculate the power of the production machinery, the rated power of the selected motor should be slightly greater than the power of the production machinery, and in many cases can be directly selected by analogy and statistical analysis through experience; For motors that run for a short time in production, overload is allowed during operation due to heating inertia, which is usually selected according to the overload coefficient.

To select the motor, you should also select the type of motor according to the requirements. According to the working conditions and funds, from the AC, DC, mechanical characteristics, speed regulation and starting performance, maintenance and price. The structure of the motor should be selected according to the working environment of the motor, with the open type in a dry and dust-free environment, the closed type in a humid environment, and the explosion-proof type in an explosive gas environment. Finally, the voltage and speed of the motor should be selected, the voltage should be selected according to the type of motor, power and the power supply voltage of the use place, the motor speed is selected according to the production machinery requirements, and the speed is generally not less than 500 r/min.

The stator of the synchronous motor is the same as that of the three-phase asynchronous motor, and the rotor is a magnetic pole and is excited by direct current. When the speed of the motor is close to the synchronous speed n_0, the rotor is excited. The speed n of the synchronous motor is constant, and it can only be used for long-term continuous work and to keep the speed unchanged in actual use, such as driving pumps, fans, compressors, etc.

DC motor is a rotating mechanical device that converts mechanical energy and direct current energy into each other. Although DC motors play an indispensable role in some aspects, they have been gradually replaced by semiconductor rectified power supplies due to their complex structure, high cost and poor reliability. However, due to its better speed regulation performance and large starting torque, the production machinery with high speed regulation requirements and large starting torque demand is often driven by DC motors.

The motor mainly consists of a magnetic pole, an armature, and a commutator (Fig. 3.7). Magnetic poles produce magnetic fields, and permanent magnets can also be used as magnetic poles in small DC motors; The rotating armature generates induced electromotive force, which is composed of silicon steel sheets. The commutator is a special device of DC motor. The brush is pressed by a spring on the surface of the commutator to connect the rotating armature winding to the external circuit.

The DC motor armature coil is forced to rotate in the magnetic field after being energized, and the induced electromotive force is also generated in the coil, and the direction is always opposite to the direction of the current or applied voltage, which is called the reverse electromotive force. The armature current I interacts with the magnetic flux Φ to produce electromagnetic force and electromagnetic torque

$$T = K_{\mathrm{T}}\Phi I \tag{3.13}$$

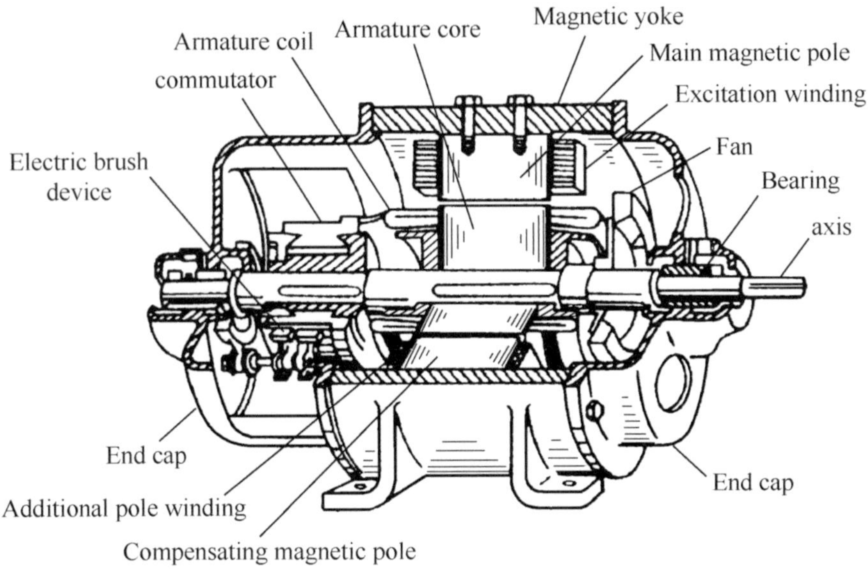

Fig. 3.7 DC motor structure diagram

In the formula, K_T is a parameter related to the motor structure. The electromagnetic torque of a motor is the driving torque that causes the armature to turn. Therefore, the electromagnetic torque T of the motor must be balanced with the mechanical load torque and no-load loss torque.

According to the operation of the excitation mode, the DC motor is divided into four kinds of separate excitation, shunt excitation, series excitation and compound excitation, as shown in Fig. 3.8, and only two kinds of separate excitation and shunt excitation are commonly used. The field winding and armature of the separately excited motor are separated and powered by two DC power sources respectively, while the two are in parallel in the shunt excited motor, but the mechanical characteristics, starting, reversing and speed regulation of the separately excited and shunt excited motor are the same.

In separately excited motor, the excitation coil is separated from the power supply of the rotor armature. In shunt excited motor, the exciting coil and the rotor armature

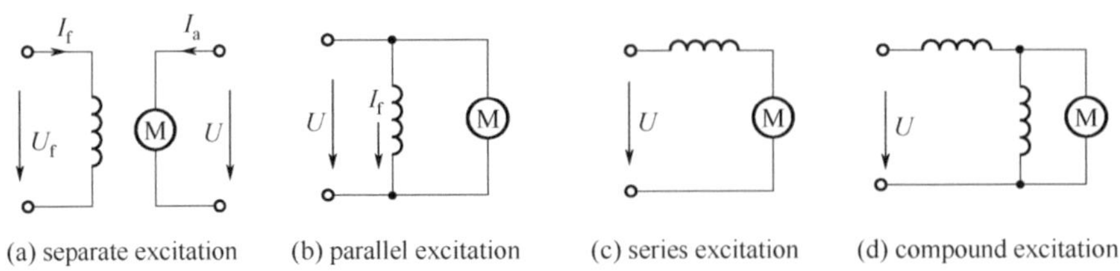

Fig. 3.8 DC motor excitation method

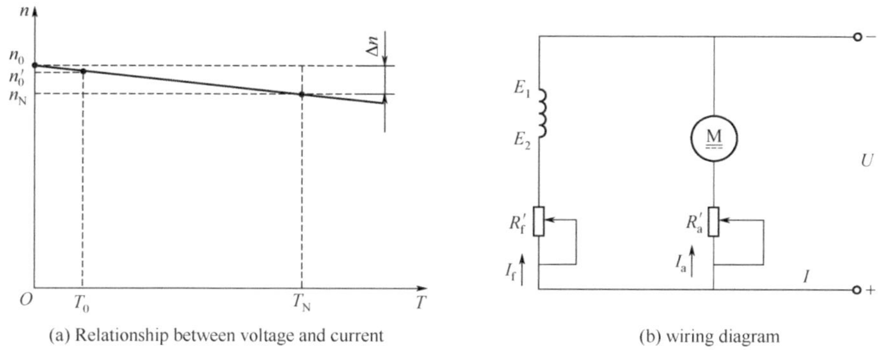

Fig. 3.9 Relationship between voltage and current of parallel excitation motor and wiring diagram

are connected in parallel to the same power supply. In a series excited motor, the exciting coil and the rotor armature are connected in series to the same power supply. In the compound motor, the exciting coil and the rotor armature are connected in series and parallel, and are connected to the same power supply.

The relationship between voltage and current is shown in Fig. 3.9. When the power supply voltage U and the excitation resistance Rt are unchanged, the excitation current and flux Φ remain unchanged. Therefore, the motor torque is proportional to the armature current. Under the condition that the power supply voltage U and the excitation resistance Rt are constant, the curve representing the relationship between the motor speed n and the torque T is called the mechanical characteristic curve of the motor, where n_0 is the ideal no-load speed, Δn is the speed drop, caused by the armature resistance.

When the shunt motor is started, the starting current is very large because the armature resistance is very small. The motor torque is proportional to the armature current. Excessive torque will cause mechanical impact and damage the transmission mechanism. Therefore, the current needs to be limited when starting. As shown in Fig. 3.9b, when the starting resistor is connected in series in the armature circuit, the starting resistance is the largest. After starting, as the motor speed increases, the starting resistance gradually becomes smaller. When the DC motor is started or working, the excitation circuit must be connected, otherwise the armature winding and commutator are in danger of burning out, and may also lead to a "speeding" accident and serious mechanical damage. If you want to change the direction of rotation of the DC motor, you can fix the direction of the magnetic field and change the direction of the armature current; you can also keep the direction of the armature current unchanged and change the direction of the excitation current.

Compared with AC asynchronous motors, shunt motors have unique advantages in speed regulation performance. They can achieve stepless speed regulation and the mechanical transmission gearbox can be greatly simplified. The output speed can be changed by changing the voltage or changing the magnetic flux. The speed regulation method of changing the magnetic flux has the advantages of smooth speed regulation, convenient control, good stability, and good speed regulation amplitude.

It is suitable for situations where the torque and speed are approximately inversely proportional and the output power is basically unchanged, such as in cutting machine tools. The voltage-changing speed regulation method has relatively hard mechanical properties, good stability, and a certain output torque, so it is generally used in lifting equipment.

Permanent magnet synchronous motor (PMSM) does not have brushes and slip rings, has a simple structure, small size, high reliability, low maintenance workload, no windings on the rotor, low heat dissipation requirements, high operating efficiency, high torque-to-current ratio, and low rotational speed. It has small inertia and is easy to realize high-performance vector control, so it has been widely used in aviation, aerospace, CNC machine tools, machining centers, robots and other fields [4].

(4) Stepper motor drive system

A stepper motor is a motor that converts electrical pulse signals into corresponding angular displacement or linear displacement. According to the output torque, it can be divided into two types: power stepper motors and fast stepper motors. Power stepper motors are mostly used in electromechanical equipment. Every time a pulse signal is input, the rotor rotates an angle or moves forward one step. Its output angular displacement or linear displacement is proportional to the number of input pulses, and the rotational speed is proportional to the pulse frequency. Therefore, stepper motors, also known as pulse motors, have the characteristics of easy control and small size, so they are widely used in intelligent manufacturing equipment such as CNC systems, automatic production lines, and automated instruments.

With the rapid development of microelectronics and the popularization and application of microcomputers, the large and complex controllers of stepper motors that used to be composed of hardware circuits can be realized with software, which not only reduces the hardware cost, but also improves the flexibility and reliability of the control sex and versatility. Figure 3.10 shows a stepper motor.

Fig. 3.10 Stepping motor

The driving power supply of the stepper motor is composed of a variable frequency pulse signal source, a pulse distributor and a pulse amplifier, from which the driving power supply provides pulse current to the motor windings. The operating performance of the stepper motor depends on the good cooperation between the motor and the driving power supply.

The advantages of stepper motors are simple structure, no cumulative error, easy use and maintenance, low manufacturing cost, and large ability to drive load inertia. They are suitable for small and medium-sized machine tools and situations where speed accuracy is not high. The disadvantages of stepper motors are low efficiency, high heat generation, and sometimes "out of step".

There are many structural forms and classification methods of stepper motors. They are generally divided into three types according to the excitation method: reactive type, permanent magnet type and hybrid type; according to the number of phases, they can be divided into single-phase, two-phase, three-phase and multi-phase forms. Permanent magnet stepper motors are generally two-phase, with small torque and volume, and the step angle is generally 7.5° or 15°; reactive stepper motors are generally three-phase, which can achieve large torque output, and the step angle is average is 1.5°, but the noise and vibration are very large, and it is gradually withdrawing from the market; the hybrid stepper motor combines the advantages of the permanent magnet type and the reactive type, and is divided into two phases and five phases: the two-phase step angle is generally 1.8°, and the five-phase step angle is generally 0.72°. Hybrid stepper motors are the most widely used. The main stepper motors used in our country are reactive stepper motors. The operating performance of stepper motors is closely related to the control method. From the perspective of its control method, stepper motor control systems can be divided into the following three categories: open-loop control systems, closed-loop control systems, and semi-closed-loop control systems.

(5) Linear motor drive system

After the rotary motor is cut in the radial direction, it is straightened and unfolded to form a linear motor. It eliminates the coupling, ball screw nut pair and other transmission links and directly drives the worktable to move, as shown in Fig. 3.11. When linear motion is required to be accurate and fast, linear motors should be used. A linear motor is equivalent to an AC synchronous motor spread out on a plane [5].

Advantages of linear motors:

① Simple structure. The tubular linear motor directly generates linear motion without the need for an intermediate conversion mechanism, Simplified structure, small motion inertia, improved dynamic response performance and positioning accuracy, and high reliability.

② Suitable for high-speed straight motion. Because there is no constraint of centrifugal force, ordinary materials can also achieve better performance high speed. Low wear of transmission components, low mechanical loss, and high efficiency.

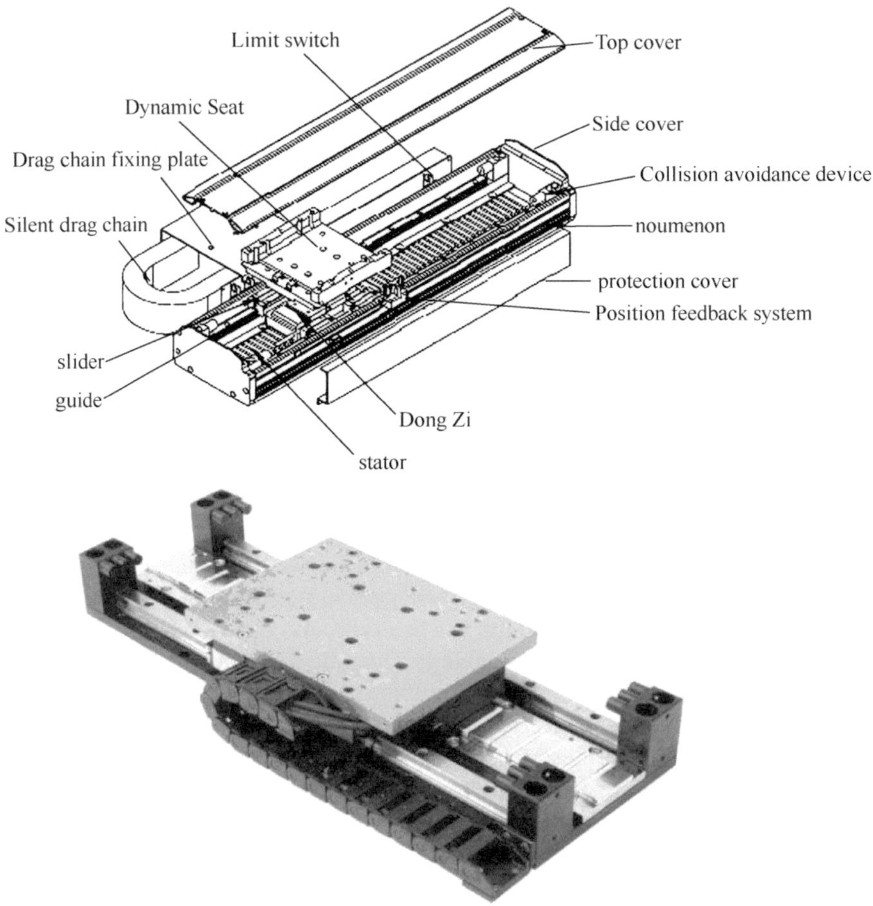

Fig. 3.11 Linear motor

③ High utilization rate of primary winding. In tubular linear induction motors, the primary winding is pie shaped and does not end winding, therefore the winding utilization rate is high.

④ Easy to adjust and control. By adjusting the voltage or frequency, different speeds and electromagnetic thrusts can be obtained force, suitable for low speed reciprocating operation.

⑤ High acceleration. This is a linear motor drive compared to a screw, synchronous belt, and gear rack drive significant advantages.

3.2.2 *Selection of Stepper Motors*

When designing electromechanical equipment and selecting motors, it is usually hoped that the output torque, starting frequency, and operation of the motor will be high operating frequency, small step error, and high cost-effectiveness. Therefore, when selecting a motor, it is necessary to comprehensively consider torque and speed, performance and cost to make a reasonable choice.

① Selection of output torque. The output torque needs to be selected based on the maximum holding torque and torque-frequency characteristics of the stepper motor. The maximum output torque of the motor at low frequency can reach 70 ~ 80% of the maximum value, and then drops to 10 ~ 70%. Generally, the output torque of the stepper motor is selected to be 20 ~ 30% of the maximum static torque.

② Selection of step angle. The step angle is an important parameter that determines the pulse equivalent of the open-loop servo system. The design should meet the control accuracy and operating speed requirements of the equipment. For equipment with low positioning accuracy and low operating frequency, a stepper motor with a larger step angle and lower frequency can be selected; on the contrary, a motor with a small step angle and a higher operating frequency can be selected. Requirements can also be achieved through a variable speed system or subdivided step angle.

③ Selection of startup frequency and operating frequency. Select the starting frequency and operating frequency of the stepper motor according to the load operating speed requirements. If the load torque is known, check the starting frequency from the starting torque frequency characteristic curve. The starting frequency should be less than or equal to the checked value. If the continuous operating frequency of the stepper motor is known, the maximum dynamic torque can be known from the motor's torque-frequency characteristic curve, and the load carried by the motor should be less than this value.

④ Static torque. The dynamic torque of a stepper motor is difficult to determine, but the static torque can be determined based on the load the motor is working with. The loads mainly include inertia load and friction load. These two loads need to be considered when starting directly. Inertia load is mainly considered during acceleration, and friction load is mainly considered at constant speed. Just choose the static torque to be 2–3 times the friction load.

3.2.3 *Servo Motor Selection Principles*

Intelligent manufacturing equipment often encounters some complex movements, which have a great impact on the power load of the motor. Servo drives are the core of many electromechanical systems, so the selection of servo motors becomes particularly important. In the selection process of servo motors for intelligent manufacturing equipment, we must first select the motor that meets the given load requirements, and then select the most suitable motor based on technical and economic indicators such as price, weight, and volume.

(1) Traditional selection principles

Only the power of the motor is considered here. Linear motion is represented by speed v(t), acceleration a(t) and required external force F(t). Rotary motion is represented by angular velocity ω(t), angular acceleration α(t) and required external force F(t). Torque T(t) indicates that they are all functions of time and have nothing to do with other factors. Obviously, the maximum power P of the motor should be greater than the peak power P peak required by the working load, but this alone is not enough. Power in the physical sense includes two parts: torque and speed, but in the actual transmission mechanism they are Restricted. Use ω peak value and T peak value to represent the maximum value or peak value. The maximum speed of the motor determines the upper limit of the reduction ratio of the reducer, n upper limit = ω peak value, maximum/ω peak value. Similarly, the maximum torque of the motor determines the lower limit of the reduction ratio, n lower limit = T peak value/T motor, maximum. If the n lower limit is greater than the n upper limit, the selected motor is inappropriate. Instead, the feasible range of gear ratios between upper and lower limits can be determined by a broad analogy for each type of motor. Only using peak power as the principle for selecting motors is insufficient, and the accurate calculation of the transmission ratio is very cumbersome. The T-ω curve of the motor is shown in Fig. 3.12.

(2) New selection principles

The new selection principle is to separate the motor characteristics from the load characteristics and express them in the form of diagrams, which facilitates the feasibility check of the drive device and the comparison between different systems, and also provides a possible range of transmission ratios. This approach works for a

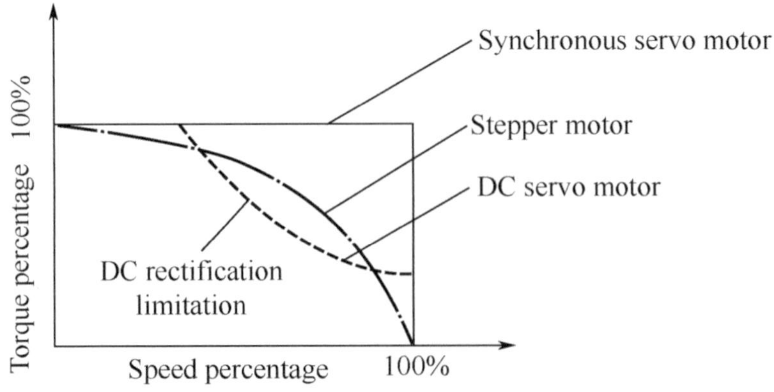

Fig. 3.12 T-ω curve of motor

variety of load situations. By separating the characteristics of the load and the motor, various parameters related to the power can be expressed in the form of diagrams and are applicable to various motors.

Intelligent electromechanical equipment motion control systems mostly use stepper motors or fully digital AC servo motors as execution motors. The two are similar in control methods (pulse train and direction signal), but there are big differences in performance and application situations. When designing the electromechanical system, the selection should be based on the situation.

(1) Accuracy

The step angle of two-phase hybrid stepper motors is generally 1.8°, 0.9°, and the step angle of five-phase hybrid stepper motors is generally 0.72°, 0.36°. Some high-performance stepper motors are subdivided by drivers or software. The step angle is smaller.

The control accuracy of the servo motor is guaranteed by the encoder. For a motor with a 17-bit encoder, every time the driver receives 131,072 pulses and the motor rotates once, its pulse equivalent is 360°/131,072 = 0.0027466°, which is the pulse equivalent of a stepper motor with a step angle of 1.8°, 1/655.

(2) Low-frequency characteristics

Stepper motors are prone to low-frequency vibration at low speeds. The vibration frequency is related to the load condition and driver performance, which is determined by the working principle of the stepper motor. Low-frequency vibration is detrimental to the operation of smart manufacturing equipment. This phenomenon can be improved by using driver subdivisions or installing dampers.

The servo motor runs smoothly, has no vibration at low speed, and has a resonance suppression function. The system has a frequency analysis function (FFT) inside the system, which can detect the mechanical resonance point and facilitate system adjustment.

(3) Moment frequency characteristic

The output torque of a stepper motor decreases as the speed increases, and drops sharply at higher speeds, so the maximum operating speed of a stepper motor is generally 300 ~ 600 r/min. The AC servo motor has a constant torque output and can output rated torque within the rated speed.

(4) Overload capacity

Stepper motors generally do not have overload capability, while AC servo motors have strong overload capability. When choosing a stepper motor as a driver, a motor with a larger torque is often needed to overcome the inertia moment, which will cause a waste of torque during normal operation.

(5) Operational performance

The control of the stepper motor is an open-loop control. If the starting frequency is too high or the load is too large, it is easy to lose steps or stall. When stopping, the speed is too high and overshoot is easy to occur. The problem of speed increase and decrease should be handled well to avoid Affect control accuracy. The AC servo drive system is a closed-loop control. The driver can directly sample the motor encoder feedback signal. It internally forms a position loop and a speed loop. There will be no loss of steps or overshoot of the stepper motor, and the control performance is more reliable.

(6) Different response performance

It takes 200 ~ 400 ms for a stepper motor to accelerate from standstill to working speed. The AC servo system has good acceleration performance and only takes a few milliseconds from standstill to rated speed, so it can be used in control situations that require quick start and stop.

(7) Cost

Compared with servo motors, the structure and driver of stepper motors are relatively simple, so stepper motors are more economical and servo motors are more expensive.

In the design process of the control system, various factors such as control requirements and cost must be comprehensively considered to select an appropriate motor.

3.2.4 Precautions for Selecting Servo Motors

(1) Maximum speed of the motor

Motor selection is based on the maximum operating speed of the load. The motor speed required for the highest speed operation of electromechanical equipment should be strictly controlled within the rated speed of the motor.

$$n = \frac{V_{\mathrm{max}} u}{P_{\mathrm{h}}} \times 10^3 \le n_{\mathrm{nom}} \tag{3.14}$$

In the formula, n_{nom} is the rated speed of the motor, r/min; n is the speed of the motor during rapid travel, r/min; V_{max} is the linear running speed, m/min; u is the system transmission ratio, $u = n\ motor/n\ screw$; P_h is the screw lead, mm.

(2) No-load acceleration torque

No-load acceleration torque refers to the torque when the executing part is accelerated from rest to step command to fast operation. The no-load acceleration torque should be limited to within 80% of the maximum output torque of the variable frequency drive system.

$$T_{\mathrm{max}} = \frac{2\pi n(J_{\mathrm{L}} + J_{\mathrm{M}})}{60 t_{\mathrm{ac}}} T_{\mathrm{F}} \le T_{\mathrm{Amax}} \times 80\%$$

In the formula, T_{Amax} is the maximum output torque of the variable frequency drive system matched with the motor, N m, T_{max} is the acceleration torque at no load, N m, T_F is the load torque transferred to the motor shaft at fast travel, N m, t_{ac} is the acceleration and deceleration time constant for fast travel, ms, J_L is the load inertia, kg m^2, J_M is the motor inertia, kg m^2.

(3) Inertia matching and load inertia

In order for the system to respond sensitively, sufficient angular acceleration should be ensured to meet the stability requirements of the system. The load inertia J_L should generally be limited to 2.5 times the motor inertia J_M, that is, $J_L < 2.5J_M$.

$$J_L = \sum_{j=1}^{M} J_j\left(\frac{\omega_j}{\omega}\right)^2 + \sum_{j=1}^{N} m_j\left(\frac{V_j}{\omega}\right)^2$$

In the formula, J_j is the rotating inertia of each rotating part, kg m^2, ω_j is the angular velocity of each rotating part, rad/min; m_j is the mass of each moving part, kg; V_j is the speed of each moving part, m/min, and ω is the angular velocity of the servo motor, rad/min.

(4) Load torque

In the normal working state, the load torque T_{ms} does not exceed 80% of the motor rated torque T_{MS}.

In the formula, T_c is the maximum load torque, N m; D is the maximum load ratio.

$$T_{ms} = T_c D^{\frac{1}{2}} \leq T_{MS} \times 80\%$$

(5) Continuous overload time

The continuous overload time tLon of mechanical and electrical equipment should be limited to the overload time tMon specified in the motor specification, otherwise the motor will overheat or even burn out.

3.2.5 *Select Servo Motor According to Load Torque*

According to the working curve of the servo motor, when the equipment is running with no load, the load torque added to the shaft of the servo motor in the entire speed range should be within the range of the continuous rated torque of the motor, that is, in the continuous working area of the working curve; The maximum load torque, loading period and overload time should be within the allowable range of the characteristic curve. The load torque added to the motor shaft can be converted to the load torque added to the motor shaft.

Taking cutting machine tools as an example, when choosing a servo motor according to the load torque:

① The friction torque of equipment moving parts such as guide rail needs to be considered.
② Due to the preloading of the bearing and nut of the equipment, as well as the friction of the ball contact surface of the preload force of the lead screw, the torque generated needs to be considered according to the situation.
③ The reaction force of the cutting force will increase the friction of the table. In the moment of bearing the large cutting reaction force, the load on the surface of the slide block also increases.
④ The friction torque is greatly affected by the feed rate, it is necessary to study and measure the change of friction caused by the change of speed, table support (slide, ball, pressure), slide surface material and lubrication conditions.

In addition, should also consider the use of equipment environment may affect the motor torque.

3.2.6 Select the Servo Motor According to the Load Inertia

In some applications, it is required to have good fast response characteristics, with the change of the control signal, the motor should complete the necessary action in a short time. For example, numerical control machine tools need to have good fast response characteristics to ensure processing accuracy and surface quality. The load inertia is related to the response of the motor and the fast moving time. With large inertia load, when the speed instruction changes, the motor needs a longer response time. Therefore, the size of the load inertia added to the motor shaft will directly affect the sensitivity of the motor and the accuracy of the whole servo system. When the load inertia is greater than 5 times the motor inertia, the sensitivity of the rotor will be affected, the motor inertia J_M and the negative load inertia J_L must meet:

$$1 \leq \frac{J_L}{J_M} < 5$$

All moving parts on the gear drive chain driven by the motor, whether rotating moving parts, or straight line moving parts, are the load inertia of the motor. The total load inertia on the motor shaft can be obtained by adding the inertia of the individual parts being driven.

(1) Inertia of the cylinder

Such as ball screw, gear and other inertia around its central axis can be calculated according to the following formula:

$$J = \frac{\pi \gamma}{32} \times D^4 L \left(\text{kg cm}^2\right)$$

In the formula, γ is the density of the material, kg/cm^3; D is the diameter of the cylinder, cm; L is the length of the cylinder, cm.

(2) Inertia of an axially moving object

The inertia of the workpiece, the table and the isoaxial moving object can be obtained from the following formula:

$$J = W\left(\frac{L}{2\pi}\right)^2 \left(\text{kg cm}^2\right)$$

(3) The inertia of the cylinder as it moves around the center

Such as a large diameter gear, in order to reduce the inertia, often dig out evenly distributed holes in the disc, or do a spoke type gear:

$$J = J_0 + WR^2 \left(\text{kg cm}^2\right)$$

In the formula, J_0 is the inertia of the cylinder when it rotates around its center line, kg cm^2; W is the mass of the cylinder, kg; R is the radius of rotation, cm.

(4) Calculation of inertia relative to motor shaft mechanical speed change

The calculation method of load inertia J_0 is converted to the motor shaft:

$$J = \frac{N_1}{N_2} J_0 \left(\text{kg cm}^2\right)$$

In the formula, N_1 and N_2 are the number of teeth of the gear.

3.2.7 *Select the Servo Motor According to the Torque When the Motor Is Accelerating and Decelerating*

(1) According to the acceleration torque of linear acceleration and deceleration

Figure 3.13 shows the torque curve of the motor during acceleration or deceleration. The acceleration torque at linear acceleration and deceleration is calculated as follows:

$$T_a = \frac{2\pi n_m}{60 \times 10^4} \times \frac{1}{t_a}(J_M + J_L)\left(1 - e^{-K_s t_a}\right) (\text{N m})$$

In the formula, n_m is the steady speed of the motor, t_a is the acceleration time, J_M is the motor rotor inertia, kg cm^2, J_L is the load inertia converted to the motor shaft, kg cm^2, and K_s is the position servo open-loop gain.

The speed at which the acceleration torque begins to decrease is as follows:

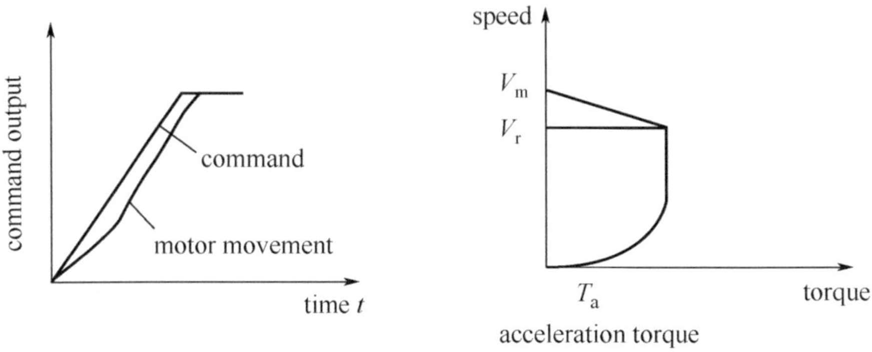

Fig. 3.13 Torque during motor acceleration or deceleration

$$n_r = n_m\left[1 - \frac{1}{t_a K_s}\left(1 - e^{-K_s t_a}\right)\right]$$

(2) Accelerate on an exponential curve

The torque curve of the motor accelerated by exponential curve is shown in Fig. 3.14. At this time, the torque T_0 at zero speed can be given by the following formula:

$$T_0 = \frac{2\pi n_m}{60 \times 10^4} \times \frac{1}{t_e}(J_M + J_L)\ (\mathrm{N\,m})$$

In the formula, t_e is the acceleration time constant of the exponential curve.

(3) Enter the phased speed instruction

In this case, the acceleration torque T_a is equivalent to T_0, which can be obtained from the following formula (where $t_s = K_s$).

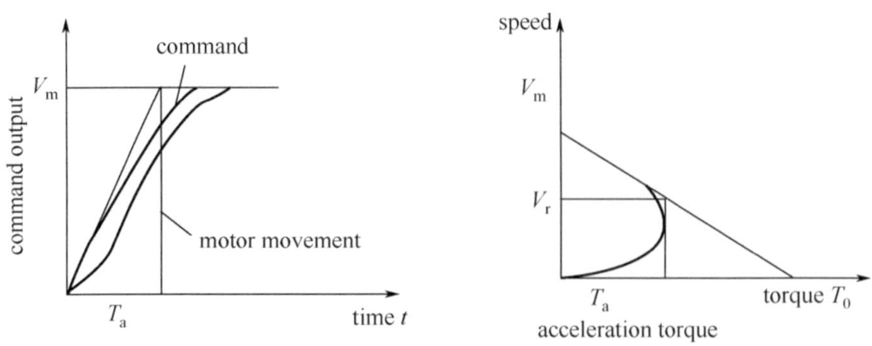

Fig. 3.14 Torque curve of the motor when accelerating along an exponential curve

$$T_a = \frac{2\pi n_m}{60 \times 10^4} \times \frac{1}{t_s}(J_M + J_L)\ (\text{N m})$$

3.2.8 Select Servo Motor According to the Root Mean Square of Motor Torque

The working machinery needs torque when starting and braking frequently. When the working machinery starts and brakes frequently, it is necessary to check whether the motor is overheated, so it is necessary to calculate the root mean square value of the motor torque in a cycle, and the root mean square value should be less than the continuous torque of the motor. The root mean square of the motor is given by the following formula:

$$T_{rms} = \sqrt{\frac{(T_a + T_f)^2 t_1 + T_f^2 t_2 + (T_a - T_f)^2 t_1 + T_0^2 t_3}{T_{周}}}$$

In the formula, T_a is the acceleration torque, N m; T_f is the friction torque, N m; the torque of T_0 during the stop period, N m; T_1, T_2, T_3, T cycle is shown in Fig. 3.15.

For the calculation of the periodically varying torque of the load (Fig. 3.16), it is also necessary to calculate the root mean square value of the torque in one cycle, which is less than the rated torque. So that the motor will not overheat and work normally.

The selection principle of the feed servo motor in the design is as follows: first, check whether the load torque and acceleration and deceleration torque meet the requirements according to the torque-speed characteristic curve, and then check

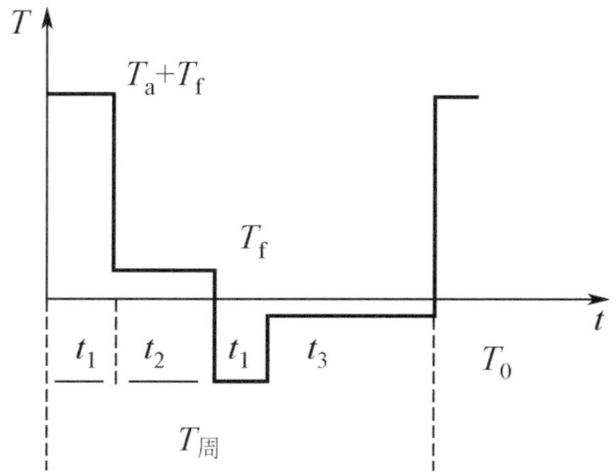

Fig. 3.15 Torque curve of $t_1, t_2, t_3, T_{period}$

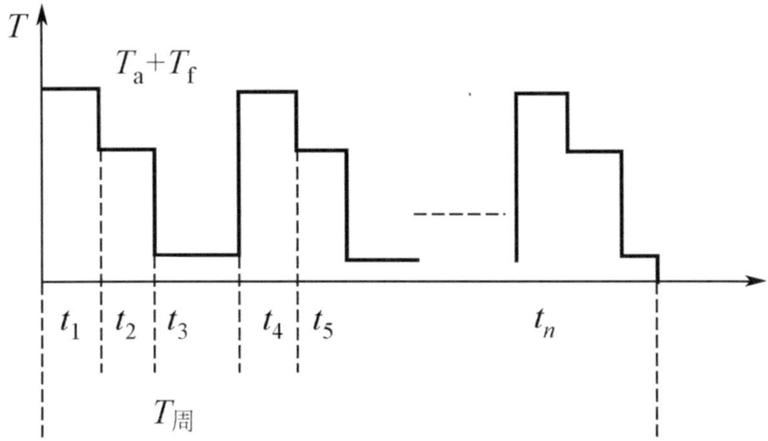

Fig. 3.16 Torque with periodic load variation

the load inertia. The motor that requires frequent start-up and braking should also check its root mean square torque, so that the selected motor can not only meet the requirements, but also avoid the problems caused by the excessive selection of motor rated torque.

3.2.9 Servo Motor Selection Steps, Methods, Formulas

(1) Determine the mode of operation

According to the control content of the mechanical system, determine the operation mode of the servo motor, start time t_a, deceleration time t_d according to the actual situation.

(2) Calculate the load converted to the moment of inertia GD_m^2 on the motor shaft.

In order to calculate the starting torque TP, we must first find the moment of inertia of the load:

$$GD_l^2 = \frac{\pi}{8}\rho LD^4 \times 10^4 \ (\text{kg m}^2)$$

In the formula, L is the length of the cylinder, cm; D is the diameter of the cylinder, cm; ρ is the material density, kg/m^3.

$$GD_m^2 = \left(\frac{N_l}{N_m}\right)^2 GD_l^2 + \left(\frac{1}{R}\right)^2 \times \frac{\pi}{8}\rho l_2 d_2^4 + \frac{\pi}{8}\rho l_1 d_1^4 \ (\text{kg m}^2)$$

In the formula, l_2 is the thickness of the load-side gear; d_2 is the diameter of the load-side gear; l_1 is the thickness of the motor-side gear; d_1 is the diameter of the motor-side gear; ρ is the material density; GD_l^2 is the load moment of inertia, kg m^2; N_l is the load shaft speed, r/min; N_m is the motor shaft speed, r/min; 1/R is the reduction ratio.

(3) Primary motor

Calculate the power P_O and torque T_L when the motor is running stably. T_L is the load torque converted to the motor shaft:

$$T_L = \frac{N_l}{N_m \eta} T_l$$

In the formula, η is the efficiency of the mechanical system.

$$P_O = \frac{T_l N_l}{9535.4\eta}$$

In the formula, T_l is the load shaft torque.

(4) Calculate acceleration and deceleration time or acceleration and deceleration power

For the preliminary selection motor, the acceleration and deceleration time must be calculated according to the requirements of the mechanical system, which must be less than the required value of the mechanical system.

Acceleration time:

$$t_a = \frac{(GD_m^2 + GD_l^2)N_m}{38.3(T_P - T_l)}$$

Deceleration time:

$$t_d = \frac{(GD_m^2 + GD_l^2)N_m}{38.3(T_P + T_l)}$$

The mechanical value of the motor is used in the above two formulas, so after finding the time after adding the start signal, it must be added as the control circuit lag time 5 ~ 10 ms. Load acceleration torque T_P can be calculated from the starting time, if T_P is greater than the rated torque of the primary motor, but less than the instantaneous maximum torque of the motor (5–10 times the rated torque), it can also be considered appropriate for the motor primary.

(5) Effective torque calculation considering working cycle and duty cycle factors

In the workplace where the movement speed is relatively fast, such as robots, the influence of acceleration and deceleration exceeding the rated current cannot be ignored, and the effective torque needs to be obtained by the duty cycle factor. If the value is below the rated torque of the primary motor, the motor is suitable for selection.

$$T_{\mathrm{rms}} = \sqrt{\frac{T_{\mathrm{P}}^2 t_{\mathrm{a}} + T_{\mathrm{l}}^2 t_{\mathrm{c}} + T_{\mathrm{P}}^2 t_{\mathrm{d}}}{t}} \times f_{\mathrm{w}}$$

In the formula, t_a is the start time, s; t_c is the uptime, s; t_d is deceleration time, s; f_w indicates the waveform coefficient. If the T*rms* does not meet the rated torque type, it is necessary to increase the motor capacity and calculate again.

3.3 Hydraulic Drive System

Due to the many advantages of electric transmission and the ease with which electric motors convert electrical energy into mechanical energy, some designers of electromechanical systems may think that hydraulic or pneumatic systems are no longer necessary, but this is not the case. In many cases, reducing the weight of the system is important, and in this respect hydraulic transmission has outstanding advantages over electric transmission. Because the typical value of the power to weight ratio of the hydraulic pump and the motor is 168 W/N, and the power to weight ratio of the motor is 16.8 W/N. Due to the saturation effect of magnetic materials, the force or torque output of the motor is limited to a certain extent. In the hydraulic system, higher force or torque can be obtained by increasing the working pressure. Generally speaking, the force-mass ratio of linear motors is 130 N/kg; The force-mass ratio of the linear hydraulic motor is 13,000 N/kg, which is 100 times higher. The torque inertia ratio of rotary hydraulic motor is generally 10–20 times that of the equivalent capacity motor, and only the slotless DC torque motor can be equivalent to the hydraulic transmission. In addition, the output stiffness of the open-loop hydraulic system is large, while that of the motor system is small [6].

3.3.1 Overview

The hydraulic system uses oil as the working medium and transmits power through the pressure inside the oil. The complete hydraulic system consists of five parts: power components, executive components, control components, auxiliary components and hydraulic oil. Hydraulic system can be divided into hydraulic transmission system and hydraulic control system two categories. Hydraulic transmission system to transfer power and movement as the main function, mainly the use of various components composed of a certain function of the basic control loop, and then the various basic control loop integrated to complete the special task of transmission

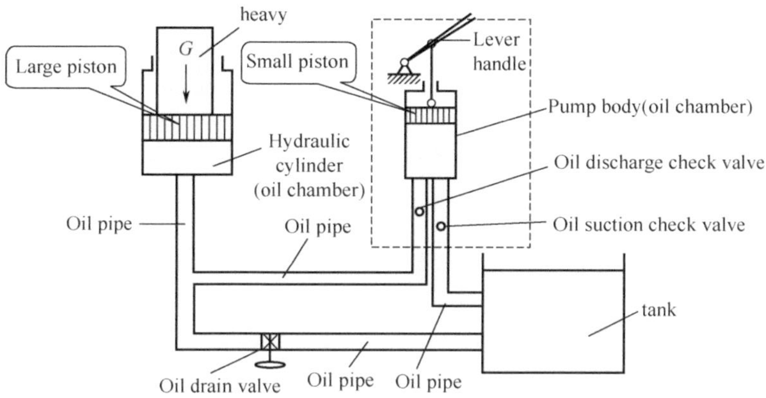

Fig. 3.17 Hydraulic transmission schematic diagram

and control system, to achieve the conversion between energy. Figure 3.17 shows the hydraulic transmission diagram, Fig. 3.18 shows the hydraulic system diagram.

The role of the power element is to convert the mechanical energy of the prime mover into the pressure energy of the liquid, generally the oil pump, which is responsible for providing power to the entire hydraulic system. The actuator converts the pressure energy of the liquid into mechanical energy to drive the load for linear reciprocating movement or rotary movement, generally for hydraulic cylinders and hydraulic motors. Control elements control and regulate the pressure, flow and direction of liquids in a hydraulic system.

Auxiliary components include fuel tank, oil filter, cooler, heater, accumulator, tubing and fittings, sealing ring, quick change fittings, high pressure ball valve, hose assembly, pressure gauge, oil level gauge, oil temperature gauge, etc.

Hydraulic oil is the working medium of energy transfer in hydraulic system.

Hydraulic system in machine tools, construction machinery, metallurgy, petrochemical, aviation, ships and other aspects have a wide range of applications, in the intelligent equipment manufacturing industry also has a great space for development. At present, the hydraulic system is developing in the direction of high pressure, high speed, high power, high efficiency, low noise, high integration and digitalization.

① Low power consumption. In recent years, the energy conversion rate and utilization rate of hydraulic technology have been significantly improved, but there is still a lot of room for improvement in loss. The power loss of the hydraulic system is mainly the volume and mechanical loss inside the hydraulic system. If the pressure energy utilization rate in the hydraulic system is improved, the energy loss will be greatly reduced. Therefore, the future should focus on reducing the internal pressure loss of components and systems to reduce power loss.

② Active maintenance. Current passive maintenance of hydraulic drives reduces productivity. With the progress of science and technology, the precision of hydraulic system and the development of modern hydraulic system fault diagnosis methods, hydraulic transmission is developing towards active maintenance.

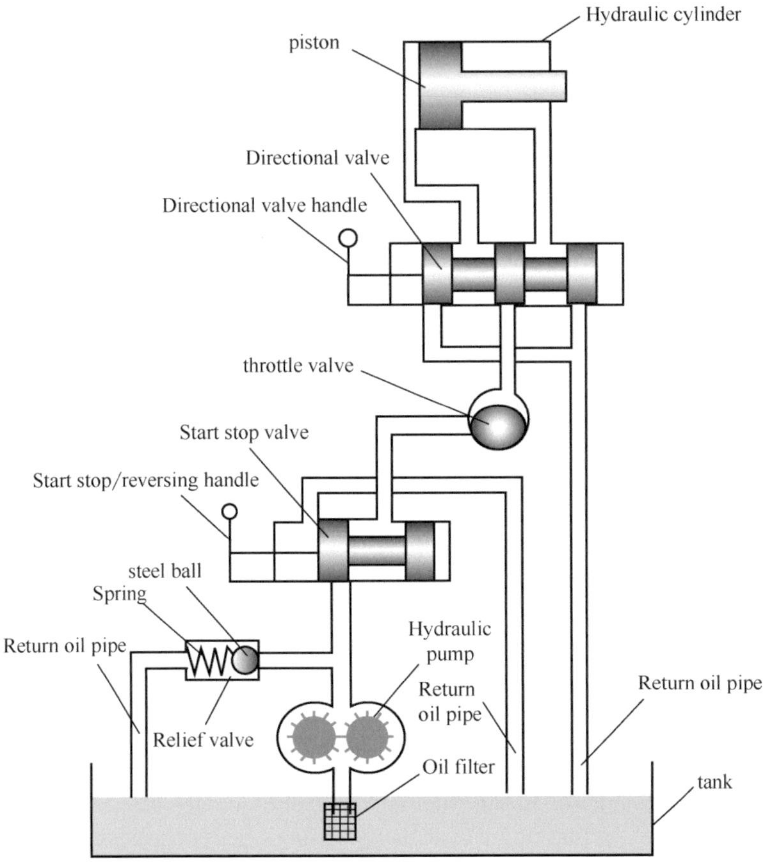

Fig. 3.18 Hydraulic system diagram

Traditional fault diagnosis methods have been unable to meet the maintenance needs of the hydraulic system, so it is necessary to gradually establish and improve the hydraulic system fault database, the use of computer technology and Internet of things technology, combined with expert systems, quickly and efficiently diagnose faults and formulate reasonable and accurate maintenance plans, take the corresponding proactive maintenance and preventive measures to achieve the purpose of improving production efficiency.

③ Integration. The rapid development of microelectronics technology has injected new vitality into the innovation of hydraulic technology. The progress of science and technology has greatly promoted the development of electro-hydraulic valves, sensing technology, etc., so that the hydraulic system gradually has the dual characteristics of electrical and hydraulic technology. Modern hydraulic technology is gradually developing towards integration, intelligence and automation.

④ Intelligent control. Intelligent control is essentially formed on the basis of electrification, compared with electrical control, intelligent control is a more advanced mode. Intelligent control can reduce the problems that may occur in construction machinery, so that construction machinery can always work in the best condition, and extend its service life. To realize intelligent control, we must first establish a perfect database, input the optimal state data of the hydraulic system, and then collect the relevant data through the sensor in real time. Systematically compare the collected data with the database data, and ensure that it is always in the best working state through system regulation, so as to reduce the system failure frequency and improve the system work efficiency.

In general, the hydraulic system design process is shown in Fig. 3.19.

① Determine the requirements of the hydraulic system. Determine the work completed by the hydraulic transmission according to the existing production conditions and costs, and then determine the work sequence of each movement, and determine the automatic working cycle for the machine with a high degree of automation; For the design of complex actions, it is necessary to draw the action cycle diagram. Secondly, according to the speed cycle diagram drawn, determine the main performance parameters of the hydraulic system, including empty stroke speed, working stroke speed and speed regulation range, and the rest of the performance parameters, such as work smoothness, reliability, conversion accuracy, residence time, etc., should also be considered.
② Schematic diagram of hydraulic system. First of all, the basic circuit of the hydraulic system is selected, and the basic circuit information such as speed

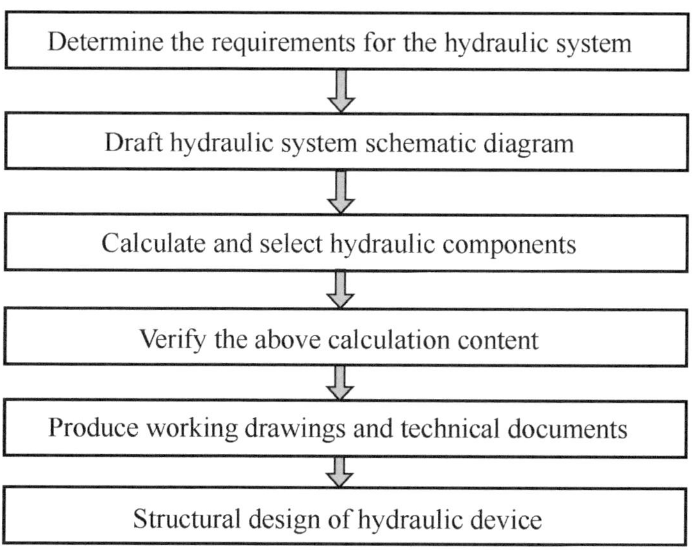

Fig. 3.19 The design process of hydraulic system

regulation mode, flow direction control and sequential action control mode is obtained through the comparative analysis of similar products, and then the auxiliary circuit and components such as the control circuit, lubricating oil circuit and filter are configured. The actions in the cycle can not affect each other, and the system structure is as simple and economical as possible, which is easy to integrate block design, manufacturing, installation and maintenance.

③ Calculation and selection of hydraulic components. First of all, the working load and the speed of the actuator must be calculated, and then the area of the hydraulic cylinder, that is, the hydraulic motor displacement, is calculated to obtain the required oil pressure and flow. Calculate the working pressure and flow of the hydraulic pump, select the appropriate hydraulic pump, and determine the drive motor power. According to the pressure and flow of the system oil, select the valve type element, determine the pipe size and tank capacity. It is worth noting that when choosing hydraulic components, choose standard components as much as possible. Table 3.2 lists the common hydraulic pump types.

④ Check the content of the preliminary calculation of the hydraulic system. After determining the specifications and dimensions of the components, the pressure loss of the loop should be estimated, and then the oil supply pressure of the system should be determined. In order to ensure the normal operation of the system, the heating and heat dissipation should also be checked. For the hydraulic system requiring high precision, it is necessary to check the hydraulic shock, reversing performance and so on.

⑤ Draw formal working drawings and prepare technical documents. The official working drawing includes hydraulic system schematic drawing, machine pipeline assembly drawing, non-standard hydraulic components parts drawing and assembly drawing. The preparation of technical documents generally includes parts and parts catalog, standard parts, general parts and purchased parts general table, trial operation requirements and technical specifications.

⑥ Design the structure of the hydraulic device. The structure of the hydraulic device has centralized type and distributed type. The centralized type is to design the hydraulic pump station independently of the oil source and the control and adjustment device of the hydraulic system. The distributed mode disperses the oil

Table 3.2 Common types of hydraulic pumps

External gear pump	Double acting vane pump	Axial piston pump
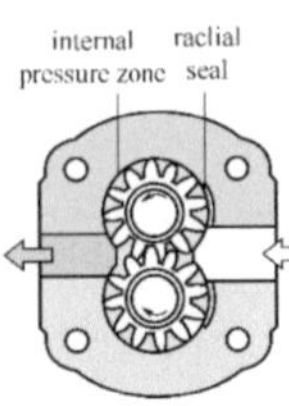		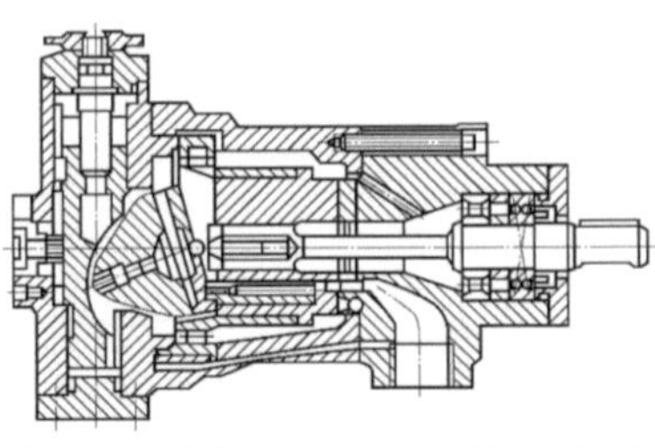

source and control adjustment device of the hydraulic system throughout the machine. Both methods have their own advantages and disadvantages, which need to be considered in comprehensive design requirements. The configuration of components in the hydraulic system has two types: plate configuration and integrated configuration. The plate configuration is to screw the components and the bottom plate to the upright plate, and the integrated configuration is to combine the hydraulic components with some auxiliary components. According to the different forms of auxiliary components, it is divided into box type, integrated block type and superimposed valve type.

3.3.2 *Form of Hydraulic System*

(1) Classification by oil circulation mode

① Open system. Figure 3.20 shows the simple structure of the open system. The open system refers to the hydraulic pump suction oil from the tank, through the reversing valve to the hydraulic cylinder (or hydraulic motor) oil supply to drive the working mechanism, the hydraulic cylinder (or hydraulic motor) oil return through the reversing valve back to the tank. Because the oil after the system is finished returns to the tank, it can play the role of heat dissipation and precipitation of impurities. However, because the oil is often in contact with the air, the air is easy to penetrate the system, resulting in unstable movement of the working mechanism and other adverse consequences. In order to ensure the smooth motion of the working mechanism, the back pressure valve can be set on the oil return circuit of the system, which will cause additional energy loss and make the oil temperature rise.

In the open system, the hydraulic pump used is a quantitative pump or a one-way variable pump, considering the self-priming capacity of the pump and avoiding the suction phenomenon, the hydraulic pump with poor self-priming capacity is usually limited to its working speed within 75% of the

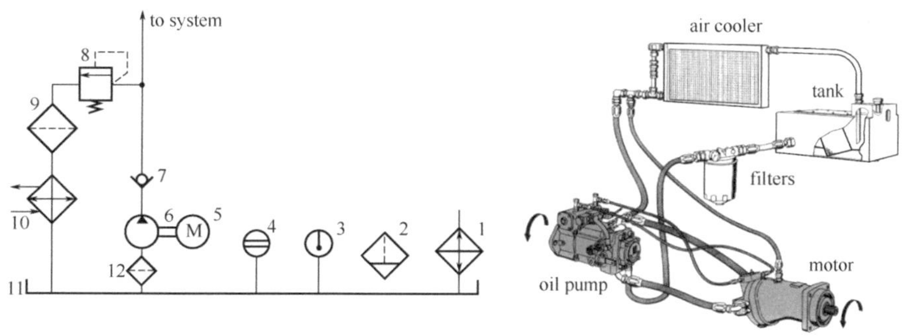

Fig. 3.20 Open hydraulic system

rated speed, or an auxiliary pump is added for perfusion, and the reversing of the working mechanism is aided by the reversing valve. When the reversing valve reverses, in addition to producing hydraulic shock, the inertia energy of the moving parts will be converted into heat energy, and the temperature of the hydraulic oil will be increased. Because of its simple structure and low cost, the open system has been widely used in construction machinery.

② Closed system. As shown in Fig. 3.21, in a closed system, the inlet pipe of the hydraulic pump is directly connected to the return pipe of the actuator, and the working liquid is closed in the pipeline of the system. Closed system structure is more compact, less contact with air, air is not easy to penetrate the system, so the smooth transmission is good. The variable speed and commutation of the working mechanism are realized by adjusting the variable mechanism of the pump or motor, which avoids the hydraulic shock and energy loss during the commutation of the open system. However, the closed system is more complex than the open system, because the oil after the work of the closed system does not return to the tank, the heat dissipation and filtration conditions of the oil are worse than the open system.

When the actuator in the closed system uses a double acting single piston rod hydraulic cylinder, the power utilization rate will be reduced during the working process due to the unequal flow rate of the size cavity, so the actuator in the closed system is generally a hydraulic motor. For example, the rotary system in large hydraulic excavators and hydraulic cranes, and the actuator elements in the walking system and vibration system of full hydraulic road rollers are hydraulic motors. Another advantage of the hydraulic motor as

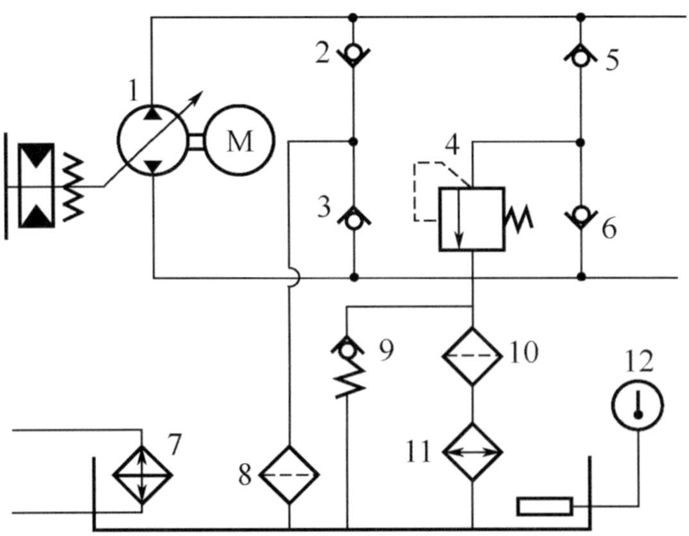

Fig. 3.21 Closed hydraulic system, 1 variable displacement pump; 2, 3, 5, 6 one-way valves; 4 overflow valve; 7 heater; 8, 10 filter; 9 bypass check valve; 11 cooling device; 12 thermometer

the actuator in the closed system is that the maximum starting torque and braking torque value are equal during starting and braking.

(2) Classified according to the number of hydraulic pumps in the system

① Single pump system. A hydraulic system that supplies oil by a hydraulic pump to one or a group of actuator elements is a single-pump hydraulic system, as shown in Fig. 3.22. The single pump system is suitable for construction machinery that does not need to carry out a variety of compound actions, such as the hydraulic system of shovel transport machinery such as bulldozers and scrapers. In the work cycle of some construction machinery (such as hydraulic excavators, hydraulic cranes), it is necessary to realize compound actions, and it is necessary to be able to adjust these actions individually, and the use of a single pump system is obviously not ideal. In order to make more efficient use of engine power and improve operating performance, a dual pump system or a multi-pump system must be used (Fig. 3.23).

② Dual pump system. Figure 3.24 shows the dual pump hydraulic system. The dual pump hydraulic system is actually a combination of two single pump hydraulic systems. Each pump can separately supply fuel to the actuator in its own loop. The power of each pump is determined according to the power required in the respective circuit, which ensures compound action.

When only a single action is needed in the system and the engine power is fully utilized, the confluence oil supply method can be used, that is, the flow of two hydraulic pumps is supplied to an actuator at the same time, so that the movement of the working mechanism can be accelerated. This two-pump hydraulic system has been widely used in small and medium-sized hydraulic excavators and cranes.

③ Multi-pump system. In order to further improve the performance of hydraulic excavators and hydraulic cranes, in recent years, a three-pump system has been adopted in large hydraulic excavators and hydraulic cranes. Figure 3.25 shows the schematic diagram of the three-pump hydraulic system. The three-pump hydraulic system is characterized by an independent closed system for the rotary mechanism and an open system for the other two circuits. According to the working condition of the host, the different loops can be combined together to obtain the best working performance of the host.

(3) According to the use of hydraulic pump form classification

① A quantitative system. A hydraulic system using a quantitative pump is called a quantitative system. The hydraulic pumps used in the dosing system are gear pumps, vane pumps or plunger pumps. When the engine speed is certain, the flow rate Q of the quantitative pump is also certain, and the pressure is determined according to the maximum resistance that needs to be overcome in the working cycle, so when the hydraulic system is working,

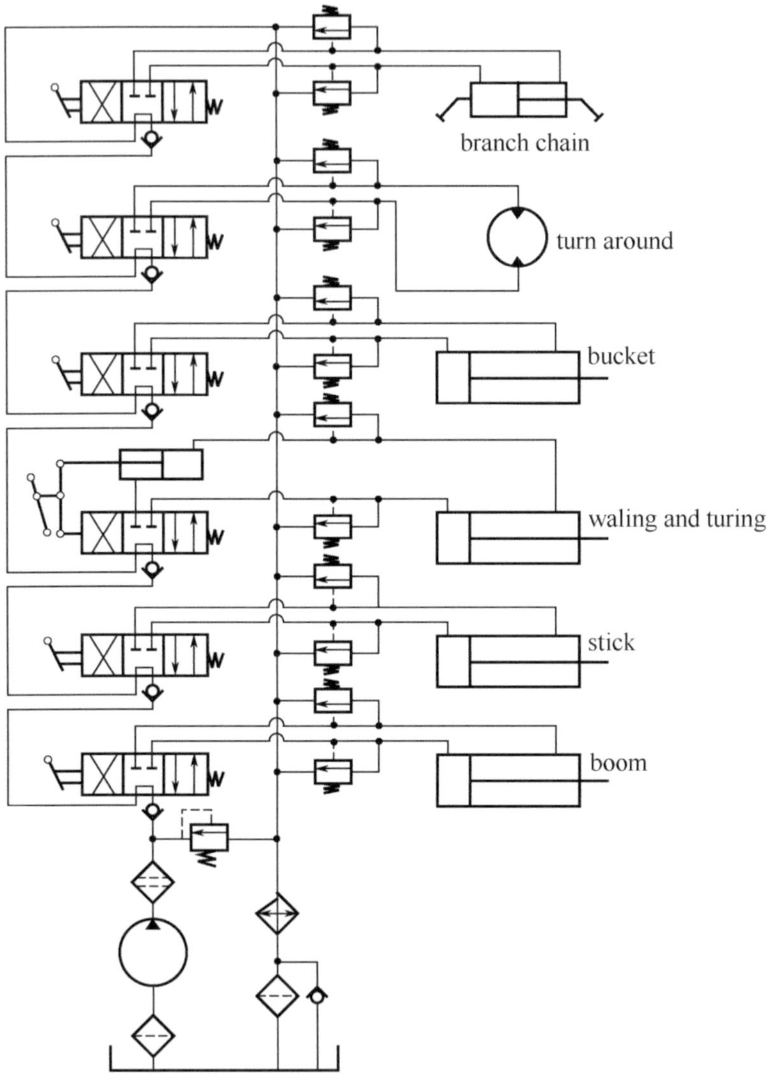

Fig. 3.22 Single pump hydraulic system

the power of the hydraulic pump changes with the change of the working resistance.

② Variable system. The hydraulic pump used in the variable system is the axial piston pump with constant power control. In the power regulator, there is pressure oil action on the right side of the control piston, and spring force action on the left side of the control piston. When the outlet pressure of the pump is lower than the pre-tightening pressure of the spring device, the spring device is not compressed, and the swing Angle of the hydraulic

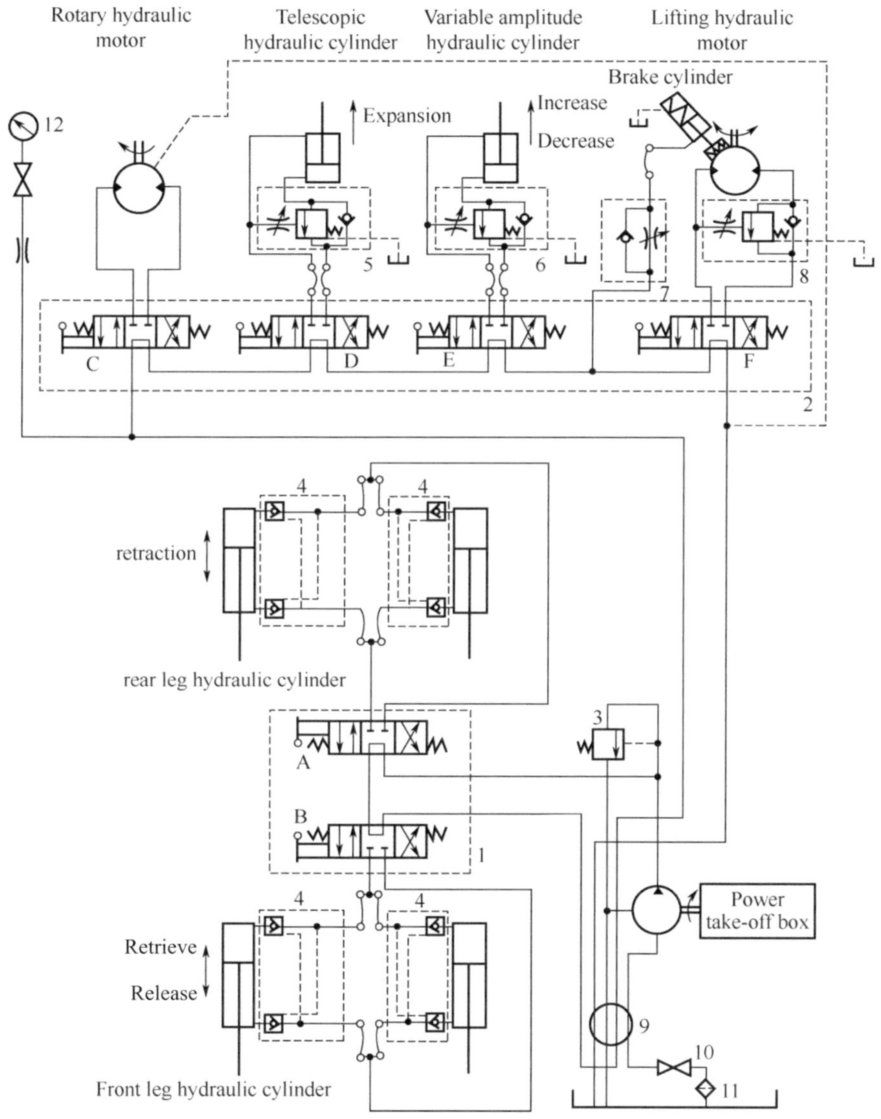

Fig. 3.23 Hydraulic system of a certain automobile crane

pump is at the maximum Angle, and the displacement of the pump is also the maximum. With the increase of the hydraulic pump outlet pressure, the spring is compressed, and the swing Angle of the hydraulic pump is also reduced, and the displacement is also reduced. The position of the hydraulic pump when the outlet pressure and the pre-compression force

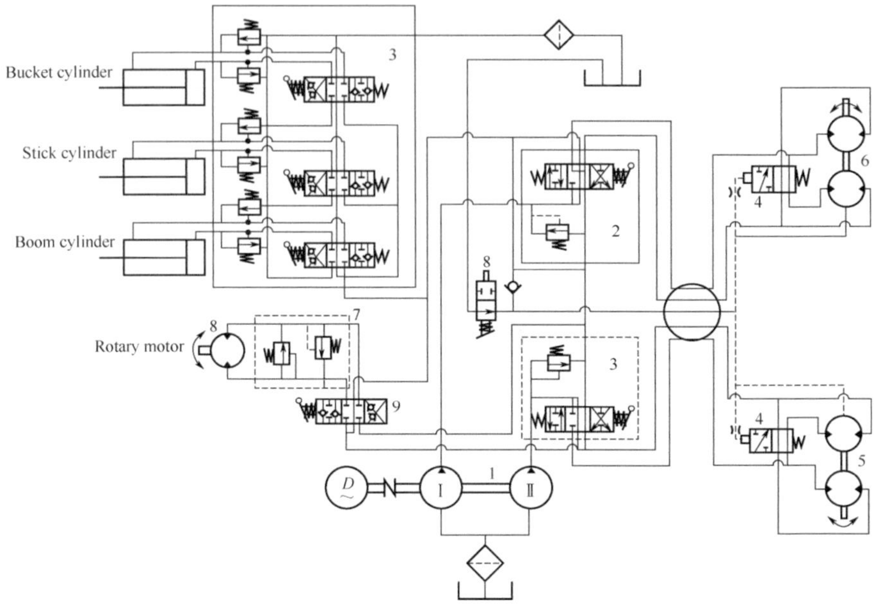

Fig. 3.24 Dual pump hydraulic system, 1 a dual hydraulic pump; 2 directional valve; 3 multi way directional valve; 4 variable speed valve; 5 pilot valves; 6 walking motor; 7 buffer brake valve; 8 rotary motors; 9 rotary motor reversing valves

of the spring device are balanced is called the adjustment starting position. When adjusting the starting position, the liquid pressure acting on the control piston in the power regulator is called the starting pressure. When the outlet pressure of the hydraulic pump is greater than the starting pressure, because the compression force of the spring in the regulator has an approximate hyperbolic change relationship with its stroke, so when the speed is constant, the outlet pressure and flow of the hydraulic pump also show an approximate hyperbolic change, and the hydraulic pump always maintains a constant power working characteristic within the adjustment range. Figure 3.26 shows a comparison between the power utilization of a quantitative system and a variable system. Figure 3.27 shows the power characteristic curve of the constant power control variable pump. Because the working pressure of the hydraulic pump changes with the size of the external load, the speed of the working mechanism can be reduced with the increase of the external load, or with the decrease of the external load, so that the engine power can be fully utilized within the adjustment range of the hydraulic pump.

The starting pressure of the variable pump is determined by the stiffness of the spring device and the requirements of the hydraulic system. The maximum pressure is determined by the hydraulic system and set by the safety valve. The swing Angle corresponding to the starting pressure is the

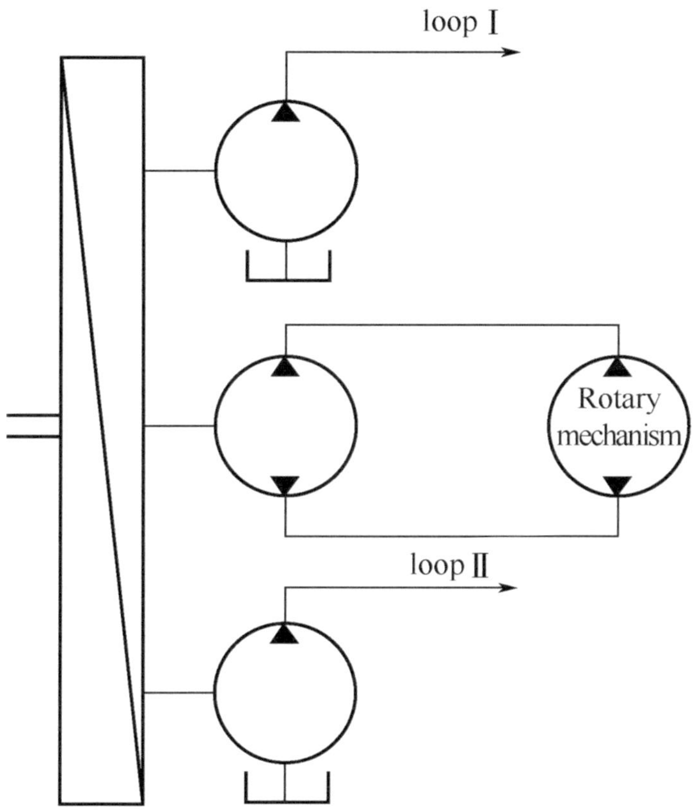

Fig. 3.25 Schematic diagram of three pump hydraulic system

Fig. 3.26 Comparison of power utilization between quantitative system and variable system

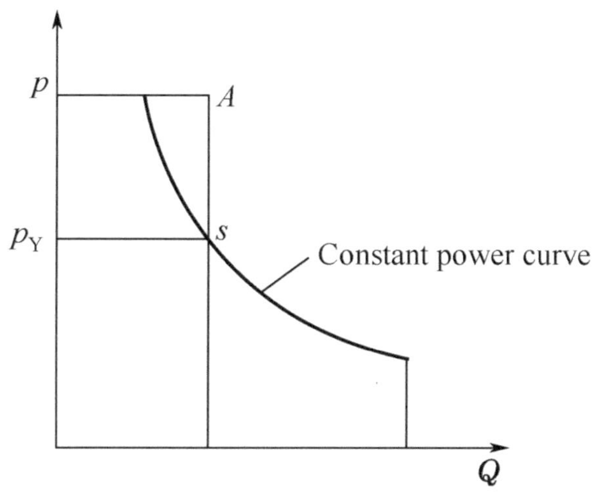

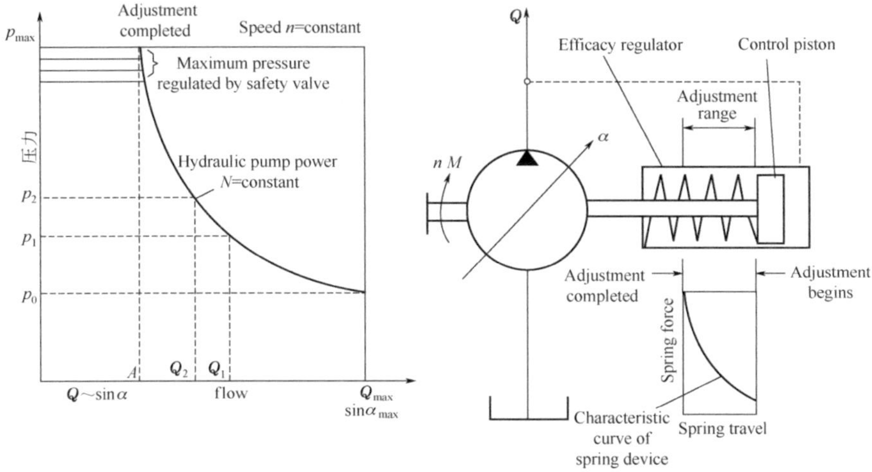

Fig. 3.27 Power characteristic curve of constant power control variable pump

largest, and the swing Angle corresponding to the end of the adjustment is the smallest. The advantage of variable pump is that it is within the adjustment range and can make full use of the power of the engine, but the disadvantage is that the structure and manufacturing process are complex and the cost is high.

(4) Classified according to the way of oil supply to the actuator

According to the way of oil supply to the actuator, it can be divided into series system and parallel system.

① Series system. When a hydraulic pump supplies oil to a group of actuators, the hydraulic system in which the oil return of the previous actuator is the oil input of the next actuator is called a series system, as shown in Fig. 3.28.

In the series system, the pressure oil output by the hydraulic pump enters the first actuator with pressure p_1 and flow Q_1, and enters the second actuator with pressure p_2 and flow Q_2. Without considering the energy loss, $Q_1 \neq Q_2$ is used for the double acting single piston rod hydraulic cylinder. Q_1 and Q_2 are directly proportional to the effective area S_1 and S_2 of the hydraulic cylinder piston, that is

$$Q_2 = \frac{Q_1 S_2}{S_1}$$

Without considering the energy loss in the pipeline and the actuator, the working pressure p_1 in the first actuator depends on the pressure p' required to overcome the load on the actuator and the working pressure p_2 of the second actuator, that is

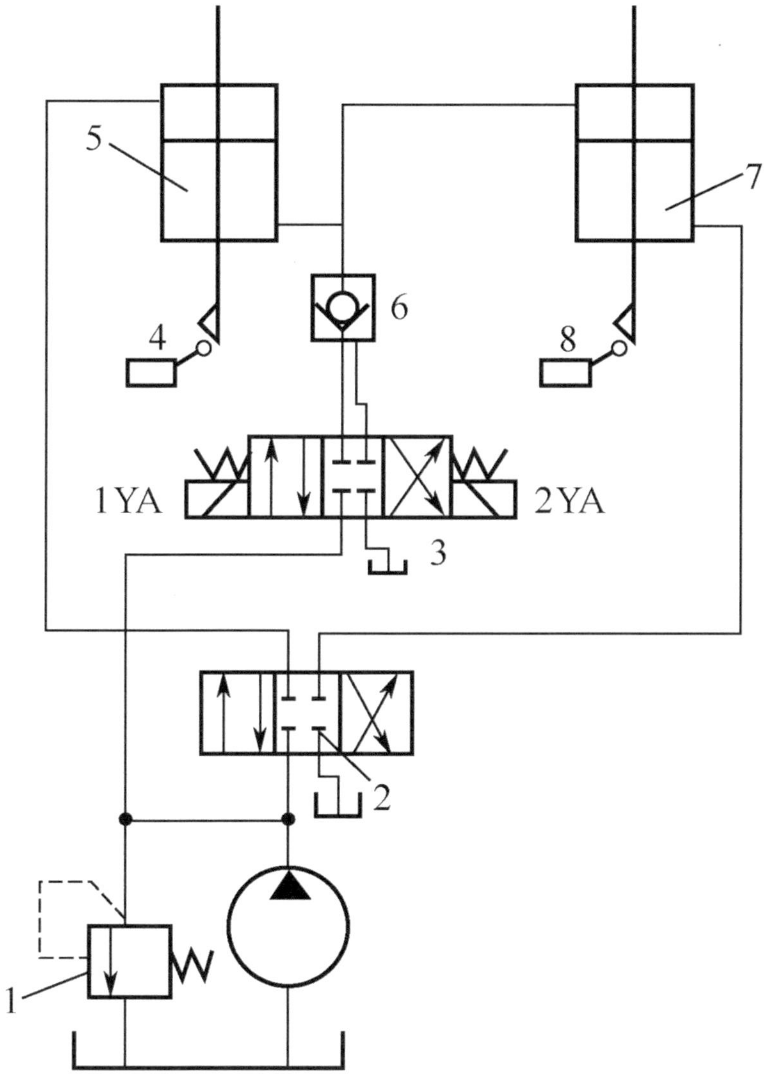

Fig. 3.28 Series system

$$p_1 = p' + p_2$$

In a series system, the working pressure is reduced once for each passing actuator. When the main pump supplies oil to the actuator controlled by the multi-way valve, as long as the outlet pressure of the hydraulic pump is sufficient, the movement of the actuator can be realized. However, because the pressure of the actuator is superimposed, the external load that can be overcome will decrease with the increase of the number of actuator elements.

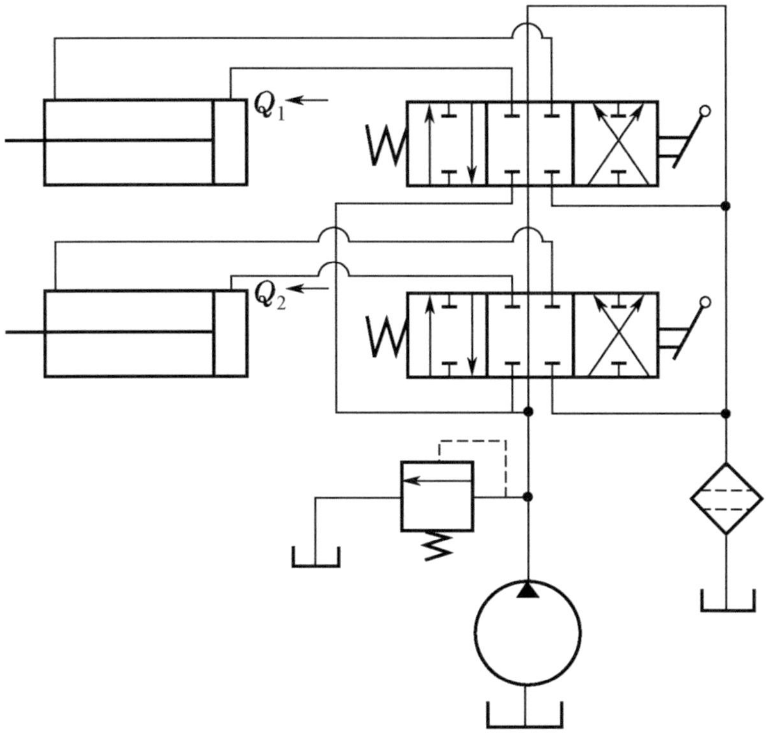

Fig. 3.29 Parallel system

② Parallel system. In a parallel system, when a hydraulic pump simultaneously supplies oil to a group of actuator, the flow into each actuator is only part of the output flow of the hydraulic pump, as shown in Fig. 3.29. In parallel system, when the main pump supplies oil to each actuator controlled by the multi-way valve, the flow distribution changes with the different load on each actuator, and the pressure oil first enters the actuator with small external load. Simultaneous action can be realized only when the upper and outer loads of each actuator are equal. The outlet pressure of the hydraulic pump depends on the sum of the pressure on the actuator with a small external load and the pressure loss on the oil circuit. Because the parallel system only needs to overcome the external load once in the working process, the ability to overcome the external load is large.

3.3.3 Hydraulic System Performance Evaluation

The hydraulic drive system has many advantages, such as small volume, light weight, high stiffness, high precision, fast response, wide speed range and so on. With the development of hydraulic and related technologies, hydraulic transmission is more and more widely used in construction machinery and machine tools at home and abroad. The performance of equipment using hydraulic drive depends on the performance of the hydraulic system. Any kind of mechanical hydraulic transmission system should meet the requirements of light weight, small size, simple structure, easy to use, high efficiency and good quality. In particular, good quality and high efficiency are emphasized, and advanced technology should be adopted as far as possible on the basis of ensuring good quality and high efficiency. The performance of the hydraulic system is based on the quality of the components used in the system and the suitability of the selected basic circuit. Generally speaking, it can be analyzed and compared from the efficiency, power utilization, speed regulation range and fine tuning characteristics, vibration and noise of the hydraulic system.

(1) Efficiency of the hydraulic system

The energy problem in today's world is becoming more and more prominent, improving mechanical efficiency is of great significance, under the premise of ensuring the performance requirements of the main engine, we should try our best to improve the efficiency of the hydraulic system. The efficiency of the hydraulic system reflects the utilization rate of the system energy, and the energy loss of the hydraulic system is expressed in the form of higher heat of the system oil temperature rise. The factors that cause the energy loss of hydraulic system mainly have the following aspects.

① The energy loss of the reversing valve during the reversing braking process. When the inertia of the actuator and its external load is large, the inertia influence of the pressure oil and the moving mechanism during the braking process increases the pressure of the oil return chamber, and the oil is extrude from the opening gap of the reversing valve or the brake valve, so that the inertia energy of the moving mechanism becomes heat consumption, and the oil temperature of the system rises. In some systems with frequent reversals and large inertia, such as the rotary system of excavators, the heating problem is particularly prominent.

② The energy loss of the hydraulic component itself. The energy loss of hydraulic components includes the energy loss of hydraulic pump, hydraulic motor, hydraulic cylinder and control components, among which the loss of pump and hydraulic motor is the largest. The structure of pipes and control elements can also affect the amount of energy lost. In the structure of the control element, the transition between two different sections should be smooth to minimize friction losses.

③ Overflow loss. When the hydraulic system is working, the working pressure exceeds the opening pressure of the relief valve (safety valve or overload valve), the relief valve is opened, and the flow output of the hydraulic pump is fully or partially overflow through the relief valve. When the system is working, measures can be taken from both design and operational factors to minimize overflow losses.

④ Back pressure loss. In order to ensure the smooth motion of the working mechanism, the back pressure valve is often set on the oil return circuit of the actuator. The greater the back pressure, the greater the energy loss. Under normal circumstances, the back pressure of the hydraulic motor is larger than that of the hydraulic cylinder; The back pressure of low speed hydraulic motors is greater than that of high speed motors. In order to reduce the heat caused by the back pressure of the oil return, the back pressure of the oil return should be reduced as much as possible under the condition of ensuring the motion stability of the working mechanism.

To sum up, in order to ensure the high efficiency of the hydraulic system, it is necessary to control and reduce the energy loss of the system and components, that is, to control and reduce the total heat of the system and improve the power utilization rate of the hydraulic system.

(2) Speed range and fine tuning characteristics

Most of the working mechanism of hydraulic machinery has a large range of load and speed change, which requires the hydraulic system to have a large speed regulation range. The speed range of different machinery is different, even in the same machinery, the speed range of different working mechanisms is not the same.

The fine-tuning characteristics reflect the sensitivity of the speed adjustment of the working mechanism. Different construction machinery has different requirements for fine tuning characteristics. For example, the requirements for fine tuning characteristics of shovel transportation machinery and digging machinery are not high, and some machinery, such as lifting engineering cranes, have strict requirements for fine tuning characteristics.

(3) Vibration and noise

Vibration and noise should be considered in the design of any machine, and hydraulic systems are no exception. The vibration and noise of the hydraulic system are caused by the vibration and noise of the components of the system, among which the pump and valve are the most serious. Vibration and noise bring a series of adverse consequences to the hydraulic system, and the hydraulic system will not work in serious cases, so the vibration and noise must be controlled. The key to reduce the vibration and noise of the hydraulic system is to control the vibration and noise of each component in the system, reduce the flow pulsation and pressure pulsation of the hydraulic pump, and reduce the impact of hydraulic oil in the pipeline.

3.3.4 Hydraulic Power System

(1) Hydraulic motor

Hydraulic motor is an executive element of the hydraulic system, is an important part of the hydraulic transmission system, it will be the hydraulic pump to provide the liquid pressure energy into its output shaft mechanical energy (torque and speed), liquid (generally hydraulic oil) is the medium of transmission force and movement. According to the rated speed, the hydraulic motor can be divided into high speed and low speed two categories. The high-speed hydraulic motor has the characteristics of high speed, small rotation inertia, easy to start and brake, speed regulation and high reversing sensitivity, and its basic forms are gear type, screw type, blade type and axial plunger type. Low speed hydraulic motor has the characteristics of large displacement, large volume, low speed, simple transmission mechanism, etc. Its basic form is radial plunger type, which can be divided into blade type, axial plunger type, swing type according to the structure type. The blade motor has the characteristics of small volume, small rotation inertia, sensitive action and high frequency commutation, but the leakage is large and can not work at low speed. Figure 3.30 shows the circuit of the double-acting vane type hydraulic motor. The axial piston motor has the characteristic of low output torque. Hydraulic motors are widely used in injection molding machinery, ships, lifting machines, engineering machinery, construction machinery, coal mining machinery, mining machinery, metallurgical machinery, Marine machinery, petrochemical industry, port machinery and so on. According to the displacement can be adjusted, the hydraulic motor can be divided into two categories: quantitative motor and variable motor, and the variable motor can be divided into one-way variable motor and two-way variable motor. In addition, there are oscillating hydraulic motors [7] (Table 3.3).

(2) Hydraulic cylinder

Hydraulic cylinder is a hydraulic actuator that converts hydraulic energy into mechanical energy and does linear reciprocating movement (or swinging movement). It has simple structure and reliable operation. Using hydraulic cylinder to achieve reciprocating movement, can eliminate the deceleration device, and there is no transmission gap, the movement is smooth, so it is widely used in various mechanical hydraulic systems. The output force of the hydraulic cylinder is proportional to the effective area of the piston and the pressure difference between the two sides. The hydraulic cylinder is basically composed of cylinder and cylinder head, piston and piston rod, sealing device, buffer device and exhaust device. Buffer device and exhaust device depending on the specific application, other devices are essential. Figure 3.31 shows the structure diagram of the double-acting single-rod piston hydraulic cylinder.

(3) Electro-hydraulic servo

Electro-hydraulic servo system is a kind of feedback control system which is composed of electric signal processing device and hydraulic power mechanism. The most common ones are electro-hydraulic position servo system, electro-hydraulic speed control system and electro-hydraulic force (or torque) control system. Electro-hydraulic servo system is a kind of feedback control system, which is mainly composed of electrical signal processing device and hydraulic power mechanism. The principle of the electro-hydraulic position servo system is shown in Fig. 3.32. A given element can be a mechanical device that provides a displacement signal,

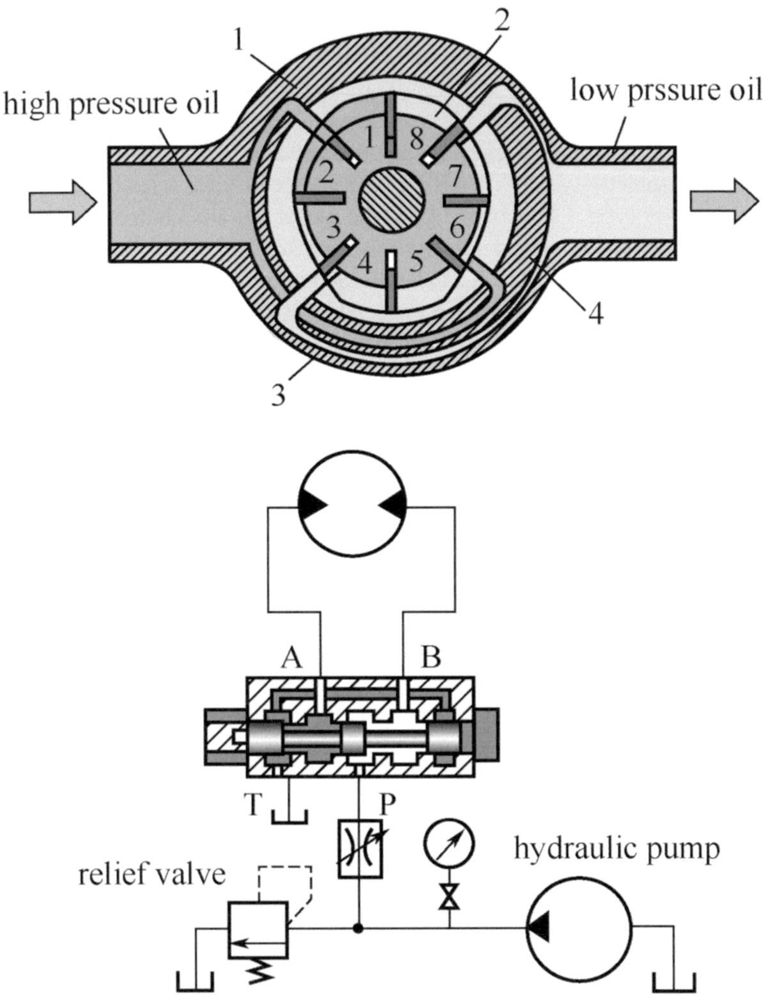

Fig. 3.30 Double acting blade hydraulic motor circuit, 1 stator; 2 rotors; 3 leaves; 4 shell; A. B valve; P speed regulation width; T fuel tank

such as a cam, a connecting rod, etc., or an electrical component that provides a voltage signal, such as a potentiometer. The feedback detection element is used to detect the actual output of the executive element and convert it into feedback signals, which can be mechanical devices such as gear pairs and connecting rods, as well as electrical components such as potentiometers and tachometer generators. The comparison element is used to compare the instruction signal and the feedback signal and to obtain the error signal. The amplifying and converting element amplifies the error signal obtained by the comparing element and converts it into an electrical signal or a hydraulic signal. The actuator converts hydraulic energy

Table 3.3 Applicable working conditions and application examples of hydraulic motors

Type			Applicable working conditions	Application examples
High speed low torque motor	Geared motor	External meshing type	Suitable for situations with low torque at high speeds, low requirements for speed stability, and minimal noise restrictions	Suitable for hydraulic systems of rotary mechanisms in drilling machines, fans, engineering machinery, agricultural machinery, and forestry machinery
		Internal meshing type	Suitable for high-speed, low torque applications that require minimal noise	
	Blade motor		Suitable for situations with low load torque, low noise requirements, and wide speed range	Suitable for machine tools (such as grinding machine rotary worktable) and other equipment
	Axial piston motor		Suitable for situations with high load speed, variable speed requirements, low load torque, and high low-speed stability requirements, i.e. medium to high speed and low torque	Suitable for equipment such as cranes, winches, forklifts, diesel locomotives, CNC machine tools, etc.
Low speed high torque motor	径向马达	Crankshaft connecting rod type	Suitable for high torque and low-speed working conditions, with poor starting performance	Suitable for equipment such as plastic machinery, walking machinery, excavators, tractors, cranes, and coal mining machine traction components
		Internal curve type	Suitable for situations with high load torque, wide speed range, good starting performance, and low speed. When the torque is relatively high, the system pressure is high (such as greater than 16 MPa), and the output bearing is subjected to radial force, it is advisable to use a crossbeam type internal curve hydraulic motor	
		Swing cylinder type	Suitable for high torque and low-speed working conditions	

(continued)

Table 3.3 (continued)

Type		Applicable working conditions	Application examples
Medium speed medium torque motor	Double inclined disc axial direction piston motor	Good low-speed performance, can be used as a servo motor	Good low-speed performance, can be used as a servo motor
	Orbit motor	Suitable for situations with low to medium load speeds and small volume requirements	Suitable for situations with low to medium load speeds and small volume requirements

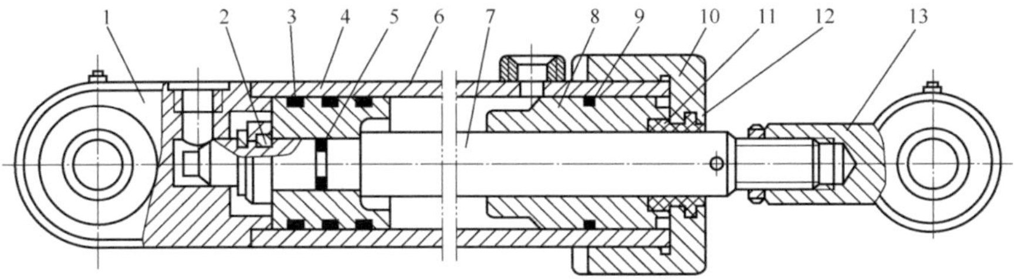

Fig. 3.31 Structural diagram of double acting single rod piston hydraulic cylinder, 1 cylinder bottom, 2 key, 3–5–9–11 sealing ring, 4 piston, 6 cylinder barrel, 7 piston rod, 8 guide sleeve, 10 cylinder head, 12 dust ring, 13 ear shaft

into mechanical energy, produces linear or rotational motion, and directly controls the controlled object, which generally refers to the hydraulic cylinder or hydraulic motor; the controlled object refers to the load of the system, such as worktable, etc. [8–10].

3.3.5 *Hydraulic System Design*

The general hydraulic system design process is divided into the following five steps.

(1) Clear design requirements, working condition analysis

The action and performance requirements of hydraulic system mainly include movement mode, stroke, speed range, load conditions, motion stationarity, precision, working cycle and so on. In addition, ambient temperature, humidity, dust, fire rating, space and so on need to be considered. In the analysis of the working conditions of the actuator, in order to make clear the variation law of the speed and load of each actuator, we can make the speed and load (take the hydraulic cylinder as an example,

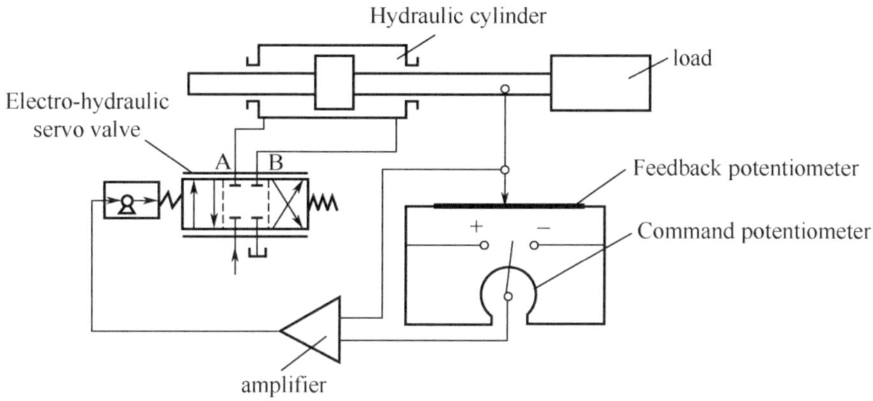

Fig. 3.32 Principle of electro-hydraulic position servo system

the load has the curve of workload, guide friction load, inertia load, gravity, seal, back pressure load, etc.) with time and displacement.

(2) Formulate the hydraulic system schematic diagram

The schematic diagram of hydraulic system is an important technical document showing the composition and working principle of hydraulic system. Drawing up the schematic diagram of the hydraulic system is the first step in designing the hydraulic system, which has a decisive influence on the performance of the system and the rationality and economy of the design scheme. First of all, it is necessary to determine the type of oil circuit: generally, systems with large space for storing fuel tanks use open oil circuit; on the contrary, systems that allow the use of auxiliary pumps to replenish oil and carry out cooling exchange to achieve cooling purposes, use a closed oil circuit. Usually, the throttle speed regulation system adopts open oil circuit, and the volume speed regulation system adopts closed oil circuit. Secondly, select the hydraulic circuit: according to the working characteristics, load properties and performance requirements of all kinds of mainframe, first determine the main circuit which has a decisive impact on the main performance of the mainframe, and then consider other auxiliary circuits. In the hydraulic system of the machine tool, the speed regulation and speed switching circuit are the main circuits; in the hydraulic system of the press, the pressure regulating circuit is the main circuit; the system with vertical moving parts should consider the balance circuit; the system with large inertia load should consider the buffer braking circuit; the system with multiple actuators may consider the sequential action and synchronous circuit; the system with no-load operation requirements should consider the unloading loop and so on.

Finally, the selected typical circuits are merged, arranged and optimized, and necessary components or auxiliary circuits are added and synthesized to form a hydraulic system with simple structure, safe and reliable operation, stable action, high efficiency, convenient adjustment and maintenance, and form the system schematic diagram.

(3) Calculation and selection of hydraulic components

① Determine the structure form and parameters of the actuator. It refers to determine the displacement or cylinder area of the actuator according to the working pressure and maximum flow of the actuator. Table 3.4 shows the form of the executing element.

The working pressure is the main basis to determine the structural parameters of the actuator, and its size affects the size and cost of the actuator, and even the performance of the whole system. The working pressure is selected high, the structure of the actuator and the system is compact, but the strength, stiffness and sealing requirements of the component are high, and the hydraulic pump with higher pressure should be used. If the working pressure is selected low, it will increase the size of the actuator and the entire system, so that the structure becomes huge, so the appropriate working pressure should be selected according to the actual situation.

② Determine the main structural parameters of the actuator. Taking the hydraulic cylinder as an example, the main structural dimensions refer to the inner diameter D of the cylinder and the diameter d of the piston rod, and D and d are determined according to a series of standard values after calculation. For the system with low speed movement requirements, it is also necessary to check the effective working area of the hydraulic cylinder. When the main size D and d of the hydraulic cylinder are calculated, it is necessary to round according to the series standard, and it is necessary to double calculate the working pressure according to the round value.

The working pressure determined by the above method has not yet calculated the back pressure of the return oil circuit, and the working pressure determined is only the part of the pressure required by the actuator to overcome the total mechanical load. After the structural parameters D and d are determined, the pressure of the working chamber of the actuator can be obtained by taking the appropriate estimated value of the back pressure.

Table 3.4 Form of executing components

Motion form	Reciprocating linear motion		Rotary motion		Reciprocating oscillation
	Short journey	Long journey	High speed	Low speed	
Possible forms of execution components	Piston hydraulic cylinder	Plunger hydraulic cylinder	High speed hydraulic motor	Low speed hydraulic motor (high torque)	Swing hydraulic cylinder

③ Determine the operating diagram of the actuator. That is, the pressure, flow, power of the executing element in a working cycle to the change of time or displacement curve. The working diagram of the whole system can be obtained by combining the working diagram of each actuator in the system. The working condition diagram of the hydraulic system can show the maximum value and distribution of the system pressure, flow rate and power in the whole working cycle, which provides a reasonable basis for selecting components, selecting circuits or revising the design in the subsequent design steps. Simple system operating diagram can be omitted.

④ Select hydraulic pump. First determine the type of pump according to the design requirements and system conditions, and then choose the specifications of the hydraulic pump according to the maximum oil supply of the hydraulic pump and the system working pressure.

⑤ Determine the maximum oil supply of the hydraulic pump. Hydraulic pump specifications model according to the calculated value in the product sample selection, in order to make the hydraulic pump work safe and reliable, hydraulic pump should have a certain amount of pressure reserve, usually the rated pressure of the pump is 25–60% higher than the working pressure. The rated flow of the pump should be comparable to the working flow, not more than too much, so as not to cause excessive power loss.

⑥ Select the motor that drives the hydraulic pump. The motor driving the pump is selected according to the drive power and the speed of the pump. The driving power of the pressure limiting variable vane pump can be calculated according to the power at the inflection point of the actual pressure flow characteristic curve of the pump. When the pressure and flow of the pump change greatly in the work, the drive power required at each stage can be calculated separately, and then the root mean square value is obtained.

⑦ Select valve components. The specifications and models of various valve components are selected from the product samples according to the hydraulic system schematic diagram and system working conditions, and the rated pressure and rated flow of various valves should generally be close to their working pressure and maximum through flow. When selecting, it should be noted that the relief valve is selected according to the maximum flow rate of the hydraulic pump; The flow valve also needs to consider the minimum stable flow rate to meet the low speed stability requirements.

⑧ Select hydraulic auxiliary components. The specification size of the tubing is mostly determined by the interface size of the connected hydraulic components, and the inner diameter and wall thickness of some important pipelines need to be checked. For fixed hydraulic equipment, the power source and valve components of the hydraulic system are often centrally installed on the hydraulic station outside the main engine, which can make installation and maintenance convenient, and eliminate the influence of vibration and oil temperature changes of the power source on the working accuracy of the main engine.

(4) Check calculation of heating and system pressure loss

After the completion of the preliminary design of the hydraulic system, it is necessary to check its main performance in order to evaluate its design quality and improve and perfect the hydraulic system. After the pipeline assembly sketch is drawn, the process pressure loss and local pressure loss of the pipeline can be calculated. For their calculation formulas, see the related books or design manuals on fluid mechanics. The total pressure loss of the pipeline is the sum of the process loss and local loss. Before the specific pipe layout of the system is not clear, the local pressure loss of the liquid flow through the valve element is usually used to estimate the pressure loss of the pipeline. When the hydraulic system is working, there is pressure loss, volume loss and mechanical loss, and most of the energy lost is converted into heat energy, which makes the oil temperature rise, resulting in the decline of the viscosity of the oil, the deterioration of the oil, the deformation of the machine parts, affecting the normal work. For this reason, temperature rise must be kept within permissible limits. The heat generated per unit time is the difference between the input power of the hydraulic pump and the output power of the actuator. Under normal circumstances, the working cycle of the hydraulic system often has several stages, and its average heat output is the average value of the heat output of each working cycle.

(5) Drawing engineering drawings and writing technical documents

Hydraulic system official working drawing includes hydraulic system schematic drawing, hydraulic system assembly drawing, hydraulic cylinder and other non-standard components assembly drawing and parts drawing. The schematic diagram of the hydraulic system should be accompanied by a detailed list of hydraulic components, indicating the model specifications, pressure and flow parameters of each hydraulic component, and generally should draw the working cycle diagram of each actuator and the action sequence table of the electromagnet. Hydraulic system assembly drawing is the installation and construction drawing of hydraulic system, including fuel tank assembly drawing, pipeline installation drawing, etc. Technical documents generally include hydraulic system design and calculation manual, hydraulic system use and maintenance technical manual, zero, component list and standard parts, general parts, purchased parts table.

After the engineering drawings and technical documents are completed, the hydraulic system design process is basically completed, and the construction design stage is followed.

3.4 Pneumatic Transmission System Design

Pneumatic transmission refers to a technology that uses compressed air as the power source to drive and control various mechanical equipment to realize the mechanization and automation of the production process. With the development of industrial machinery automation, pneumatic technology is more and more widely used in various fields, is an important means to achieve a variety of production control and automatic control, and has a very important position in the automatic production of industrial enterprises. Pneumatic transmission system and hydraulic transmission system performance has its own characteristics, in the actual design, should be according to the specific requirements to choose the appropriate drive, promote the advantages and avoid the shortcomings. Table 3.5 compares the features of the two drivers.

According to the different functions of pneumatic components and installations, the pneumatic pressure transmission system can be divided into the following four components.

① Air source device. The air source device is a device that converts the mechanical energy provided by the prime mover into the pressure energy of the gas and provides compressed air for the system. The air source device is mainly composed of air compressors, and also includes compressed air purification storage equipment (aftercooler, oil–water separator, gas storage tank, dryer and conveying pipeline).
② Executive elements. The actuator converts the pressure energy of the compressed air into the mechanical energy of the working device and plays the role of energy conversion. Its main forms are: cylinder output linear reciprocating mechanical energy, swing cylinder and gas motor output rotary swing and rotary mechanical energy respectively. For the system with vacuum pressure as the power source, vacuum suction cups are used to complete various lifting operations.
③ Control elements. The control element is used to adjust and control the pressure, flow and flow direction of the compressed air, so that the system actuator can work according to the program and performance required by the function. According to the completion of different functions, the types of control components are

Table 3.5 Comparison between pneumatic transmission system and hydraulic transmission system

Pneumatic transmission system	Hydraulic transmission system
Has a significant impact on load changes, responds quickly to speed changes, generates moderate thrust, has relatively easy signal transmission, and is easy to achieve medium distance control	Has little impact on load changes, slow transmission speed, can achieve high thrust, difficult to transmit signals, commonly used for short distance control
The working medium is air, which is inexpensive and has a long service life. It requires a separate lubrication device to lubricate the system	The working medium is hydraulic oil, which has a relatively short service life and is expensive. It can achieve self-lubricating function
The pneumatic transmission system has a simple structure, is easy to manufacture, and is easy to maintain	The hydraulic transmission system has a complex structure, relatively difficult manufacturing, difficult maintenance, and complex troubleshooting
Good flame resistance, explosion resistance, and impact resistance, with minimal pollution and unaffected by temperature	Easy to leak, pollute the environment, flammable, sensitive to temperature pollution
High noise during operation	Low noise during operation

divided into many kinds, the pneumatic transmission system generally includes pressure, flow, direction and logic four categories of control components.

④ Auxiliary components. Auxiliary components are used for internal lubrication, exhaust, noise reduction, connection between components and signal conversion, display, amplification, detection and other necessary pneumatic components, such as oil mist, water filter, pressure reducing valve, muffler, pipe fittings and fittings, converter, display, sensor, gas storage tank, air source purification device, automatic drainage device, buffer and so on.

The commonly used actuator elements of the pneumatic system are the cylinder and the motor, which convert the pressure energy of the gas into mechanical energy, the cylinder is used for linear motion, and the motor is used for continuous rotary motion. The components and functions of the pneumatic transmission system are shown in Table 3.6. The cylinder is divided into ordinary cylinder, film cylinder and rodless cylinder. Ordinary cylinder is mainly composed of cylinder, piston, piston rod, front and back end cover and seal, etc., which is the most widely used; The diaphragm cylinder is composed of cylinder body, diaphragm, die plate and piston rod, etc. The piston rod is driven by compressed air through the diaphragm for reciprocating linear movement; The rodless cylinder has no rigid piston rod and uses the piston directly or indirectly to achieve reciprocating linear motion. The structure saves installation space and is widely used in automation systems.

Table 3.6 Composition and function of pneumatic transmission system

Name	Common components	Effect
Air source device	 (a) 气泵 (b) gas station	Compress air into high-pressure air and process it to provide clean and dry compressed air to the system
Execution element	 (a) 气缸 (b) 气动马达	Realize reciprocating linear motion, rotational motion, and swinging motion under the action of compressed air
Control components	 (a) directional valve (b) sequence valve (c) pressure control valve (d) speed regulating valve	Used to control the direction, speed, time, sequence, stroke, and system pressure of the actuator
Auxiliary components	 (a) trachea (b) filters (c) muffler	Connect pneumatic components, perform noise reduction, cooling, measurement, etc. on the pneumatic system

The selection of cylinders should pay attention to the selection of cylinder structure and installation mode according to the task requirements, and determine the thrust and tension of the piston rod, should avoid the frequent impact of the piston and the cylinder head, the working speed should not be too fast, and anti-freezing measures should be taken at low temperature, and the assembly needs to be greased on the surface of the workpiece with relative movement.

Pneumatic control components are used in the system to control and adjust the compressed air pressure, flow, flow direction and signal important components, can be divided into direction control valve, pressure control valve, flow control valve three categories. Direction control valve has two kinds of one-way type and reversing type, the spool structure mainly has the end type and the slide valve type. The one-way type includes a one-way valve, or a door shuttle valve, and a quick exhaust valve. The commutation mode changes the direction of gas flow by changing the channel of gas flow, and then changes the direction of movement of the actuator. Control methods are divided into pneumatic control, electromagnetic control, mechanical control, manual control, time control. Pressure control valves are used to control the pressure of compressed gas in the system, mainly including pressure reducing valves, relief valves and sequence valves. Pressure reducing valve is an essential part of the pneumatic system, according to the different ways of adjusting pressure, divided into direct acting type and pilot type two; The relief valve plays the role of a safety valve, when the system pressure exceeds a certain value after automatic exhaust, there are direct acting type and pilot type two; Sequence valve according to the pressure in the system to control the mechanism in order to work. Flow control valves mainly include throttle valves, one-way throttle valves and exhaust throttle valves. The throttle valve adjusts the flow rate by changing the flow area of the valve; The one-way throttle valve is a combined control valve composed of a one-way valve and a throttle valve.

Figure 3.33 shows the control structure of the pneumatic transmission system. The logic control part of the pneumatic system is mostly PLC control, usually using some pneumatic logic components, which takes compressed air as the working medium, through the action of the movable parts inside the components, to change the direction of air flow, to achieve a certain logic function, pneumatic logic components according to the working pressure is divided into high pressure, low pressure, micro pressure three kinds, according to the structure can be divided into cut-off type, diaphragm type, slide valve type and ball valve type, etc.

The design of the pneumatic system also needs to be equipped with air source devices and accessories. Air source device is a set of devices used to produce compressed air with a certain pressure and flow rate and purify, treat and store it. The whole device is composed of air compressor, aftercooler, oil remover, dryer, air filter and gas storage tank. Air compressor is the power source, generally have piston type, diaphragm type, blade type, screw type several types, its rated pressure should be slightly higher than the working pressure; The aftercooler is installed at the outlet of the compressor to cool the high-temperature gas, and its structure forms are tubular, radiator, sleeve, snake tube and plate; The oil remover separates the condensed water and oil in the compressed air, and purifies the air. There are ring rotary type, impact

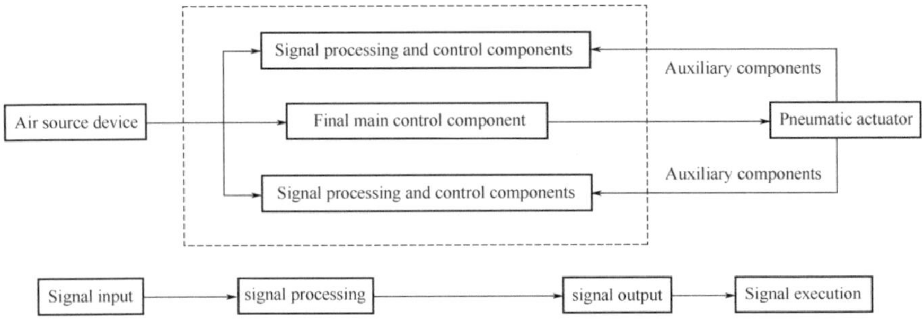

Fig. 3.33 Control structure of pneumatic transmission system

folding type, centrifugal rotary type and water bath type. The role of the dryer is to purify the air that has been preliminarily purified twice, so that the wet air becomes dry air, in the form of adsorption, heating, freezing, etc. The function of the air filter is to filter out the water, oil droplets and impurities in the compressed air to achieve the purity required by the system. When selecting, attention should be paid to the parameters such as the required flow rate, filtration accuracy and allowable pressure of the system. The air filter and the pressure reducing valve and the oil mist constitute the air source regulating device. The gas reservoir is used to regulate the air flow to reduce the fluctuation of the output air pressure. The design of pneumatic accessories is also important, mainly oil mist, muffler and converter. The oil mist is an oil injection device of the air pressure system. After atomizing the lubricating oil, the compressed air is brought into the part of the system that needs lubrication. The installation should be as close as possible to the reversing valve and installed vertically; The function of the muffler is to eliminate or reduce the noise caused by the compressed gas passing through the pneumatic component at high speed, and the exhaust resistance should not be too large when selecting; The converter can convert electrical, liquid and gas signals to each other, and should not be tilted and inverted when installed.

The basic circuit of the pneumatic system is composed of pneumatic components, which are divided into direction control loop, pressure control loop, speed control loop and multi-cylinder motion loop. The direction control loop is divided into single control reversing loop, double control reversing loop and self-locking reversing loop. The pressure control circuit is divided into pressure regulating circuit and booster circuit. The speed control loop is divided into throttle speed control loop, buffer loop and gas/liquid speed control loop, and the system also needs to be equipped with synchronization loop and safety protection loop.

Based on the above basic structure, the basic model of the pneumatic transmission system can be built (Fig. 3.34), such as the gas-hydraulic power slide, pneumatic manipulator and pneumatic servo positioning system. During installation and debugging, attention should be paid to the installation of pipelines and components in strict accordance with the manual.

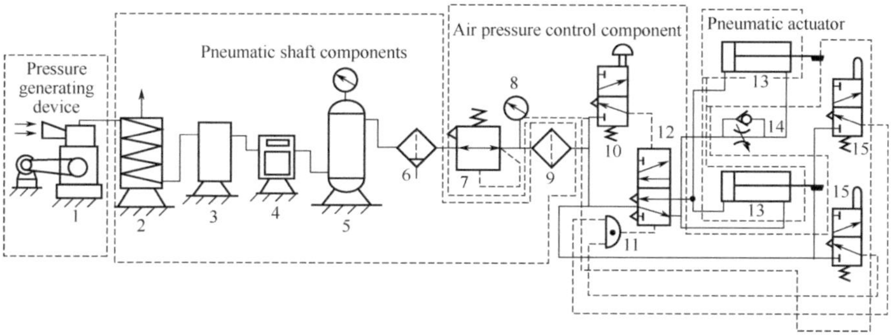

Fig. 3.34 Working principle of pneumatic transmission system, 1 compressor; 2 aftercooler; 3 oil remover; 4 drying machine; 5 gas storage tanks; 6 filter; 7 pressure regulator (pressure reducer); 8 indicator table; 9-oil mist device; 10–12, 14, 15 air pressure control components; 13 pneumatic actuator

With the development of related disciplines, pneumatic technology has formed a multi-disciplinary cross-integration model. The specific innovation of pneumatic technology can be explained from the aspects of structure function, material, intelligent control, etc.

(1) Structural and functional innovation

The human pursuit of structural design is endless. Structural and functional innovation is an important approach to product improvement and innovative design. On the basis of the improvement and perfection of the structure and function of the existing pneumatic products, the new structure and function of the pneumatic products are developed according to the social demand, and the replacement of the pneumatic products is accelerated. Structural and functional innovation includes the structure of the mechanical part of the pneumatic device, the interface form of the surrounding facilities, etc. Through innovative design, we can create a unique and novel structure and excellent performance of the pneumatic products.

(2) Material innovation

The progress of science and technology largely depends on the change of materials. The continuous emergence of new materials promotes the development of pneumatic components in the direction of miniaturization and high performance. At present, the development and selection of pneumatic products tend to miniaturization and high performance, and the material is the selection of new materials such as corrosion resistance, wear resistance, high temperature resistance, and earthquake resistance. New alloys, metal matrix composite materials, ceramic materials, polymer materials, high-performance rare earth materials, nanomaterials, smart materials are in the development and development, the application of new materials will make pneumatic products take on a new look.

(3) Intelligent innovation

Intelligence is the development trend of equipment products, and the drive mechanism is no exception. The development of high-performance sensing components and intelligent control systems is very important for pneumatic product innovation. With the development of computer technology and network technology, adaptive control, fuzzy control and neural network control have been widely used, and the research of intelligent control system with self-learning, condition monitoring and fault self-diagnosis has broad prospects in the field of pneumatics. The performance of the control system is the most important link related to the performance and quality of the pneumatic system, which will directly affect the quality of the product.

Figure 3.35 shows the pneumatic control diagram of the fully automatic drilling machine. Automatic drilling machine action sequence for starting the machine, feeding action, tightening mechanism action, feeding mechanism back, drilling, drill bit back, tightening mechanism release.

The design process of the pneumatic transmission system is as follows:

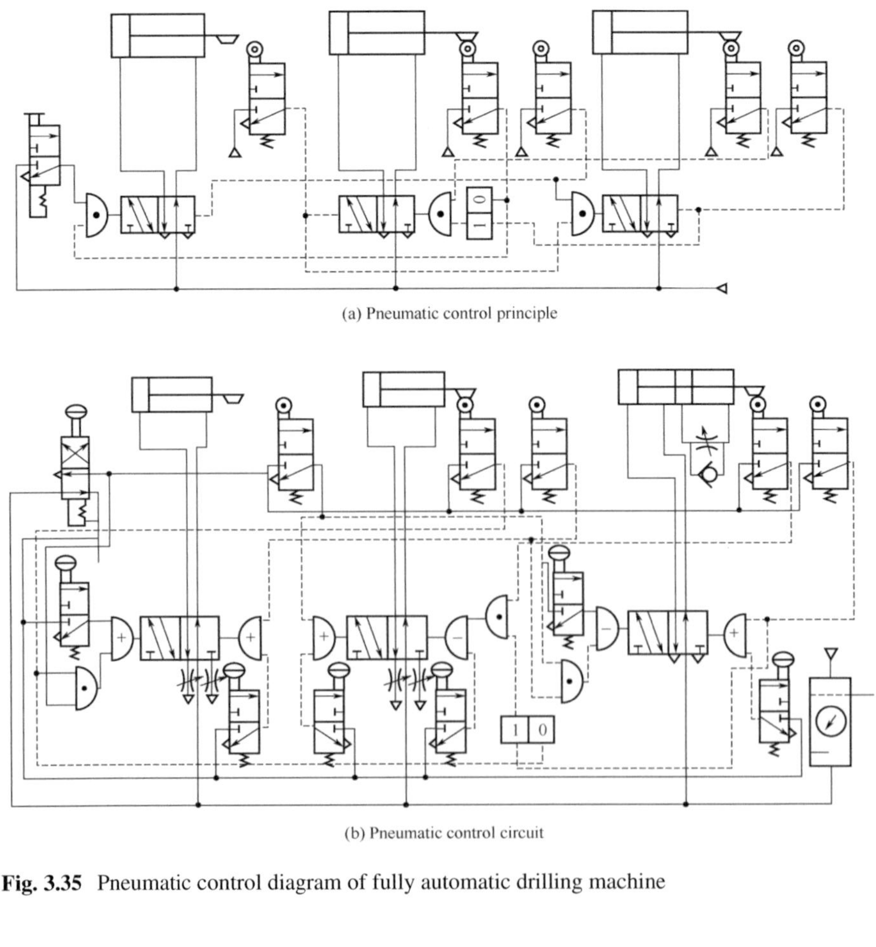

(a) Pneumatic control principle

(b) Pneumatic control circuit

Fig. 3.35 Pneumatic control diagram of fully automatic drilling machine

① Clear the working condition requirements of the system and clear the control object;
② Determine the control scheme and design the pneumatic circuit;
③ Select and calculate the execution elements;
④ Select pneumatic control components;
⑤ Select pneumatic auxiliary components;
⑥ According to the gas consumption of the actuator, determine the capacity and number of compressors;
⑦ Draw the pneumatic system diagram and list the required standard component purchase list.

3.5 Development Direction of Driving Mechanism

The driving mechanism is a very important part of intelligent manufacturing equipment. With the application of computer control technology in the actuator, the driving mechanism is developing in the direction of integration and fieldbus.

Motion controller, servo driver and servo motor will be combined in pairs to form a drive and control integration technology, and the motor ALLinONE integration method will be one of the development focuses. Micro high power density drive technology depends on the breakthrough of a new generation of semiconductor device technology, Yaskawa company in 2017 for the first time launched GaN power semiconductor driver built-in servo motor commercial products, to achieve high efficiency, servo system miniaturization, quiet, energy saving, provincial distribution and so on.

Now the servo drive and servo motor are independent, with the development of robotics and other industries, in the case of high physical space requirements need servo drive and servo motor to do integration, drive—control integration with small size, light weight, high flexibility, low cost, high reliability and other advantages, through the shared memory can transmit more control, status information. The disadvantage is that the development of high integration is difficult. It is suitable for applications with relatively high physical space integration.

The actuator, like the transmitter, has also developed rapidly in recent years, especially some foreign manufacturers have launched fieldbus actuators. It can be seen from these products that the fieldbus actuator is the development trend of the actuator in the future: the mechatronics structure will gradually replace the combined structure; Fieldbus digital communication will gradually replace analog 4 ~ 20 mA signals; The infrared remote control non-contact debugging technology will gradually replace the contact manual debugging technology; Digital control will gradually replace analog control.

Fieldbus technology is first applied in all kinds of transmitters, and then it is applied in the electric actuator, that is, the remote control of the electric actuator is realized through the fieldbus, and the status and position signal of the electric actuator is uploaded to the supreme control device, and displayed on the CRT, and even some parameters of the electric actuator can be configured and fault diagnosis in the distance Fuck.

Nowadays, many companies in the world begin to sell fieldbus actuator products, and these products are more applied to the fieldbus of PROFIBUS, FF, HART and other protocols.

With the adoption of the Field Control bus system (FCS), the control function (PID) is also integrated into the actuator, and the actuator will eventually become a separate control unit.

The intelligent valve positioner can not only easily change the flow characteristics of the control valve, but also improve the control quality of the control system. Therefore, the requirements for control valve flow characteristics can be simplified and standardized (for example, only linear control valves are produced). The intelligent function module is used to match the characteristics of the controlled object, so that the types and varieties of control valve products are greatly reduced, and the manufacturing process of control valve is also simplified.

Fieldbus will be widely used in the actuator, the output signal of some controllers, valve position signal in the same transmission line transmission, control valve and valve positioner, PID control function module combination, so that the control function is realized in the field, so that the risk is dispersed, so that the control is more timely, more rapid, more reliable. Together with other industrial automation instruments and computer control devices, it makes the function of industrial production process control more perfect, the control precision is higher, and the control effect is more obvious. Table 3.7 lists the field bus driver products.

The servo drive system will be updated every 5 years, or even less. The general development trend of equipped servo drive system is:

① High efficiency of drive system. It includes the high efficiency of the motor itself and the drive device, such as the improvement of the performance of the permanent magnet material, the optimization of the drive circuit, the improvement of the software algorithm, and the optimization of the mechanical structure.
② High speed. The servo drive system can continuously improve the speed and accuracy of the drive system by using high-performance motors, encoders, data processing modules, etc.
③ Generalization. The versatile drive system menu is rich in features, and users can easily switch forms without changing the hardware, and can drive stepper motors, servo motors, etc.
④ Intelligent. With the development of artificial intelligence, modern drive systems are becoming more and more intelligent, and can carry out fault self-diagnosis and analysis, load inertia self-adaptation, automatic gain adjustment, etc.
⑤ Networking. The rapid development of fieldbus and the Internet of Things makes the servo drive system also become a member of the Internet of Things. The subsequent development is how to adapt to the requirements of high-performance motion control for real-time, reliability and synchronization of data transmission.

Table 3.7 Fieldbus driver products

Company name	Product and type	Bus type
EIM (USA)	MOV1224 (Electric actuator)	MODBUS
Keyst one (USA)	Electrical actuators (Electric actuator)	MODBUS
ROTORK (UK)	PakscanIE (Intelligent electric actuator) FF-01 (Valve positioner) FF-01 Network interface (Electric actuator)	MODBUS FF FF
Limit orque (USA)	DDC-100T (Electric actuator)	BITBUS
AUMA (Germany)	Matic (Electric actuator)	Profibus
Sienens (Germany)	SIPART P32 (Valve positioner)	HART
Valtek (USA)	Starpac (Intelligent control valve)	HART
Masaneilan (USA)	Smart valve positioner (Valve positioner)	HART
Neles (USA)	ND800 (Control valve)	HART
Jordan (USA)	Electrical actuators (Electric actuator)	HART
Elsag Bailey (USA)	Contract (Electric actuator)	HART
ABB (USA)	EAN823 (Electric actuator)	HART
ABB (USA)	TZD-C120/220 (Valve positioner)	FF
Fisher (UK)	DVC5000	FF
Flower Rosnourt (USA)	DMC5000f series digital (Control valve)	FF
Flow serve (USA)	Logix14XX (Valve positioner) BUSwitch (Discrete control valve) MxActuator (Valve positioner)	FF
Yokog ara (Japan)	YVP (Valve positioner)	FF
Yamat ake (Japan)	SVP3000 Alphaplus AVP303 (Valve positioner)	FF
SMAR (USA)	FY302 (Valve positioner) FP302 (H1/20 ~ 100 kPa interface)	FF
Emerson (USA)	EI-0-Matic0990 (1/4 turn electric actuator) EI-0-Matic22CO (Electric actuator) EI-0-Matic7630 (Electric valve positioner) Field Q (Pneumatic valve positioner)	FF

⑥ Modularity. Modularization is not only limited to the combination of servo drive module, power module, communication module, etc., but also covers the modularization of internal software and hardware of servo drive.

⑦ Predictive maintenance. The drive system is embedded with predictive maintenance technology, which can instantly understand the equipment operating status data through the network.

⑧ Radicalization. Whether it is a motor or a hydraulic or pneumatic drive element, there are two extremes of development—large and small. Large aspects such as the current has a power of 500 kW permanent magnet servo motor.

⑨ Direct drive. Including disk motor turntable servo drive, linear servo drive of linear motor, abandon the intermediate link, accuracy, speed is greatly improved, the total mass is reduced.

3.6 Summary of This Chapter

This chapter briefly introduces the classification and characteristics of intelligent manufacturing equipment drive mechanism, introduces servo motor, stepper motor, variable frequency motor, linear motor and other motor drive systems in detail, taking servo motor drive system as the main object, introduces the selection principle and method of servo motor. In addition, this chapter also introduces the characteristics and design methods of hydraulic drive system and pneumatic drive system. In practical applications, readers can combine relevant mechanical design manuals to develop practical projects.

References

1. Li Shuquan. Electrical Machinery [M]. Beijing: China Machine Press, 2015.
2. Wu Rui. Research on Linear Motor Servo Control Technology [D]. Xinxiang: Henan Normal University, 2012.
3. Hong Naigang. Motor Motion Control System [M]. Beijing: China Machine Press, 2015.
4. Bao Xiping, Ji Zhi, Zhu Tao. Research Status and Development of High-Performance Permanent Magnet Synchronous Servo Systems [J]. Micromotors, 2014, 47(07):84–88.
5. Joseph Dillinger. Fundamentals of Mechanical Manufacturing Engineering [M]. Translated by Yang Zuqun. Changsha: Hunan Science and Technology Press, 2007.
6. Hu Hong, Yao Bowei. Principle and Application of Mechatronics [M]. Beijing: National Defense Industry Press, 2002.
7. Zhao Bo, Wang Hongyuan. Hydraulic and Pneumatic Technology [M]. Beijing: China Machine Press, 2017.
8. Wang Junzheng, Zhao Jiangbo, Wang Shoukun. Development and Prospect of Electro-Hydraulic Servo Technology [J]. Chinese Hydraulics & Pneumatics, 2014(05):1–12.
9. Shen Chan, Lu Bo, Hui Weian. Development and Innovation of Pneumatic Technology [J]. Fluid Power Transmission and Control, 2011(04):7–10.
10. Song Yusheng. Current Situation and Development Trend of China's Pneumatic Tool Industry [J]. Rock Drilling Machinery & Pneumatic Tools, 2020(01):40–46.

Chapter 4
Design of Perception System for Intelligent Manufacturing Equipment

4.1 Concept of Sensors

Sensor technology, communication technology and computer technology are the three pillars of information technology, in which sensor technology is the "sense" and the main means and way to obtain information in the information process. At present, the information technology represented by mobile Internet, Internet of things, cloud computing, big data and artificial intelligence is changing with each passing day, and the intelligent era of the Internet of everything is coming. Perceptual information technology, with sensors as the core, combined with radio frequency, microprocessor, micro-energy and other technologies, is one of the basic and decisive core technologies to realize the Internet of everything. Intelligent manufacturing equipment should have not only a good control system, but also a good perception system. The realization of sensing function of intelligent manufacturing equipment mainly depends on sensors. Sensors are important senses of intelligent manufacturing equipment and can be used as autonomous input devices of intelligent manufacturing equipment.

The architecture of the Internet of Things is divided into perception layer, network layer and application layer, and the related technologies involved in the perception layer are collectively referred to as perception technology. Perception technology is the basis of the Internet of Things, and it can be combined with some of the current basic network facilities to provide omnipresent and comprehensive perception services for the future society, and truly realize the so-called "physical world omnipresent". The objects of IoT connected technology include smart devices and the entire physical world perceived through sensors. The perception technology of the Internet of Things involves the monitoring technology of the perception terminal, the perception network technology, the perception information service technology, the perception detection technology and the network security and other key technologies. Perception technology is the basis for the construction of the Internet of Things system, and the combination with basic network facilities can enable the future society to realize information sharing [1].

M. Xu et al., *Design and Case of Intelligent Electromechanical Equipment System*,
Advanced and Intelligent Manufacturing in China,
https://doi.org/10.1007/978-981-95-3833-1_4

4.1.1 Concept and Composition of Sensors

Modern intelligent manufacturing equipment needs to test electrical parameters such as voltage, current, resistance, power, as well as non-electrical parameters such as mechanical quantity (displacement, speed, acceleration, force, strain, etc.), chemical quantity (concentration, pH value, etc.), biomass (mold, bacteria, etc.). A sensor is a device or device that can feel the specified measured and convert the available signal according to a certain law (function), usually composed of a sensitive element and a conversion element. Sensor technology is a comprehensive technology involving sensing principle, sensor design, development and application, sensor is the specific measured information (physical quantity, chemical quantity, biomass) according to a certain law into a device or device for easy processing, transmission of available signal output, is able to convert external non-electrical signals into electrical signals or optical signal output devices. Sensors make things have the senses of touch, taste and smell, and become the main way and means to obtain information in the field of nature and production, so that things have "life".

The sensing (detection) principle is the physical, chemical and biological effects on which the sensor works, and is governed by the corresponding laws and regulations. The sensor is generally composed of three parts: sensing element, conversion element and measurement circuit (Fig. 4.1). Sensitive materials are the material basis for the development of sensing technology, and processing technology and means are also essential components of sensing technology. The processing technology of modern sensors mainly includes micro-machining technology and lithography technology.

The conversion element is the core of the sensor, and its function is to convert non-electrical information into electrical signals. The sensor is a sensor that converts the measured non-electric quantity into another non-electric quantity that is easy to be converted into electricity, and then converts it into electricity. Not all sensors contain these two parts, for physical sensors, generally only the conversion element; The structural sensor includes two parts: the sensor element and the conversion element. The measuring circuit usually adopts special circuits such as bridge circuit, high impedance input circuit, pulse width modulation circuit, oscillation circuit, etc., which converts the output power of the conversion element into a useful electrical signal circuit that is easy to display, record, control and process.

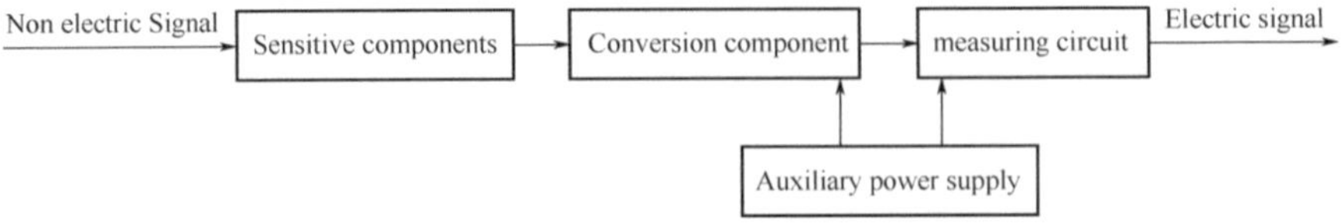

Fig. 4.1 Composition of sensor

4.1.2 Characteristics of the Sensor

(1) Sensor Static Model and Static Characteristics [2, 3]

Static model refers to the mathematical model obtained by the sensor under static conditions, which can be expressed by algebraic equations (without considering hysterescence and creep):

In the formula

Y	Output quantity;
X	input quantity;
a_0	Zero position output;
a_1	sensitivity of the sensor, usually expressed in K or S;
$a_2, a_3, \ldots, a_n$	constants to be determined for nonlinear terms.

The static characteristic indicators of the sensor are:

$$Y = a_0 + a_1 X + a_2 X^2 + \cdots + a_n X^n \tag{4.1}$$

① Linearity: It indicates the percentage of the maximum deviation between the sensor curve and the fitting straight line to the full scale (FS) output value.

$$\delta_{\mathrm{L}} = \frac{\pm\Delta_{\mathrm{max}}}{Y_{\mathrm{FS}}} \times 100\% \tag{4.2}$$

where

Δ_{max}	is the maximum deviation between the calibration curve and the fitting straight line;
Y_{FS}	Sensor full-scale output, $Y_{FS} = Y_{max} - Y_0$.

② Sensitivity. The sensitivity of a sensor is the ratio of the change in sensor output to the input that causes the change when the sensor is in a stable working state.

③ Accuracy. The accuracy of a sensor refers to the percentage of the maximum absolute error relative to the full-scale output of the sensor under specified conditions, which represents the reliability of the measurement results.

$$A = \frac{\Delta A}{Y_{\mathrm{FS}}} \times 100\% \tag{4.3}$$

④ Resolution. The resolution of the sensor refers to the minimum change in the measured input that can be detected within the specified measurement range, and can also be expressed as a percentage of the value relative to the full scale input value.

⑤ Hysteresis. Under the same working conditions, the degree of non-coincidence of the input–output curve of the sensor in the positive and reverse stroke is the ratio of the maximum deviation of the positive and reverse stroke to the full scale.

⑥ Repeatability. The repeatability of the sensor is the inconsistency of the characteristic curve obtained for several consecutive times in the full range of the input in the same direction under the same working conditions. In terms of value, it is expressed by the percentage of the maximum standard deviation of each measurement value of 2 or 3 times and the full scale, reflecting the size of the accidental error of the measurement result, but not the error with the true value.

⑦ Zero drift. In the absence of input, the sensor output deviates from the magnitude of the zero value to the ratio of the full scale.

⑧ Warm bleaching. The degree to which the output value of the sensor deviates when the temperature changes. It is generally expressed as the percentage of maximum output deviation and full scale for every 1 °C change in temperature.

⑨ Indicates the threshold. The minimum input value that produces a measurable change in the sensor output.

(2) Dynamic model and dynamic characteristics of the sensor

The dynamic characteristic of the sensor is the response characteristic of the sensor to the input quantity changing with time. The signals detected by the sensor are mostly a function of time, and the dynamic characteristics reflect the ability of the output of the sensor to truly reproduce the changing input quantity.

1. Dynamic model. Dynamic models are generally expressed in the form of differential equations with constant coefficients and transfer functions. The transfer function is the ratio of the output Laplace transform to the input Laplace transform.

The ordinary differential equation has the form:

$$\begin{aligned}&\left(a_nD^n + a_{n-1}D^{n-1} + \cdots + a_1D + a_0\right)Y(t)\\&= \left(b_mD^m + b_{m-1}D^{m-1} + \cdots + b_1D + b_0\right)X(t)\end{aligned}$$

The transfer function is of the form

$$W(s) = \frac{Y(s)}{X(s)} = \frac{b_ms^m + b_{m-1}s^{m-1} + \cdots + b_1s + b_0}{a_ns^n + a_{n-1}s^{n-1} + \cdots + a_1s + a_0}$$

The sensor dynamic model can generally be described in three forms: zero-order sensor, first-order sensor and second-order sensor, often expressed in the form of transfer function.

① Zero-order sensor.

Differential equation form:

$$y = b_0/a_0 = Kx$$

Transfer function form:

$$W(s) = \frac{Y(s)}{X(s)} = K$$

in the formula

K Static sensitivity.

② First-order sensor.

$$a_1 \mathrm{d}y/\mathrm{d}t + a_0 y = b_0 x$$

Differential equation form:

$$W(s) = \frac{Y(s)}{X(s)} = \frac{K}{\tau s + 1}$$

Transfer function form:
in the formula

K static sensitivity, $K = bo/a$;
r Time constant, $\tau = a1/a$.

③ Second-order sensor.

$$a_2 \mathrm{d}^2 y/\mathrm{d}t^2 + a_1 \mathrm{d}y/\mathrm{d}t + a_0 y = b_0 x$$

Differential equation form:

$$W(s) = \frac{Y(s)}{X(s)} = \frac{K}{\frac{s^2}{\omega_0^2} + \frac{2\xi s}{\omega_0} + 1}$$

Transfer function form:
in the formula

$$K = b_0/a_0;$$

K Static sensitivity,

$$\omega_0 = (a_0/a_2)^{0.5};$$

w_o natural frequency,

$$\xi = a_0/2(a_0a_2)^{0.5}.$$

ε Damping ratio,

2. Dynamic characteristics. The dynamic characteristics of the sensor: first, the difference between the output and the ideal output after the output reaches a stable state; The second is the error of the transition state of the output from one steady state to another when the input quantity jumps. Any periodic function can be decomposed into various harmonic components by Fourier series, and it is approximately expressed as the sum of these sinusoidal quantities, so the method of input signal functions such as sine function and step function is commonly used for analysis in engineering.

① Zero-order sensor.

a. Frequency response characteristics: independent of frequency, no amplitude and phase distortion problems, so it can be called proportional or inertial-free links.
b. Step response characteristics: The step response is proportional to the input and has ideal dynamic characteristics.

② First-order sensor.

a. Frequency response characteristics. The frequency transfer function is:

$$W(\mathrm{j}\omega) = \frac{K}{\mathrm{j}\omega\tau + 1}$$

Assume the input is $x = X \sin \omega t$, the output is $y = Y \sin(wt + \psi)$, then the amplitude-frequency characteristic is:

Phase frequency characteristics:

$$|W(\mathrm{j}\omega)| = \frac{K}{\sqrt{1 + \omega^2\tau^2}}$$

According to the frequency characteristic, the smaller the time constant, the better the frequency response characteristic.

b. Step response characteristics.

The step response index includes: time constant τ—the time required for the output value to rise to the steady state value of 63.2%; Rise time Tr—the time required from 10 to 90%; Response time Ts—the time it takes for the output value to reach the error range of ± Δ%; Overshoot σ%—expressed as a percentage of the ratio of the maximum value ΔA(overshoot) exceeding the steady-state value during the transition process to the steady-state value; Attenuation rate ψ—the percentage decrease in the height of two adjacent peaks; Steady-state error ess—the difference between steady-state output value and target value.

Step response:

$$Y(t) = 1 - e^{-t/\tau} \tag{4.15}$$

From the above formula, we can see that the dynamic characteristics of a first-order sensor depend on the time constant τ. The smaller the time constant τ is, the faster the response is. There is no overshoot and attenuation rate. When $t > 5\ \tau$, the output is close to the steady-state value.

③ Second-order sensor.

a. Frequency response characteristics. The frequency transfer function is:

$$W(j\omega) = \frac{K}{\left(\frac{j\omega}{\omega_0}\right)^2 + \frac{2\xi j\omega}{\omega_0} + 1} \tag{4.16}$$

Width characteristics:

$$|W(j\omega)| = \frac{K}{\sqrt{\left[1 - \left(\frac{\omega}{\omega_0}\right)^2\right]^2 + 4\xi^2\left(\frac{\omega}{\omega_0}\right)^2}} \tag{4.17}$$

Phase frequency characteristics:

$$\psi = -\arctan\frac{2\xi\left(\frac{\omega}{\omega_0}\right)}{1 - \left(\frac{\omega}{\omega_0}\right)^2} \tag{4.18}$$

According to the amplitude-frequency and phase-frequency characteristics: ω/ω 0 ≪ 1, approximate zero-order segment, A(ω) ≈ 1, ψ ≈ 0; When ω/ω0, A(ω) ≈ 0, ψ ≈ 180°, that is, the sensor has no response, and the frequency of the measured parameter is much higher than its natural frequency; When ω/ω0 = 1 and ξ → 0, the sensor appears resonance, A(ω) ≈ ∞, and the amplitude and phase of the output signal are seriously distorted.

The damping ratio ξ has a great influence on the frequency characteristics. When ξ increases, the maximum value of the amplitude-frequency characteristic decreases. When ξ > 0.707, resonance will not occur; when ξ = 0.707, the flat segment of the amplitude-frequency characteristic is the widest, which is the best damping.

b. Step response characteristics. The higher the natural frequency, the faster the response curve rises. Otherwise, the slower it is. Under damping (ξ < 1), attenuation oscillation occurs; When overdamping (ξ > 1) and critical damping (ξ

= 1), there is no oscillation and no overshoot. When $\xi = 0$, constant amplitude oscillation is formed.

4.1.3 Characteristics and Functions of Intelligent Sensor

The smart sensor is reflected in the "intelligence", the sensor itself has a microprocessor (chip), with the ability of data acquisition, processing and analysis, is the product of modern microelectronics technology, information technology, material technology and processing technology, and has been widely used in intelligent manufacturing equipment. Intelligent sensor is the function integration of analog interface circuit, integrated analog-to-digital converter microcontroller, through the simulation of human senses and brain coordination action, combined with long-term testing technology research and practical experience and put forward, is a relatively independent intelligent unit, its appearance makes the original hardware performance of the harsh requirements to reduce, and rely on software help can make sensingThe performance of the device is greatly improved. Compared with ordinary sensors, smart sensors have the following significant characteristics:

① High precision. The intelligent sensor has the function of information processing, and the software can not only correct various deterministic system errors, but also properly compensate random errors, reduce noise, and greatly improve the accuracy of the sensor.

② High repetition accuracy. The repetition accuracy reflects the stability of the sensor between multiple measurement outputs. Multiple measurements of the same quantity can determine a range that can include all measurements around the nominal value, and this range is repeated accuracy.

③ High reliability. The miniaturization of the intelligent integrated sensor system eliminates some unreliability factors of the traditional structure and improves the anti-interference performance of the whole system. At the same time, the smart sensor also has diagnostic, calibration, data storage and adaptive functions, and has good stability.

④ High cost performance. Under the demand of the same accuracy, the multifunctional intelligent sensor has a significantly improved cost performance compared with the ordinary sensor with a single function, especially after the adoption of a cheaper single-chip microcomputer.

⑤ Function diversification. The intelligent sensor can realize the comprehensive measurement of multi-sensor and multi-parameter, and expand the measurement and application range through programming; It has a certain self-adaptation ability, and changes the form of the inverse output data of the range accordingly according to the change of the detection object or condition. The intelligent sensor has the function of digital communication interface, which is directly sent to the remote computer for processing, and has a variety of data output forms, which is suitable for various application systems.

Generally speaking, smart sensors have self-calibration, self-calibration, self-calibration functions; With automatic compensation function; It can automatically collect data and preprocess the data. Can automatically test, self-select the range, self-find fault; With data storage, memory and information processing functions; With two-way communication, standardized digital output or symbol output function; It has the function of judgment and decision processing. Some smart sensors also have firmware based information processing, data verification, and multi-parameter sensing capabilities.

Intelligent sensors have been widely used in industrial automation, scientific research, astronomical exploration, earth-sea exploration, environmental protection and energy saving, medical health, automotive, national defense, biopharmaceutical and many other fields.

4.1.4 Classification of Sensors

(1) By purpose

The quantities that need to be measured in life are distance, smoke and gas, touch, motion, light, acceleration and angular momentum, electromagnetism, sound, etc. [4]. According to different uses, the sensor is divided into position sensor, liquid level sensor, energy consumption sensor, speed sensor, acceleration sensor, ray radiation sensor, thermal sensor, temperature sensor, humidity sensor, pressure sensor, flow sensor, liquid level sensor, force sensor, torque sensor, etc. (Fig. 4.2).

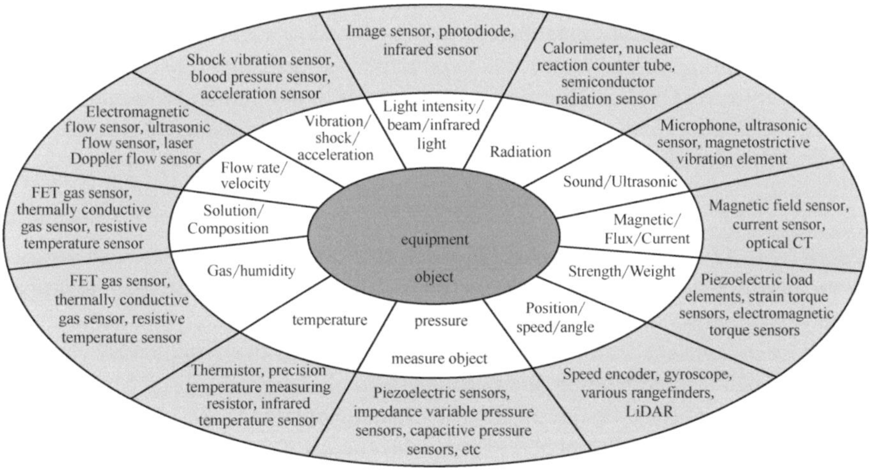

Fig. 4.2 Classification of sensors

(2) According to the detection principle

According to different sensing principles, it can be divided into vibration sensor, humidity sensor, magnetic sensor, gas sensor 181, vacuum sensor, biosensor and so on.

(3) According to the output signal

According to the different output signals, it can be divided into analog sensors, digital sensors, fake digital sensors and switch sensors. An analog sensor converts a measured non-electrical quantity into an analog electrical signal. A digital sensor converts the measured non-electrical quantity into a digital output signal (both direct and indirect). A pseudo-digital sensor converts the measured semaphore into a frequency signal or short-period signal output (including direct or indirect conversion). A switching sensor is a sensor that outputs a set low or high level signal when a measured signal reaches a certain threshold.

(4) According to its manufacturing process

According to the different manufacturing process, the sensor can be divided into integrated sensor, thin film sensor, thick film sensor, ceramic sensor. Integrated sensors are manufactured using standard process technology for the production of silicon-based semiconductor integrated circuits, usually also integrating part of the circuit used for the initial processing of the signal under test on the same chip. Thin film sensors are formed by depositing a thin film of the corresponding sensitive material on a dielectric substrate (substrate). When the hybrid process is used, part of the circuit can also be manufactured on this substrate. The thick film sensor is made by coating the paste of the corresponding material on a ceramic substrate, which is usually made of Al2O3, and then heat treatment to make the thick film form. Ceramic sensors are produced using a standard ceramic process or some variant of it (sol, gel, etc.), after the completion of appropriate preparatory operations, the formed components are sintered at high temperatures. The process of thick film and ceramic sensor has many commonalities, and in some aspects, it can be considered that the thick film process is a variant of the ceramic process.

(5) According to the measurement purpose

According to different measurement purposes, it can be divided into physical sensors, chemical sensors and biological sensors. Physical sensors are made using the properties of certain physical properties of the substance being measured to change significantly. Chemical sensors are made of sensitive elements that can convert chemical quantities such as composition and concentration of chemical substances into electrical quantities. Biosensors are made of the characteristics of various organisms or biological substances to detect and identify the chemical composition of organisms.

(6) According to its composition

According to the composition, the sensor can be divided into basic type, combined type and application type sensor. The basic sensor is one of the most basic single conversion devices. The combined sensor is a sensor composed of different single conversion devices. The applied sensor is a sensor composed of a basic sensor or a combined sensor and other mechanisms.

(7) According to the form of action

According to the different forms of action, the sensor can be divided into active and passive sensors. The active sensor also has an action type and a reaction type. The sensor can send a certain detection signal to the tested object, and can detect the change of the detection signal in the tested object, or the signal is formed by the detection signal producing some effect in the tested object. The sensor that detects the change mode of the detection signal is called the action sensor, and the sensor that detects the response and forms the signal mode is called the reaction sensor. Radar and radio frequency range detectors are examples of active sensor applications, while photoacoustic effect analysis devices and laser analyzers are examples of reactive sensor applications. Passive sensors only receive signals generated by the measured object itself, such as infrared radiation thermometers and infrared camera devices.

There are many kinds of sensors, the working principle, measurement method and the measured object are different, and the classification method is also different. All sensors can be divided into two types: passive (passive) sensors and active (active) sensors. Passive sensors do not need any additional energy source to generate electrical signals directly in response to external excitation, while active sensors require external energy sources when working [5, 6]. On a macro level, sensors can also be divided into traditional discrete sensors, analog integrated sensors and intelligent sensors.

① Traditional separate sensor. This kind of sensor is a basic sensor in the traditional sense, which is manufactured by a non-integrated process, and its function is relatively simple, and it only has the function of obtaining signals.
② Analog integrated sensor. Integrated sensor is made of silicon semiconductor integration process, also known as silicon sensor or monolithic integrated sensor. Analog integrated sensor was born in the 1980s, the sensor integrated on a chip, can complete the measurement and analog signal output function, its main characteristics are single function (only measure a certain physical quantity), small measurement error, low price, fast response, long transmission distance, small size, low power consumption, simple peripheral circuit, suitable for remote measurement, control, usedNo nonlinear calibration is required.
③ Smart sensors. Smart sensors (digital sensors) are developed with the development of microelectronics technology, computer technology and automatic testing technology, and were born in the mid-1990s. Smart sensors contain sensors, A/D converters, signal processors, memory (or registers) and interface circuits. Some products also have multiplexers, central processing units, RAM and ROM.

The smart sensor is characterized by the ability to output measurement data and related control quantities, which can be adapted to various microcontrollers. Its testing function depends on the software, so the degree of intelligence depends on the level of software development.

4.1.5 Microelectro Mechanical System (MEMS) Sensor

The term MEMS is often used to describe the type of sensor, as well as the preparation process of the sensor [7]. MEMS consists of mechanical microstructures, microsensors, microactuators, and microelectronics, all integrated on the same silicon substrate. MEMS sensor is a new type of sensor manufactured by micromachining technology, which is an important branch of MEMS. There are many kinds of MEMS sensors, which can be divided into physical type, chemical type and biological type according to their working principle. According to the measurement can be divided into acceleration, angular velocity, pressure, displacement, flow, electricity, magnetic field, infrared, temperature, gas composition, humidity, pH value, ion concentration, biological concentration and touch sensors. Table 4.1 shows the classification system of MEMS sensors based on the two classification methods.

With the development of MEMS technology, inertial sensor has become one of the most widely used MEMS devices, and micro-accelerometer is a typical representative of inertial sensor. The theoretical basis of microaccelerometers is Newton's second law of motion. The acceleration can be measured inside a system, and if the initial velocity is known, the linear velocity can be calculated by integrating, and the linear displacement can then be calculated. Combined with a gyroscope (used to measure angular velocity), the object can be precisely positioned.

The main products of MEMS include micro pressure sensors, inertial measurement devices, microflow systems, read/write heads, optical systems, printer nozzles, etc., among which products from the automotive industry and information industry account for about 80% of sales [7, 8].

(1) MEMS accelerometer.

MEMS accelerometers are miniature accelerometers, which are instruments used to measure the acceleration of objects. Compared with traditional accelerometers, MEMS accelerometers are smaller in volume and mass, and have been widely used in intelligent manufacturing equipment. The working principle is that when the accelerometer accelerates along with the external object, the mass block moves backward under the action of inertial force, and its displacement is limited by the spring and the damper. When the external acceleration is constant, the mass block has a definite displacement. When the external acceleration changes, the displacement of the mass block also changes accordingly. When the displacement of the mass block changes, the capacitance between the movable arm and the fixed arm (sensor) will change accordingly. If the change of sensor output voltage is measured, that is, the displacement of the actuator (mass block) is measured, and the displacement has

Table 4.1 Classification of MEMS sensors

MEMS sensor	MEMS physical sensors	MEMS mechanical sensor	MEMS accelerometer
			MEMS angular velocity (gyroscope)
			MEMS Inertial Measurement Unit
			MEMS pressure sensor
			MEMS flow sensor
			MEMS displacement sensor
		MEMS electrical sensor	MEMS electric field sensor
			MEMS electric field strength sensor
			MEMS current sensor
		MEMS magnetic sensor	MEMS magnetic flux sensor
			MEMS magnetic field strength sensor
		MEMS thermal sensor	MEMS temperature sensor
			MEMS heat flow sensor
			MEMS thermal conductivity
		MEMS optical sensor	MEMS infrared sensor
			MEMS visible light sensor
			MEMS laser sensor
		MEMS acoustic sensor	MEMS noise sensor
			MEMS surface acoustic wave sensor
			MEMS ultrasonic sensor
	MEMS chemical sensors	MEMS gas sensor	Combustible gas sensor
			Toxic gas sensor
			Air pollution gas sensor
			Automotive sensors
		MEMS temperature sensor	
		MEMS ion sensor	MEMS pH sensor

(continued)

Table 4.1 (continued)

			MEMS ion concentration sensor
	MEMS biosensor	MEMS physiological quantity sensor	MEMS biological concentration sensor
			MEMS tactile sensor
		MEMS biochemical sensor	

a definite corresponding relationship with the acceleration to be measured, and the output voltage and acceleration have a definite relationship [9].

In Fig. 4.3, Vm represents the input voltage signal, Vs represents the output voltage, Cs1 and Cs2 respectively represent the two capacitors between the fixed arm and the movable arm, then the relationship between the input signal and the output signal can be expressed as:

$$V_s = \frac{C_{s1} - C_{s2}}{C_{s1} + C_{s2}} V_m$$

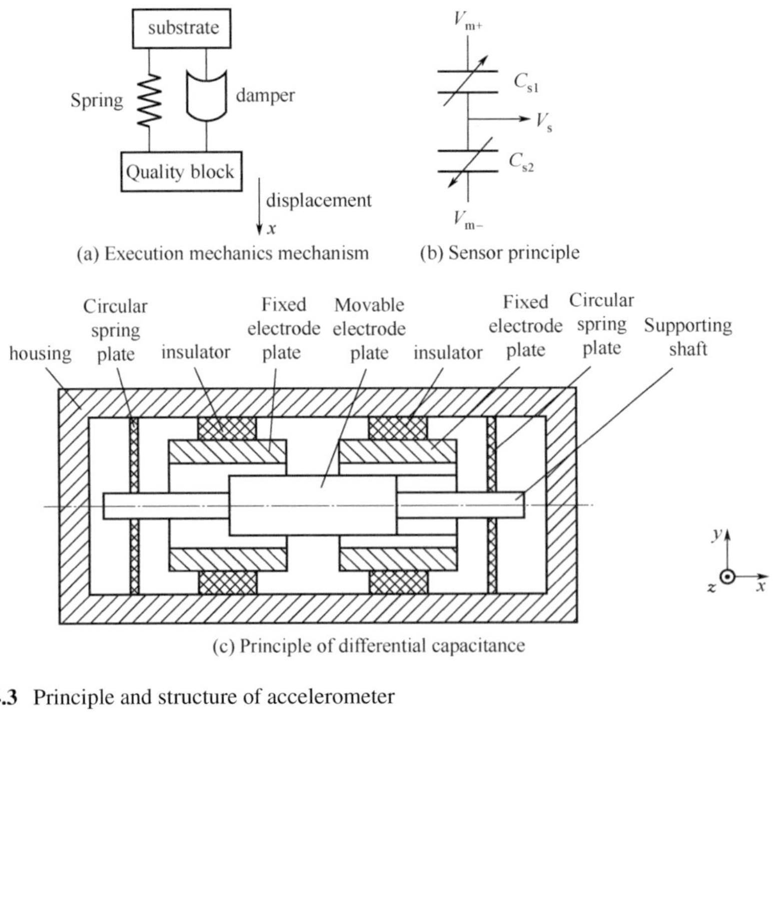

Fig. 4.3 Principle and structure of accelerometer

The relationship between capacitance and displacement is given by the definition of capacitance:

$$C_{s1} = \frac{\varepsilon_0 \varepsilon}{d - x}, \quad C_{s2} = \frac{\varepsilon_0 \varepsilon}{d + x} \tag{4.20}$$

where, x is the displacement of the movable arm (actuator); d is the distance between the fixed arm and the cantilever when there is no acceleration; ε_0 and ε are capacitance parameters. From Eqs. (4.19) and (4.20), we can get

$$V_s = \frac{x}{d} V_m \tag{4.21}$$

According to the principle of mechanics, the mechanical equation of the mass block under stable conditions is:

$$kx = -ma_{\text{ext}} \tag{4.22}$$

where k is the spring constant and m is the mass of the mass block.

Therefore, the relationship between the external acceleration and the output voltage is:

$$a_{\text{ext}} = -\frac{kx}{m} = -\frac{kdV_{\text{s}}}{mV_{\text{m}}} \tag{4.23}$$

It can be seen from formula (4.23) that when the structure and input voltage of the accelerometer are determined, the output voltage is directly proportional to the acceleration.

According to the measurement principle, MEMS accelerometers can be divided into piezoresistive micro-accelerometers, capacitive micro-accelerometers, piezoelectric micro-accelerometers, etc.

1. Piezoresistive micro-accelerometer. When a certain stress is applied to a certain axis of a semiconductor, the phenomenon that its resistivity changes is called the piezoresistive effect of a semiconductor. The principle of piezoresistive micro accelerometer is: when the external acceleration input, the mass block will be affected by an inertial force, the cantilever beam will be deformed under the action of this inertial force, resulting in the deformation of the piezoresistive film firmly connected with the cantilever beam, the resistance value of the piezoresistive film will change, and the voltage value at both ends will change. Figure 4.4 shows the principle of the silicon piezoresistive micro acceleration sensor. Through the experiment, a series of relations between voltage and inertial force are obtained, and the inertial force is caused by external acceleration, so that the relationship between voltage and acceleration can be obtained, and then the acceleration can be measured.

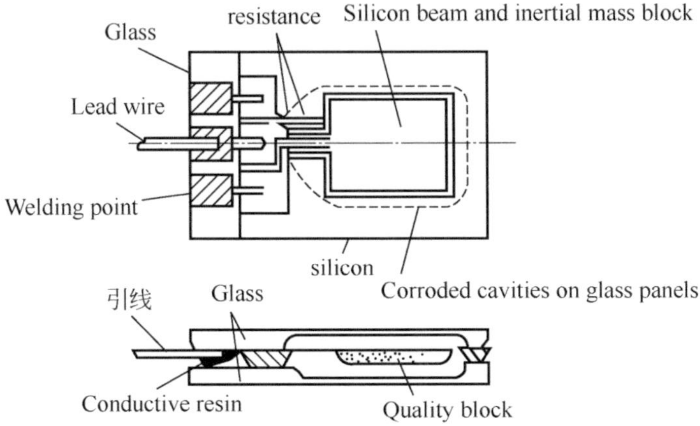

Fig. 4.4 Principle of silicon piezoresistive micro accelerometer

Piezoresistive micro-accelerometer has simple principle, structure and fabrication process, easy to implement interface and internal circuit. However, it is very sensitive to the change of temperature, which affects the measurement accuracy. The sensitivity is relatively low, and it is not easy to measure small acceleration changes.

2. Capacitive micro-accelerometer. According to the electrical engineering knowledge, the change of capacitance is related to the distance between the two plates, so the change of the distance between the plates can be reflected by the change of capacitance, and the change of displacement can be derived from the change of capacitance, and then the differential operation can get the acceleration value.

   The mass block is fixed on the matrix, and one plate of the capacitive micro-accelerometer capacitor is fixed on the moving mass block, and the other plate is fixed on the fixed matrix. When the acceleration action, the mass block displacement causes the change of the upper and lower capacitance, according to the capacitance change difference value of acceleration. Capacitive micro-accelerometer has high sensitivity and measurement accuracy, good stability, small temperature drift and low power consumption. However, the output circuit is complex and susceptible to parasitic capacitance and electromagnetic interference.
3. Piezoelectric micro-accelerometer. The piezoelectric body is electrically polarized by the external mechanical force, and leads to the bound charge of opposite symbol in the surface of the piezoelectric body at both ends, and its charge density is proportional to the external mechanical force, which is called the positive piezoelectric effect. The piezoelectric body deforms under the action of external electric field, and its shape variable is proportional to the intensity of external electric field, which is called inverse piezoelectric effect. The crystal with piezoelectric effect is called piezoelectric crystal, commonly used piezoelectric crystal quartz, piezoelectric ceramics and so on. The working principle

of the piezoelectric micro-accelerometer is: the elastic beam is covered with a layer of piezoelectric material film, when the external acceleration acts on the mass block, the elastic beam will be deformed under the action of inertial force, and the voltage will be generated between the upper and lower electrodes of the device structure, and the mathematical model transformation formula will be determined by measuring the voltage change, so as to obtain the acceleration change.

Since 1977, Stanford University in the United States first made an open-loop micro-accelerometer by micromachining technology, accelerometers with various structures and principles have been developed at home and abroad. The micro-silicon accelerometers with high resolution and large range have become the focus of research. Because the inertial mass block is relatively small, the inertial force used to measure acceleration and angular velocity is correspondingly small, and the sensitivity of the system is relatively low, so it is particularly important to develop a high-sensitivity accelerometer. Whether it is civil or military, the micro-accelerometer with high precision and large range will greatly expand its application range. Low temperature drift and low hysteresis effect have become new performance objectives. Selecting suitable materials, adopting reasonable structure and applying new low-cost temperature compensation link can greatly improve the precision of micro-accelerometers. The development of multi-axis accelerometer has become a new direction.

(2) Micro pressure sensor

Micro pressure sensor is a new type of pressure sensor made of semiconductor material and MEMS process. Compared with the traditional pressure sensor, the micro pressure sensor has the advantages of small size, high precision, high sensitivity, high frequency dynamic characteristics and good stability. The micro-pressure sensor is easy to integrate with the micro-temperature sensor to increase the precision of temperature compensation. Micro pressure sensors are widely used in aerospace, vehicle, control and other fields.

(3) MEMS gyroscope

In the process of aircraft flight, it is necessary to measure the three degrees of freedom of the aircraft pitch, yaw and roll, not only the acceleration but also the angular velocity. Acceleration can be measured using an accelerometer, while angular velocity is generally measured using a gyroscope. Conventional gyroscopes measure angular velocity by taking advantage of the fact that a high-speed rotating object has the property of maintaining its angular momentum. The precision of this gyroscope is very high, but its complex structure, short service life and high cost are generally only used for navigation. Now the gyroscope used in the aircraft due to the requirements of external conditions, its accuracy is very high, but the cost of high precision is the complex structure, short life, so that the use of costs increased significantly. Common micromechanical angular velocity sensors include double balance ring structure, cantilever beam structure, tuning fork structure, vibration ring structure, etc. Fig. 4.5 shows an example of a MEMS gyroscope.

(4) Microflow sensor

MEMS flow sensor is becoming a research hotspot in the field of microflow measurement because of its small diameter, small flow measurement and high integration degree. The microflow sensor is suitable for microfluidics precision measurement and control because of its small size, low measurement level, small dead zone capacity and short response time. The microflow sensors studied at home and abroad can be divided into thermal, mechanical and resonant types according to different working principles. The output of the thermal flow sensor is nonlinear, which is affected by the heat insulation effect of the matrix, and is suitable for the micro-flow measurement with low precision requirements. However, the flow measurement range is wide, the sensitivity is high, and the flow limit is low, which is one of the research hotspots. At present, the research of microflow sensor has been developed in the direction of further miniaturization, and it can distinguish the direction of flow. Researchers are also constantly exploring new measurement methods, such as vibration, photoelectric measurement, etc. Microflow measurement technology combining multiple measurement methods to integrate multi-source information is also an important development direction.

(5) Microgas sensor

With the development of micro-nano technology, various gas sensors with different properties have become the focus of research, and micro-gas sensors have emerged. MEMS technology is easy to make gas sensors and temperature detection components together to ensure the excellent performance of gas sensors. According to the different materials of the micro gas sensor, the micro gas sensor is divided into silicon gas sensor and silicon gas sensor. The microgas sensor can integrate various

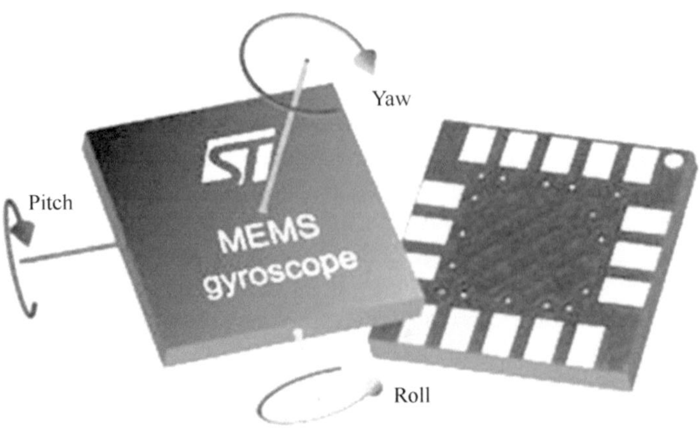

Fig. 4.5 Schematic of MEMS gyroscope

sensors in one chip to meet the needs of various measurements when measuring gas. At present, the development of micromechanical manufacturing technology is relatively perfect, and the development of micro and nano technology allows a chip to complete many different functions. The gas sensor and temperature sensor can be integrated into a chip, and the temperature can be measured at the same time to ensure the accuracy of gas measurement.

(6) Micro temperature sensor

The micro-temperature sensor is small in size, light in weight, and its inherent heat capacity is small, which has greater advantages in temperature measurement than the existing temperature sensors such as thermistors. The micro-cantilever beam temperature sensor uses the principle that the thermal expansion coefficients of silicon and silicon dioxide materials are different, and the shape variables are different at different temperatures, so the shape variables of different parts of the cantilever beam are also different. The detection circuit located at the bottom of the cantilever beam is used to measure the different deformation at different temperatures, and the corresponding relationship between temperature and deformation is obtained. The measuring method has high precision, good linearity and wide measuring range.

The electric polarization intensity of pyroelectric crystal is related to temperature, and the sensor made according to the principle of pyroelectric effect is called pyroelectric temperature sensor, which has the advantages of simple structure and small size. According to the linear relationship between voltage and temperature at both ends of transistor PN junction, a PN junction micro-temperature sensor is developed. In addition, the temperature sensing element and the subsequent amplifier are integrated into a chip, which is a functional device sensor that integrates sensing and amplification.

4.1.6 Development of Sensors

The development of sensors has roughly gone through three stages (Fig. 4.6). The first stage was in the 1950s, when structural sensors appeared, which used structural parameter changes to sense and transform signals. The second stage began in the 1970s, and solid-state sensors gradually developed, which are composed of solid components such as semiconductors, dielectrics, and magnetic materials, and are made of certain characteristics of materials. The third stage is the beginning of the end of the twentieth century, intelligent sensors appear and get rapid development. Intelligent sensor is a combination of microcomputer technology and detection technology. The sensor has the characteristics of artificial intelligence.

The world of the future Internet of Things is also the world of sensors. At present, the development and innovation of domestic sensor technology focus on three aspects: material, structure and performance improvement. Sensitive materials from liquid to semi-solid, solid direction; The structure is developing in the direction

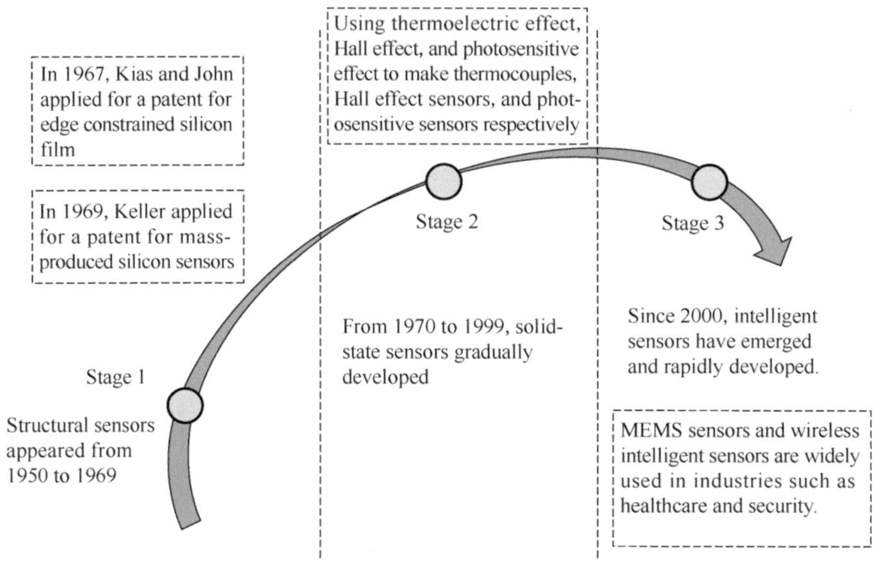

Fig. 4.6 Development stage of sensors

of miniaturization, integration, modularity and intelligence. Micro-sensors manufactured by MEMS technology have a series of advantages, such as miniaturization, integration, low cost and easy mass production, etc. The advantages presented have been paid attention to by more and more countries, and many countries have begun to invest heavily in the development of micro-sensors. The performance develops to the direction of wide detection range, high detection accuracy, strong anti-interference ability, stable performance and long life. In particular, it is worth mentioning that the development of Internet of Things technology has put forward new requirements for traditional sensing technology, and sensor products are gradually developing in the direction of MEMS technology, wireless data transmission technology, infrared technology, new material technology, nanotechnology, ceramic technology, thin film technology, optical fiber technology, laser technology, composite sensor technology, and multidisciplinary integration.

4.2 Intelligent Manufacturing Equipment Sensors

4.2.1 Overview

The sensing system of intelligent manufacturing equipment mainly relies on intelligent sensors. Intelligent sensor is a sensor with information processing function, with the ability to collect, process and exchange information, is the product of the combination of sensor integration and microprocessor, and is an important sense of intelligent manufacturing equipment. Compared with general sensors, intelligent sensors have diversified functions, high precision information acquisition can be achieved through software technology, and have certain programming automation capabilities and low cost.

As the autonomous input device of intelligent manufacturing equipment, intelligent sensor is a variety of sensory organs of intelligent manufacturing equipment and a window for intelligent manufacturing equipment to obtain external environmental information. The sense of intelligent manufacturing equipment for the external environment mainly includes vision, position sense, speed sense, force sense, touch and so on.

Intelligent sensors can store and process the detected physical quantities, and intelligent sensors can exchange information and transmit effective data independently, and automatically complete data analysis and processing. A good smart sensor is a suite of sensors and meters driven by a microprocessor, and has functions such as communication and onboard diagnosis, providing relevant information to the monitoring system or operator, improving work efficiency and reducing maintenance costs. The intelligent sensor integrates all functions and some control functions of the sensor and intelligent instrument, has high linearity and low temperature drift, reduces the complexity of the system and simplifies the system structure. Intelligent sensor through the test data transmission or receive instructions to achieve various functions, such as gain setting, compensation parameter setting, internal check parameter setting, test data output.

① Self-compensation and calculation function. The sensor's temperature drift and output nonlinearity are difficult to solve, and despite a lot of efforts by engineers, the fundamental problem has not been solved. The self-compensation and calculation function of smart sensor opens up a new way for temperature drift and nonlinear compensation of sensor, which is difficult to complete by hardware can be realized by software. Appropriately relax the sensor processing accuracy requirements, as long as the repeatability of the sensor can be ensured, the microprocessor is used to calculate the test signal through software, and the temperature drift and nonlinearity are compensated in real time by multiple fitting and difference calculation methods, so as to obtain accurate measurement results.

② Self-test, self-calibration, self-diagnosis function. In order to ensure the accuracy of the normal use of the sensor, the ordinary sensor needs regular inspection and calibration, generally need to disassemble the sensor from the use site to the laboratory or inspection department, if the online measurement sensor is abnormal, it can not be diagnosed in time. The smart sensor has obvious advantages, and the self-diagnosis function enables the smart sensor to self-test when the power is turned on, and determine whether the component is faulty according to the self-test results. In addition, calibration can be carried out online according to the use time, and the microprocessor uses the measurement characteristics data stored in the EPROM for comparison calibration.

③ Sensing compound. Intelligent sensor has a composite function, can simultaneously measure a variety of physical and chemical quantities, give a more comprehensive reflection of the law of material movement information, the original need for multiple sensors to work at the same time, now only need one. For example, a composite liquid sensor can simultaneously measure the temperature, flow rate, pressure and density of the medium. The composite mechanical sensor can simultaneously measure the three-dimensional vibration acceleration, velocity and displacement of a certain point of the object.

④ Sensor integration. The development of large-scale integrated circuit enables the sensor and the corresponding circuit to be integrated into the same chip, which is not only small in size, strong in function, but also has some "intelligence", which is an integrated intelligent sensor. The sensor with high integration has a high signal-to-noise ratio, and the weak signal of the sensor is amplified by the integrated circuit signal and then transmitted over a long distance, which can greatly improve the signal-to-noise ratio. Since the sensor and the circuit are integrated on the same chip, the sensor's zero drift, temperature drift and zero position can be automatically calibrated periodically by the self-calibration unit, and the sensor's frequency response can be improved by appropriate feedback.

4.2.2 The Role of Sensors in Intelligent Manufacturing Equipment

Sensors are the basis for intelligent manufacturing equipment information input, and they are also the basis for intelligent manufacturing equipment to obtain information, judge and act independently. Intelligent manufacturing equipment can sense the external environment, make decisions after independent analysis and judgment, and achieve autonomous feedback or action. This process is realized through three functional modules: input system, calculation system and output system. Part of the input of intelligent manufacturing equipment is the manually input setting parameters, and the other part is the information obtained by sensing the external environment through its own sensors. Manual input parameters are the Settings of intelligent manufacturing equipment, reflecting the user's purpose and expectations, and the data input by the sensor is the information that intelligent manufacturing equipment obtains through sensing the external environment and is conducive to the operation of the equipment.

As the only autonomous input of intelligent manufacturing equipment, sensors are equivalent to various sensory organs of intelligent manufacturing equipment and robots. The senses of intelligent manufacturing equipment for the external environment mainly include vision, position sense, speed sense, force sense, touch and so on. Vision can be divided into intuitive vision and environmental model vision, which is the most commonly used input system for intelligent manufacturing equipment. The data of intuitive vision is a picture composed of pixels, typical applications such as machine vision, object recognition based on high-speed cameras, camcorders, etc.

The visual data type of environmental model is the spatial model composed of point cloud data, and the typical application is the spatial modeling based on 3D LiDAR, laser scanner, etc.

Position sense refers to the perception of the distance between the surrounding objects and their own, so as to determine their own environmental position, such sensors are laser rangefinder, 2D LiDAR, magnetometer (to judge the direction), millimeter wave radar, ultrasonic sensors, etc.

Speed sense refers to the intelligent manufacturing equipment for its own running speed, acceleration, angular speed and other information, such sensors have speed encoders, acceleration sensors, gyroscopes and so on.

Force perception is used in intelligent manufacturing equipment to sense the force of external contact objects or internal mechanical mechanisms, typical applications such as force sensors installed on joint drives to achieve force feedback; A force sensor installed between the end of the robot arm and the last joint of the robot to detect the force applied by the object.

In intelligent manufacturing equipment, touch can be further divided into contact sense, pressure sense and slip sense. Such sensors include optical touch sensor, piezoresistive array touch sensor and slip sense sensor, etc. Among them, slip sense sensor is a necessary condition for realizing robot grasp function [10].

In addition to the above five senses, some physical sensors also have feelings beyond the human body, such as biological sensors can measure blood pressure, body temperature, etc., environmental sensors can measure temperature and humidity, air dust particulate matter content, ultraviolet light intensity and so on. These sensors, which transcend human senses, are now paired with wearable devices, thus being given the function of expanding human senses. Figure 4.7 shows the sensor and application on the smart device.

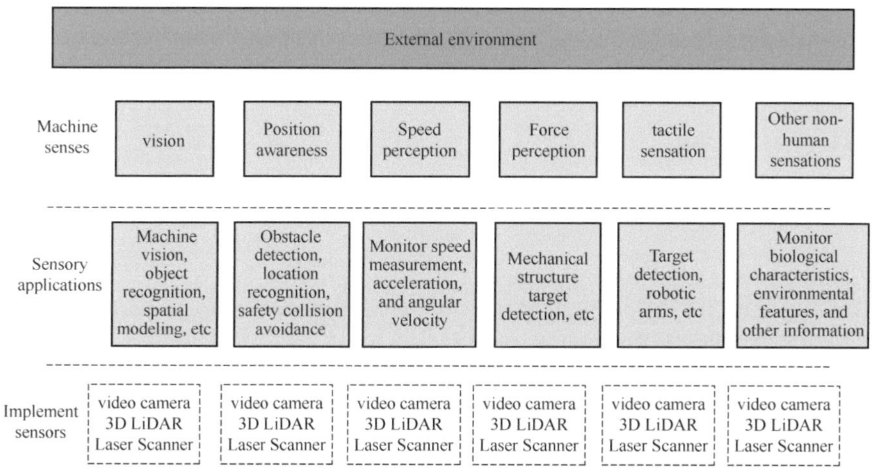

Fig. 4.7 Sensors on smart devices

4.2.3 Classification of Sensors for Intelligent Manufacturing Equipment

The sensors in intelligent manufacturing equipment can also be divided into motion sensors, biosensors, and environmental sensors according to their functions.

① Motion sensor. The functions of motion sensors are motion detection, navigation, entertainment, human–computer interaction, etc., including acceleration sensors, gyroscopes, geomagnetic sensors (electronic compass sensors), atmospheric pressure sensors, etc. Electronic compass sensors can be used to measure direction, enable or aid navigation. It is of great value to measure, record and analyze the activity of the human body anytime and anywhere through the motion sensor, and users can know the number of running steps, the number of swimming laps, the distance of cycling, energy consumption and sleep time, and even sleep quality. Now many wearable devices, including mobile phones, have such sensors, which can record people's movement information at any time, and give some health references through data analysis.

② Biosensors. The main functions of biosensors include health and medical monitoring, entertainment and so on. Biosensors mainly include blood glucose sensors, blood pressure sensors, ECG sensors, myoelectric sensors, body temperature sensors, brain wave sensors, etc. With the help of these sensors applied in wearable technology, health warning and condition monitoring can be achieved, doctors can improve the diagnosis level, and family members can also better communicate with patients. Wind speed and direction sensor.

③ Environmental sensor. Environmental sensors include temperature and humidity sensors, gas sensors, pH sensors, ultraviolet sensors, evaporation sensors, rainfall sensors, environmental light sensors, particulate sensors or dust sensors, pressure sensors, etc. These sensors mainly achieve environmental monitoring, weather forecasting, health warnings and other functions. In today's increasingly prominent environmental issues, environmental sensors will play an increasingly large role.

4.2.4 Wireless Sensor Networks

Wireless sensor network Fig. 4.8 is a multidisciplinary emerging frontier field, involving sensor technology, network communication technology, wireless transmission technology, distributed information processing technology, microelectronics technology, micro-machining technology, embedded technology, software technology, etc., linking the information world with the physical world.

Since wireless sensor networks emphasize wireless communication, distributed data monitoring and processing, and sensor networks, they have the following outstanding advantages [11]:

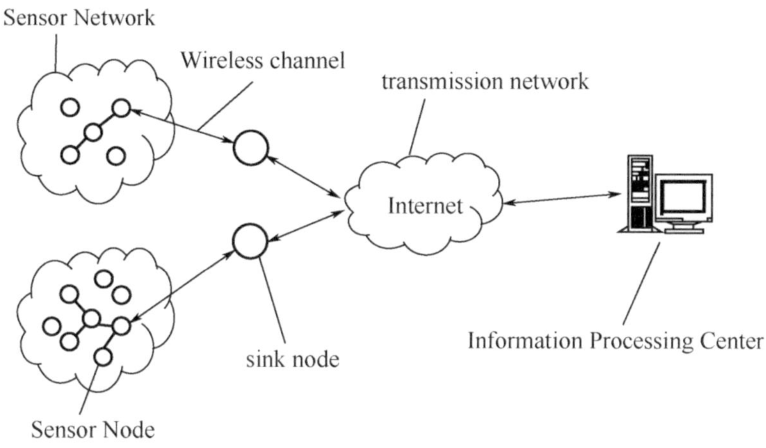

Fig. 4.8 Schematic diagram of wireless sensor network

① Due to the integration of various types of sensors, the monitoring performance is enhanced.
② It solves the problem of maintenance and replacement of sensors arranged in harsh monitoring environment.
③ Because the network system can contain a large number of sensor nodes, the fault tolerance of the system can be enhanced, which exponentially increases the reliability of the diagnosis of the entire system.
④ Through the arrangement of a large number of wireless sensors near the monitoring area, the effect of monitoring through distributed monitoring is far better than that of using only a single sensor.
⑤ Through distributed data processing, that is, the sensor node and the cluster head carry out local collaborative calculation, and the cluster head transmits the user's needs and part of the processed data through the data aggregation, reducing the amount of data transmission, especially the energy consumed when the radio signal is transmitted at a long distance.
⑥ Wireless link formation is fast, no need to wire digging trench, line opening fast.
⑦ Wireless communication coverage is large, almost free from geographical environment restrictions, and can provide 64 Kbps ~ 11 Mbps communication rate, bit error rate is less than 10–10.
⑧ Network scale and topology can be flexibly developed according to application requirements, and links can be added at any time, convenient installation and expansion.
⑨ High security of the wireless link, which can effectively prevent eavesdropping.
⑩ Compared with the wired network, the investment cost is low, the engineering cycle is short, and the cost performance is high.

Wireless sensor networks have been widely used in industrial control, smart home and consumer electronics, security, military safety, logistics, intelligent fine agriculture, environmental awareness, health monitoring, intelligent transportation, logistics management, pipeline monitoring, aviation monitoring, health monitoring and behavior monitoring and many other fields.

Wireless sensor network (WSN) collects the information collected by many sensors with wireless communication and computing capability distributed in an area through wireless means to realize the monitoring and control of the specific state in the area. WSN technology is closely related to the Internet of Things technology, WSN can be said to be the technical support of the Internet of Things. The wireless sensor network takes the signal source of unattended monitoring or measurement as the sensing object, and obtains the target attributes such as temperature, pressure and acceleration through various physical signals such as heat and electricity of the target.

Wireless sensor networks are constructed in a free manner. The construction of wireless network sensor is not limited by external conditions. The network topology structure is uncertain, the sensor nodes constituting the network topology can be increased or decreased at any time, the network topology structure can be separated or merged at any time. The control mode is not centralized, the control mode between each sensor node is distributed, and the functions of routing and host are realized by the terminal of the network. Each host runs independently and does not interfere with each other. Wireless sensor networks use wireless mode to transmit information, so sensor nodes in the process of transmitting information is easy to be invaded by the outside world, resulting in information leakage and damage to wireless sensor networks, most of the nodes of wireless sensor networks are exposed, reducing the security of wireless sensor networks. Figure 4.9 shows the wireless sensor node structure. Figure 4.10 shows the application of wireless sensor networks in smart homes.

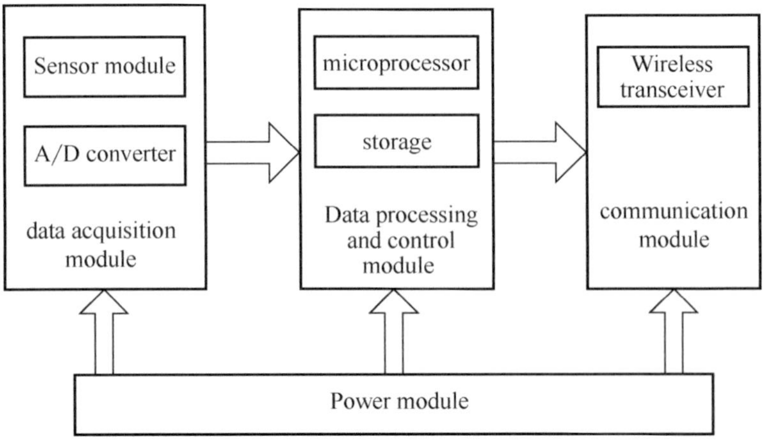

Fig. 4.9 Schematic diagram of wireless sensor node structure

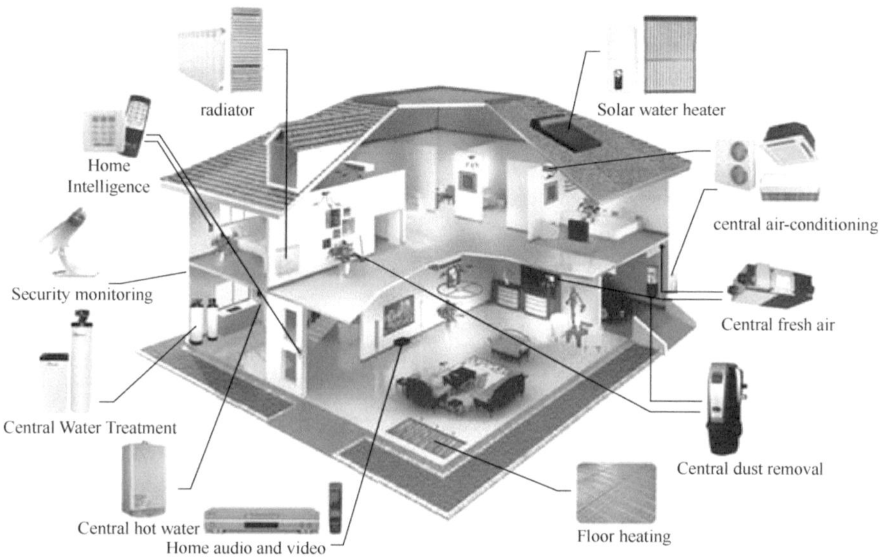

Fig. 4.10 Application of wireless sensor networks in smart homes

4.2.5 *Fuzzy Sensors*

Fuzzy sensor belongs to the category of intelligent sensor, which is an integrated symbolic measurement system composed of numerical measurement and language symbol representation. Based on the classical sensor numerical measurement, fuzzy reasoning and knowledge integration, the intelligent sensor outputs the measurement results in the form of natural language symbol description. Fuzzy sensor is a kind of sensor which can generate and process symbol information of related measurement based on numerical quantity. The basic logic structure of fuzzy sensor consists of four parts: signal extraction, signal processing, numerical conversion and fuzzy concept synthesis. Fuzzy sensor has the characteristics of general intelligent sensor, but also has its own characteristics. Fuzzy sensor has learning function, reasoning function, perception function, communication function, etc. Fig. 4.11 shows the structure of a fuzzy sensor.

Fuzzy sensor has been widely used in actual production and life, such as fuzzy control washing machine in the amount of cloth detection, water level detection, water turbidity detection, rice cooker water, meal detection, fuzzy mobile phone charger and so on. In addition, fuzzy distance sensor, fuzzy temperature sensor, fuzzy color sensor and so on are also developed by foreign experts. With the development of science and technology, the mutual integration of scientific branches, fuzzy sensors are also applied to neural networks, pattern recognition and other systems.

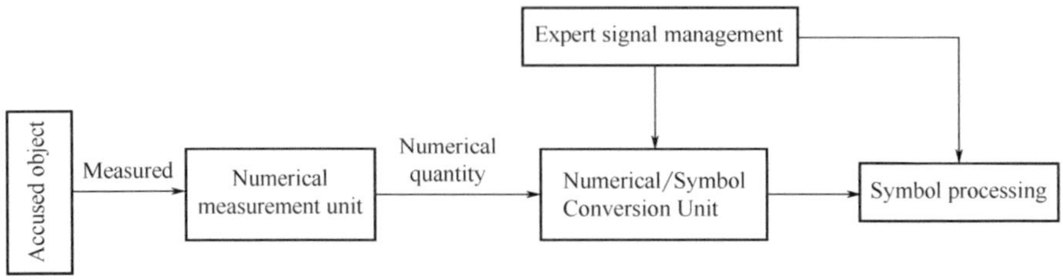

Fig. 4.11 Fuzzy sensor structure

4.3 Selection of Sensors for Intelligent Manufacturing Equipment

In the system of mechatronics products, especially intelligent manufacturing equipment, the sensor is the first in the system, its role is equivalent to the system of the sensory organ, used to detect the external environment and its own state of various physical quantities (such as force, temperature, distance, deformation, position, power, etc.) and its changes, and convert these signals into electrical signals, and then through the corresponding transformation, amplification, modulation and Demodulation, filtering, operation and other circuits will detect useful signals and feed back to the control device or display. The sensor that realizes the above functions and the corresponding signal detection and processing circuit constitute the detection system in the mechatronics product. When the sensor is applied, the corresponding test system is generally set up, especially in intelligent products, the sensor and its detection system are an essential part.

Taking CNC machine tools as an example, the sensors used are rotary encoders, Hall sensors, rotary transformers, induction synchronizers, grating displacement sensors, magnetic gate displacement sensors and other sensors, as well as a variety of sensors involving cutting force measurement and workpiece detection.

The selection of sensors is an important part of the design of intelligent manufacturing equipment, which should generally consider the use environment, sensitivity, frequency response characteristics, linear range, stability and accuracy.

(1) Determine the type of sensor according to the measurement object and the measurement environment

When measuring a specific indicator, it is necessary to analyze various factors and comprehensively consider what kind of sensor is used. Even if it is to measure the same physical quantity, there are a variety of principles of sensors to choose from. The size of the measuring range, the requirements of the measured position on the sensor volume, the measurement method is contact or non-contact, the signal extraction method, wired or non-contact measurement, etc., need to be considered. After considering the above issues, it is possible to determine what type of sensor to choose, and then consider the specific performance indicators of the sensor.

(2) Sensitivity

In general, in the linear range of the sensor, the higher the sensitivity of the sensor, the better. The higher the sensitivity, the smaller the change that the sensor can perceive, the larger the value of the output signal corresponding to the measured change, which is conducive to signal processing. However, when the sensitivity of the sensor is high, the external noise unrelated to the measurement is also easy to mix, and will be amplified by the amplification system, affecting the measurement accuracy. Therefore, the sensor itself is required to have a high signal-to-noise ratio to minimize interference signals introduced from the outside world. The sensitivity of the sensor is directional. When the measurement is a one-dimensional vector, and the requirement for its directivity is high, other sensors with small directional sensitivity should be selected; If the measured vector is multidimensional, the cross sensitivity of the sensor is required to be as small as possible. Closely related to the sensitivity is the measurement range, the maximum input should not make the sensor into the nonlinear region, let alone into the saturated region. Some test work should be carried out under strong noise interference. The input includes not only the measured amount, but also the interference amount, and the sum of the two cannot enter the nonlinear region. Excessive sensitivity will narrow its applicable measurement range.

(3) Frequency response characteristics

The frequency response characteristic of the sensor determines the frequency range to be measured, and the sensor must maintain the measurement state within the allowable frequency range. In fact, the response of the sensor always has a certain delay, and the shorter the delay time, the better the performance of the sensor. The frequency response of the sensor is high, and the frequency range of the detectable signal is wide. Due to the influence of the structural characteristics, the inertia of the mechanical system is large, so the frequency of the detectable signal of the sensor with low frequency is also low. In dynamic measurement, the response characteristics of the sensor have a direct impact on the test results, and the sensor should be selected according to the characteristics of the measured physical quantity signal (steady state, transient, random, etc.) to avoid excessive errors.

(4) Linear range

The sensor works in the linear region, which is the basic condition to ensure the measurement accuracy. Any sensor has a certain linear range, and the linear range of the sensor refers to the range where the output is proportional to the input. In theory, within this range, the sensitivity remains constant. The wider the linear range of the sensor, the larger its range, and can ensure a certain measurement accuracy. However, it is not easy for any sensor to guarantee its absolute linearity, and it can be applied in its approximate linear region within the permissible limits. When selecting a sensor, after determining the type of sensor, it is first necessary to see whether its range meets the requirements. When the required measurement accuracy is relatively low, in a certain range, the sensor with small nonlinear error can be approximated as linear.

(5) Stability

The stability of the sensor refers to the ability of its performance to remain unchanged after a period of use. In addition to its own structure, the main factor affecting the stability of the sensor is its use environment, so the sensor must have a strong ability to adapt to the environment in order to have good stability. Before selecting a sensor, it should be investigated in its use environment, and select the appropriate sensor according to the specific use environment, or take appropriate measures to reduce the impact of the environment on the sensor, such as building a special house, considering reducing the intensity of electromagnetic radiation and other measures. The stability of the sensor has a time limit, and it should be re-calibrated before being used again after the calibration period to determine whether the performance of the sensor has changed. In some cases where the sensor is required to be used for a long time and cannot be easily replaced or calibrated, it is necessary to choose a sensor with excellent stability that can withstand the test of a long time.

(6) Reliability

Reliability is the life of sensors and all measuring devices. The so-called reliability refers to the ability of instruments, devices and other products to achieve the specified functions under the specified conditions and within the specified time. Only the performance parameters of the product are within the specified error range, can be considered to complete the specified function. Therefore, in the selection process of sensors, its reliability is one of the important indicators for selecting sensors.

(7) Accuracy

The accuracy of the sensor indicates the degree to which the output of the sensor agrees with the truth value being measured. Precision is an important performance index of sensor, and it is an important parameter related to the measurement accuracy of the whole measurement system. Whether the sensor can truly reflect the measured value has a direct impact on the whole test system. The price of the sensor is proportional to its accuracy, the design and selection should consider the cost performance according to the purpose and requirements of the test, the accuracy does not have to be too high, as long as the accuracy requirements of the entire measurement system can be met in many sensors to meet the same measurement purpose to choose a relatively cheap and simple sensor. If it is for qualitative analysis, the sensor with high repetition accuracy can be selected, and the sensor with high absolute value accuracy should not be selected. If it is for quantitative analysis, it is necessary to obtain accurate measurement values, and it is necessary to choose a sensor with an accuracy level that can meet the requirements. In some special use cases, if the appropriate sensor cannot be selected, it is necessary to design and manufacture the sensor by itself. The performance of the self-made sensor should meet the requirements of use.

(8) Measurement method

The working mode of the sensor under actual conditions is also a factor that must be considered in sensor selection. Different working environments have different requirements for sensors.

(9) Others

In addition to the above factors that should be fully considered when selecting the sensor, it should also take into account the simple structure, small size, light weight, cheap price, easy maintenance, easy to replace and other factors as much as possible.

4.4 Design of Intelligent Manufacturing Equipment Perception System

4.4.1 Customized Development of Perception System

(1) Design of perception system

The design of sensing system is similar to the development of intelligent instrument, in order to complete the function of the system, it is necessary to follow the correct design principles and develop the sensing system according to the scientific design steps. First of all, it is necessary to clarify the design requirements of the sensing system, and the requirements for the design of the sensing system are:

① Function and technical indicators. The functions required by the perception system mainly include information output form, communication mode, man–machine dialogue, etc. The technical indicators of the system mainly include accuracy, measurement range, working environment conditions and stability. Small size, light weight, high precision and sensitivity, fast response, good stability and high signal-to-noise ratio are the goals to pursue.

② Reliability. The sensing system is the "organ" of intelligent manufacturing equipment. In order to ensure that each component of the sensing system can work stably and reliably for a long time, various measures should be taken to improve the reliability of the system. In terms of hardware, it is necessary to rationally select components, leave a certain margin for technical parameters such as load, speed, power consumption and working environment of components during design, carry out aging inspection and screening of components, and carry out experiments under extreme conditions, such as making the sensing system withstand low temperature, high temperature, shock, vibration, interference, smoke, etc., to ensure the adaptability of the environment. In terms of software, the modular design method is adopted, and the software is comprehensively tested to eliminate loopholes, reduce the failure rate and improve reliability.

③ Easy to operate and maintain. In the design process of the sensing system and each front-end sensing module, the convenience of on-site operation and maintenance should be considered, so that the operator can master the use of the system without special training. In addition, the main system structure should be standardized and modular as far as possible, it is best to be equipped with on-site fault diagnosis procedures, once the fault occurs, it can ensure that the fault is effectively located in order to replace the corresponding equipment module, so that the system has good maintainability.

④ Process structure. Process structure is also one of the important factors affecting system reliability. According to the working environment conditions of the system and each component, determine whether the process structure such as waterproof, dustproof, sealing, shock resistance, shock resistance, corrosion resistance is required, and seriously consider the overall structure of the system and the connection relationship between the modules. (5) Environmental adaptability. The perceptual system has a strong ability to adapt to environmental conditions. When designing, the site environment should be fully investigated, and then the sensor type should be carefully selected, which should not be affected by the measured object (such as resistance, permeability), nor affect the external environment, so as to improve the reliability of the sensor and improve the life of the electromechanical equipment.

(2) Design method of the system

When designing the system, top-down design and open design can be adopted.

① Top-down design. According to the function and design requirements of the system and each component module, the designer puts forward the general task of system design, draws the hardware and software general block diagram (overall design), and then breaks the task into a number of independently characterized sub-tasks, until each sub-task is simple enough to be directly and easily implemented. Subtasks can adopt certain generic modules and can be designed and debugged as separate entities. This modular system design method not only simplifies the design process, shorts the design cycle, but also has flexible structure, convenient maintenance, easy to expand and update, enhances the adaptability of the system and improves the reliability of the system.

② Open design. The system design can adopt the open design principle, leaving room for future update and expansion, in order to facilitate the expansion of user functions or secondary development and upgrading, to meet the requirements of different levels of users, should be in comprehensive consideration of various factors after the correct selection of reasonable design scheme.

(3) System design steps

System design is the process of designing a system that can meet the target requirements to the maximum extent according to the results of system analysis, using the ideas and methods of system science.

1. Determine the design task. Clarifying the working principle of the design object begins with the analysis of the project requirements and ends with the determination of the overall technical scheme. Comprehensively understand the design content, figure out the problems to be solved, and mainly decompose the hardware design requirements, including hardware functional requirements, performance indicators, reliability indicators, manufacturability requirements, serviceability requirements and testability requirements; Hardware requirements are quantified and evaluated for their feasibility, rationality, reliability, etc. Hardware design requirements are the basis and basis for the overall technical plan design of hardware engineers. According to the final design goal of the system, a detailed design task specification is made to clarify the function of the sensing system and the technical indicators to be achieved.
2. Formulate the overall design plan. Formulate the design scheme according to the design task description, including theoretical analysis, calculation and necessary simulation experiments, to verify whether the scheme can meet the design requirements, and then demonstrate the feasibility of the scheme. Finally, from the overall advancement, reliability, cost, production cycle, maintainability and other aspects of comparison, comprehensive formulation of the design scheme until the completion of the hardware outline design. Mainly for the design and verification of hardware unit circuit, local circuit or circuit with new technology, new device application, key process, structural assembly and other uncertain technology verification and commissioning, to provide design basis and design support for the summary design. Figure 4.12 shows the procedure for designing and debugging the sensing system.

According to the overall design scheme, determine the core components of the system and the distribution of software and hardware. The system is divided into functional modules which are easy to realize by top-down design method. The working flow charts of software and hardware of each module are drawn and debugged respectively. After each module is debugged, it is adjusted to complete the design of the sensing system.

The first step is to determine the core components of the system according to the overall scheme of the system. The perceptive components play a great role in the overall performance and price of the system, and will affect the design of hardware and software. The intelligent control components in the system are usually optional MCU or MPU.

MCU is in A chip integrated CPU, RAM, ROM, clock, timing/counter, series parallel I/O interface and many other functional components, some models of MCU including A/D converter, D/A converter, analog comparator, pulse width modulator, USB interface, with strong function, small size, low price, support software, easy to develop and so onDot. Therefore, the front-end node module of the sensing system selects MCU as an intelligent control component. When selecting a specific model, we should consider word length, instruction function, addressing

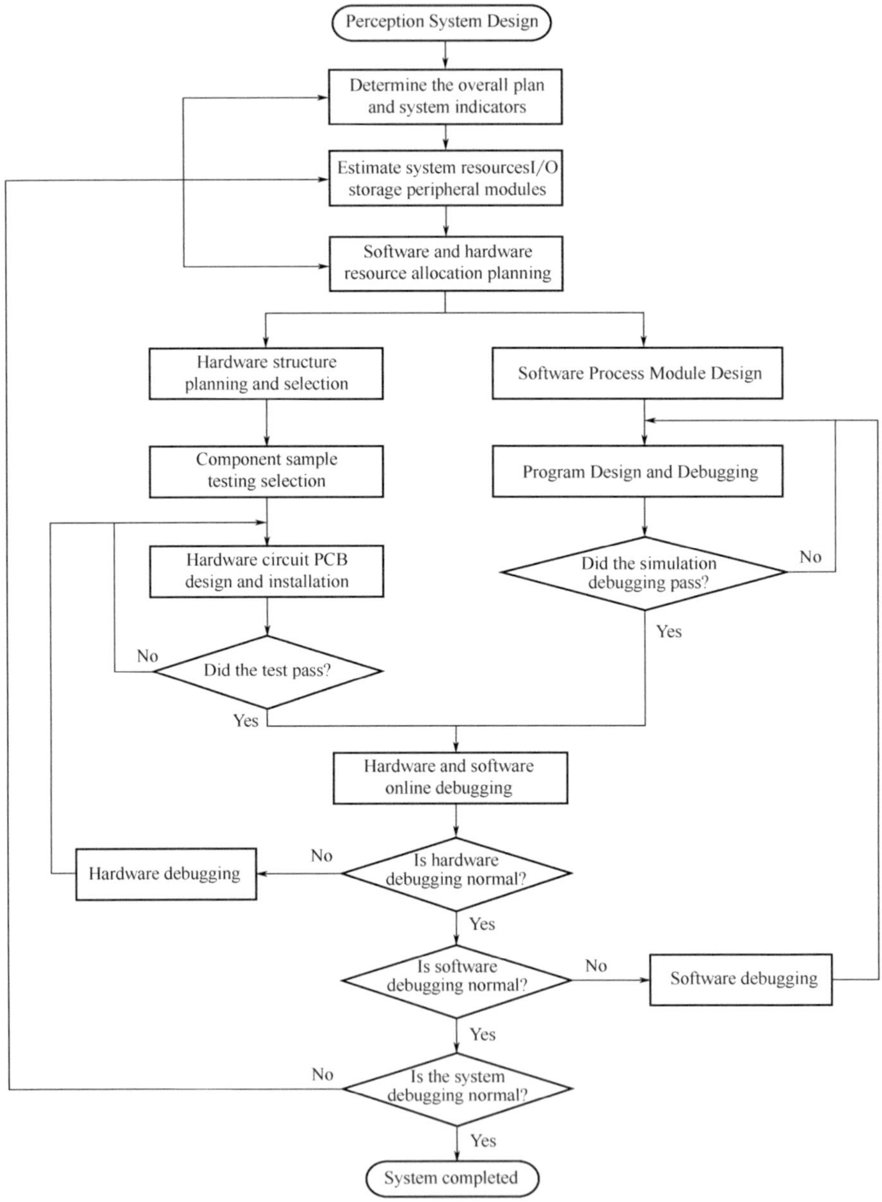

Fig. 4.12 Design and debugging steps of perception system

range, addressing mode, internal memory capacity, bit processing capacity, interrupt processing capacity, supporting hardware, chip price and development platform. At present, the commonly used MCU has ATMEL company's AT89 series, AVR series, TI company's MSP430 series, Motorola company's 68HCXX series and compatible with a variety of improved upgraded chips. The characteristics of MCU are very suitable for applications with high integration and low cost.

Second, select the sensor. First of all, according to the requirements of the use of many sensors to choose their own needs. Some sensor input/output characteristics, the theoretical analysis is more complex, but the practical application is very simple, the user only needs to use according to its main performance parameters, such as measurement range, accuracy, resolution, sensitivity, etc. The sensor performance parameter index contains a wide surface, for a specific sensor, should be based on actual needs and possibilities, in the case of ensuring its main performance indicators, appropriate relaxation of the requirements for secondary performance indicators, do not blindly pursue a variety of characteristic parameters are high indicators, in order to obtain a higher cost performance. Secondly, we should pay attention to the application environment, use conditions and maintenance requirements of different series of products. Environmental changes (such as temperature, vibration, noise, etc.) will change some characteristics of the sensor (such as sensitivity, linearity, etc.) and can cause outputs independent of the parameters being measured, such as zero drift. Therefore, sensors should be reasonably selected according to environmental requirements.

The third step is design and commissioning. The first is the design and debugging of hardware component modules and software programming. In general, the design of hardware modules and software is carried out separately. However, because the perception and recognition of intelligent manufacturing equipment are closely related to the software and hardware of the detection system or component module, they can also be crossed.

The design process of the hardware part is to design each unit circuit according to the hardware block diagram respectively, and then to synthesize the hardware to form a complete hardware circuit diagram. After completing the design, the circuit board (PCB) is drawn and printed, and then assembled and debugged.

Software design can first design the overall structure diagram, and then the overall structure is divided into multiple sub-modules according to the principle of top-down, using structured programming method, draw a detailed flow chart of each sub-module, choose the appropriate language to write programs and debug. Considering the function, cost, development cycle and expense of the system or module comprehensively, the proportion of software and hardware can be reasonably allocated to make the system achieve a higher cost performance.

The fourth step is joint debugging of hardware and software. After the software and hardware are successfully debugs, the software and hardware must be jointly debugs. If the problem in debugging is a hardware fault, the hardware circuit can be modified; If it is a software problem, modify the program; If it is a system problem, modify both the software and hardware. After commissioning, it is also necessary to test software and hardware, mainly including functional testing, pressure testing, performance testing and other professional tests, such as anti-interference testing, product life testing, anti-moisture testing, high temperature and low temperature testing.

In the process of sensing system debugging, there is an important work is sensor calibration and calibration, and according to the use of the site, the sensor still needs to be calibrated regularly. Calibration is the process of using standard equipment to generate known non-electricity and using it as a reference to determine the relationship between the output electricity and the input non-electricity of the sensor. It is worth pointing out that the sensor should be calibrated at the factory, and the main performance parameters (or indicators) listed in the manufacturer's product list are obtained through calibration. The calibration of the sensor shall be carried out in an environmental state similar to its conditions of use and under specified installation conditions. When the sensor is used before or during use or after shelving for a period of time, its performance parameters must be retested or adjusted and corrected as necessary to ensure its measurement accuracy. This retest process is "calibration".

(4) Interface and embedded communication

In recent years, wireless sensor networks have developed rapidly. In this process, various wireless network data transmission standards have emerged, and different protocol standards correspond to different application fields [12]. There are different ways of communication between the subsystems of the sensor system and between the external interfaces, in order to unify the standards as much as possible, the IEEE has developed the 1451 standard, which provides the interface standard for connecting the sensor and the transmitter to the device, and the standard defines the form of the sensor electronic number table, including the key information of different sensors generated by different manufacturers.

The digital interface of the sensor is generally a serial interface, and some digital image sensors have a large factor of data transmission and need a parallel interface. The serial interface has asynchronous mode such as RS232, RS485 and synchronous mode such as SPI. The serial interface circuit is simple, the data transmission distance is wider, widely used, the interface used to be one of the standard interfaces between PC and peripherals, and now it has been gradually replaced by USB, but it is still widely used in the sensor field. RS485 is used to configure the local network and multipoint communication link, gradually replaced by the controller local area network (CAN) Controller Area Networks, but is still widely applied in the field of factory automation, etc.

In terms of wireless communication standards, the sensor application field is mainly Bluetooth, ultra-wideband (UWB), ZigBee and Wi-Fi four key low-power communication standards, which have different standards and ranges of use. Table 4.2 lists common wireless transmission protocols.

Table 4.2 Comparison of common wireless transmission protocols

Standard/ protocol	Concept	Main advantages	Main disadvantages
Blue tooth	Bluetooth technology is a global standard for open wireless data and voice communication. It is a special short-range wireless technology connection based on low-cost short-range wireless connections, which establishes communication environments for fixed and mobile devices	Low power consumption, strong anti-interference ability, mature and stable technology, transmission rate up to 24Mbps, low cost, support for mobile phones and other terminals, intelligent Bluetooth theoretically can infinitely expand the number of communication devices	Classic Bluetooth has a limited number of nodes and is not suitable for high-speed data exchange. The device search speed is slow, the connection energy consumption is high, and the Bluetooth stack is difficult to achieve precise time synchronization
ZigBee 802.15.4	ZigBee, Also known as Purple Bee, it is a wireless network protocol for low-speed and short distance transmission, with the underlying layer consisting of the media access layer and physical layer using the IEEE802.15.4 standard specification	Simple protocol, efficient power management, support for large node mesh networks, good data security, support for identity verification and data encryption	Lack of multi hop protocol causes communication interference, difficult operation in high congestion environments, slow data transmission rate, and does not support local smartphones
Ultra Wide Band (UWB)	Ultra Wide Band (UWB) technology is a wireless carrier communication technology that does not use sine wave carriers, but instead utilizes nanosecond level non sine wave narrow pulses to transmit data, thus occupying a wide spectrum range. High speed wireless access suitable for indoor and other dense multipath environments	Low energy consumption, anti noise interference, anti multipath attenuation, strong signal penetration	High cost, low penetration rate, long signal acquisition time, and mutual interference with other wireless signals
Wi-Fi	Wif, Also known as 802.11b standard, it is a technology that can wirelessly connect terminals such as personal computers and handheld devices (such as PDAs, mobile phones) to each other	High coverage and transmission speed in high blocking environments, supporting tablets and smartphones, strong adaptability and scalability, good security, and wide application	High power requirements, single hop network, higher cost compared to Bluetooth and ZigBee

4.4.2 *Customized Development Methods and Cases of Perception Systems*

(1) Similar sensor superposition achieves longitudinal depth combination on a single function

The organic combination of similar sensors to form a redundant structure ensures the safety of the system in this function. Take a driverless car equipped with a large number of sensors as an example, Fig. 4.13 shows. The perception system of driverless cars organically combines a variety of vision and position sensors to form a complementary redundant structure. Through the perception system, road information such as obstacles, roads and traffic signals can be independently identified, so as to ensure that the system can correctly and efficiently perceive the external environment in real time and make correct driving decisions. The driverless car perception system is the key to the machine replacing the driver. The system is mainly composed of various vision, position sense and sensors, and different sensors of the same type assist and make up for each other, forming multiple security guarantees and ensuring the high security of the system. A driverless car capable of autonomous driving should have the following functions:

① Ranging support 1. The 3D LiDAR mounted on the roof of the car can actively construct a spatial model of the surrounding environment. The Google self-driving car is equipped with Velodyne's Lidar sensors, which calculate the distance of objects at a range of 200 m and use this to create a three-dimensional image of the environment. The Lidar sensor is the vision system of Google's driverless car and is the main source of information input for the driverless system.

Fig. 4.13 Autonomous vehicles

② Ranging support II. The millimeter-wave radar installed in the front and rear bumpers is not affected by weather and light, which is the second powerful guarantee for driving safety. A total of four millimeter-wave radars are installed on the front and rear bumpers of Google's driverless car, which is part of the adaptive cruise control system, which can ensure that the driverless car is in a safe distance from the car when driving on the road. The driverless car needs to maintain a safe reaction distance of 2–4 s with the front car, and the specific setting changes according to the speed, so as to maximize the guarantee of passengersThe safety of…At present, the standard vehicle radar mostly uses millimeter wave radar, and other cases also use infrared radar. However, the common disadvantage of millimeter wave radar and infrared radar is that the reflection effect of pedestrians is extremely weak, so it can only be applied to maintain the front and rear vehicle distance as an aid to 3D LiDAR.

③ Ranging support III. Ultrasonic radar has the best ranging stability, but the closest distance is the third guarantee of driving safety. Ultrasonic sensor is the reverse radar on ordinary cars, because of its excellent ranging stability, not affected by light, weather, and can detect obstacles regardless of texture is widely used, but it is limited by the measurement distance, can only measure objects within 10 m.

④ External environment identification. Front, side and rear cameras can clearly and effectively identify things. Cameras mounted on the front of the vehicle can better help the car recognize objects in front of it, including pedestrians, other vehicles, and so on. It can also recognize traffic signs and signals, as well as various speed limits, one-way, two-way, and sidewalk signs. In-vehicle cameras or other sensors capture lane information in real time, which can provide path and direction information for the car to travel.

⑤ Body positioning. High-precision Beidou navigation system or GPS for route planning. Driverless cars make full use of Beidou or GPS technology to locate themselves, and then use commercial maps to achieve optimal path planning. However, due to the influence of weather and other factors, the accuracy of GPS is generally on the order of a few meters, and can not achieve enough accuracy. In order to achieve accurate positioning, commercial maps are generally online mode, the positioning data and the real-time data collected before need to be integrated, with the movement of the vehicle, the real-time map in the car will also be updated according to the new situation, so as to display a more accurate map to facilitate positioning.

⑥ Body status monitoring. The speed encoder and acceleration sensor installed in the wheel are used to collect the real-time speed of the wheel to obtain the speed information of the autonomous vehicle such as speed, wheel speed, angular speed and inertia. By judging the speed information of the wheel, it can also indirectly determine whether the tire pressure is stable.

(2) Multiple sensors combined to achieve horizontal breadth combination in multiple functions

Multiple sensor combination applications for product innovation is the most common sensor application trend, and the innovation and combination of functions will also

Fig. 4.14 Pepper robot

give birth to various forms of new intelligent equipment in the future, especially in the fields of home applications, social services, and public services. In order to meet the multi-type and multi-level input and output needs of the system, multiple types of sensors are innovated to form a variety of senses of intelligent equipment, and intelligent feedback is formed according to a variety of senses. Take Pepper, an emotion-interactive robot developed in Japan, as an example, as shown in Fig. 4.14. The robot is a typical combination of sensors used products, Pepper is equipped with a variety of sensors to achieve vision (camera, infrared sensor), position sense (laser rangefinder), hearing (microphone), touch (contact sensor, slip sensor) and other feelings, and equipped with a special display to achieve facial expression and mood expressionRobotic arms to achieve body language, etc.

Pepper's main sensors include: cameras located in the head and mouth to recognize objects and record images; A laser transmitter and a laser receiver located in the eye to measure the distance between the target object and itself; Infrared sensors located in the head, used to identify human facial contours, to judge human emotions; The contact sensor and slip sensor located in the hand are used to realize the functions of grasping objects; Others such as microphones and mechanical sensors used to assist the internal structure of the machine.

There is no primary or secondary relationship between the sensors of the Pepper robot, and the various hardware equally serves the overall system. Its human emotion recognition system, artificial intelligence system of speech judgment and feedback are the key technologies to determine the performance of products.

The BigDog robot (Fig. 4.15) developed by Boston Dynamics, which has attracted widespread attention in recent years, is also a typical case of using multiple sensor combinations. BigDog is capable of walking, running, jumping, climbing and carrying heavy objects. BigDog's four legs can absorb shocks to recover energy, and its unique feature is that its exquisite mechanical design and various sensor applications make BigDog have ultra-high stability and coordination, can travel smoothly

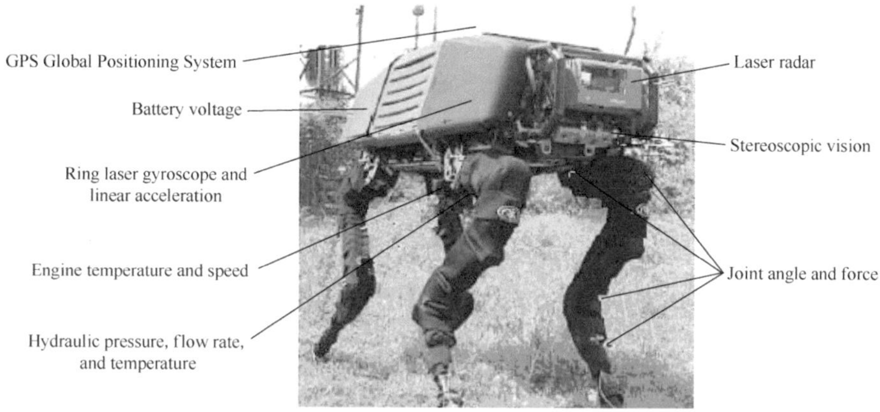

Fig. 4.15 BigDog robot

in poor road conditions, mountains, and can respond to external forces such as collisions, kicks and other impacts to prevent falls. There are four kinds of sensors in the proprioceptive module of BigDog robot: linear potentiometer, which is used to measure the displacement of BigDog robot joints to judge the change of the force direction of the joint; The force sensor is used to measure the force borne by the actuator and ankle, and the linear potentiometer is used to maintain the force balance of the body; Current sensor, used to measure whether the servo motor provides the correct current; Gyroscope is used to measure the angular velocity and linear acceleration of the robot body. The information from all these sensors is combined to maintain the force balance of the robot body. The sensor constituting the external perception module has a stereo camera, which is used to perceive the slope of the ground to adjust the force balance, and can also be used to identify obstacles for avoidance; In addition, there is Lidar, which is used to locate the guide for automatic obstacle avoidance tracking.

(3) The application of new sensors to traditional equipment makes scene innovation sprout vitality

The application of new sensors to traditional equipment is a scene innovation, the most typical case is the sweeping robot Fig. 4.16). There are two kinds of sweeping robot: random collision type and path planning type. The path planning robot will gradually replace the random collision robot because of its high cleaning efficiency, strong escape ability, convenience and so on. Path planning requires GPS technology, vision technology, laser technology.

GPS technology makes the sweeping robot clear about its position in the room, effectively avoiding repeated cleaning and improving cleaning efficiency, but it can not avoid obstacles, so the detection of obstacles is still "collision". And because of problems such as GPS positioning errors, it makes little sense to use in the smaller space of the house.

Fig. 4.16 Cleaner robot

The sweeping robot based on vision technology refers to obtaining images through the camera, planning paths and avoiding obstacles through algorithms, and has no requirements for indoor objects, which is suitable for various indoor rooms, but the positioning accuracy is lower than that of laser navigation. At present, the products on the market adopt SLAM(real-time positioning and mapping) technology, which can observe the room through the camera, identify the sign objects and main features of the room, draw the room map for navigation through the triangulation principle, so as to confirm their own position in the room for cleaning.

The sweeping robot based on laser technology and the laser radar navigation sweeping robot refer to the use of some low-cost liDAR to obtain the distance information of surrounding objects, and plan the path through intelligent algorithms to avoid obstacles. The sweeping robot guided by LiDAR has high accuracy (centimeter level) and can be applied to a large space (Lidar with a radius of 5 m can cover 80 m^2), but it cannot detect high-reflectivity objects such as floor glass and vases, and rotating lidar may also have life problems.

At present, the price of laser-guided sweeping robots is still higher than people's expectations. In the future, with the decrease of the cost of sensors such as 2D Lidar, laser-guided sweeping robots are expected to enter thousands of households.

4.5 Sensors for Industrial Robots

4.5.1 Sensory System of Industrial Robot

With the shortage of labor, "machine replacement" is a general trend. Industrial robot is a multi-joint manipulator or multi-degree of freedom mechanical device widely used in the industrial field, mainly composed of mechanical structure system, drive system, sensing system, robot-environment interaction system, human–computer interaction system and control system, with a certain degree of automatism, can rely on its own power energy and control capabilities to achieve a variety of industrial processing and manufacturing functions. Industrial robots are widely used in automotive, home appliances, electronics, logistics, chemical and other industrial fields. Industrial robots have high positioning motion requirements, many degrees of freedom, frequent movement, and long working time, so reliable sensors are necessary. The commonly used sensors on industrial robots mainly include internal sensors of industrial robots and external sensors of industrial robots. The sensory system of industrial robots is mainly vision, hearing, touch, smell, taste, balance, etc. The general requirements of its sensors are high precision, good repeatability, good stability and reliability, strong anti-interference ability, light weight, small size, easy installation. Table 4.3 lists the common sensors used by robots.

The internal sensors of industrial robots are installed on the operating machine, including displacement, speed and acceleration sensors, which are mainly used to detect the internal state of the robot operator and serve as feedback signals of the servo control system. External sensors such as vision, touch, force and distance are used to detect the working object and the communication between the environment and the robot. Sensors play a very important role in the control of robots, and it is because of sensors that robots have human-like perceptual functions and reaction capabilities.

Table 4.3 Common sensors for robots

Sensor location			Basic types
Internal sensors	Position sensor		Potentiometer, rotary transformer, code wheel
	Speed sensor		Code wheel, speed generator
	Accelerate sensor		Strain gauges, piezoelectric, MEMS accelerometers
	Tilt angle sensor		Vertical oscillator type, liquid type
	Torque sensor		Strain gauge, piezoelectric type
External sensors	Visual sensor	Measurement sensor	Optical type
		Identify sensors	Optical and ultrasonic
	Tactile sensor	Tactile sensor	Single point, distributed
		Pressure sensor	Single point, distributed, high-density integration
		Slip sensor	Point contact, line contact, surface contact
	Proximity sensor	Proximity sensor	Air type, magnetic field type, electric field type, optical type, ultrasonic type
		Distance sensor	Optical and ultrasonic

4.5.2 *Internal Sensors of Industrial Robots*

In the internal sensors of industrial robots, position sensor and speed sensor are indispensable components in the feedback control of robots.

(1) Inspection at specified positions and angles

Detect the pre-specified position or Angle, you can use the on/off two state values, mainly detect the starting origin of the robot, the over-the-limit position or determine the position. The electrical contact of the mechanical limit switch can be disconnected or switched on when a specified displacement or force is applied to the movable part of the micro switch. The limit switch is usually packed in a box to prevent the action of external forces and the erosion of water, oil and dust.

The photoelectric switch is a light-transmitting switch composed of a certain distance between an LED light source and a photosensitive diode or a photosensitive transistor. When the light passes through the gap between the light source and the photosensitive element by the reference position of the shutter, the light does not shine on the photosensitive element, and plays the role of a switch. Sometimes in order to improve the reliability of the limit, the mechanical limit and the photoelectric switch are used simultaneously.

(2) Position and Angle measurement

The sensor for measuring the joint displacement and angular displacement of robot is an essential element in the position feedback control of robot. Commonly used sensors are potentiometers, rotary transformers, encoders and so on.

(3) Velocity and angular velocity measurement

The measurement of velocity and angular velocity is an essential part of the feedback control of the driver. Sometimes the displacement sensor is also used to measure the speed, that is, to detect the displacement per unit sampling time, but this method has its limitations: the measurement is unstable at low speed; Only low measurement accuracy can be obtained at high speed. The most commonly used speed and angular velocity sensors are tachometers or sensors called tachometers or ratio generators. The tachometer generator for measuring angular velocity can be divided into DC tachometer generator, AC tachometer generator and induction type AC tachometer generator according to its structure.

(4) Acceleration measurement

With the improvement of industrial site requirements, the load and acceleration of industrial robots have new requirements, and with the high-speed ratio and high precision of robots, the vibration problem of robots is put on the agenda. In order to solve the vibration problem, it is necessary to install an acceleration sensor on the moving arm of the robot to measure the vibration acceleration and feed it back to the drive. The commonly used sensors for acceleration measurement include MEMS acceleration sensor, strain gauge acceleration sensor, servo acceleration sensor, piezoelectric induction acceleration sensor, etc. This part has been elaborated on earlier.

4.5.3 External Sensors of Industrial Robots

Industrial robot (Fig. 4.17) the role of external sensors is to detect the working object and the environment or the relationship between the robot and them. Industrial robots are equipped with tactile sensors, visual sensors, force sensors, proximity sensors, ultrasonic sensors and auditory sensors, which greatly improve the working condition of the robot and enable it to complete complex work more fully. External sensors for a variety of disciplines in one of the products, some aspects are still being explored, with the further improvement of external sensors, the function of the robot will be more and more powerful.

(1) Tactile sensor

The sense of touch is a combination of mechanical stimuli such as contact, impact and compression. The sense of touch can be used for robot grasping and can be used to further perceive physical properties such as the shape, softness and hardness of objects. Generally, the sensors that detect the sense of contact, the sense of pressure and the sense of proximity caused by direct contact with the outside are called robot tactile sensors.

(2) Force sensor

Force perception refers to the perception of forces in the movement of the robot's fingers, limbs and joints. It mainly includes: wrist force perception, joint force perception and support force perception. According to the load of the measured object, the force sensor can be divided into force sensor (single-axis force sensor), moment

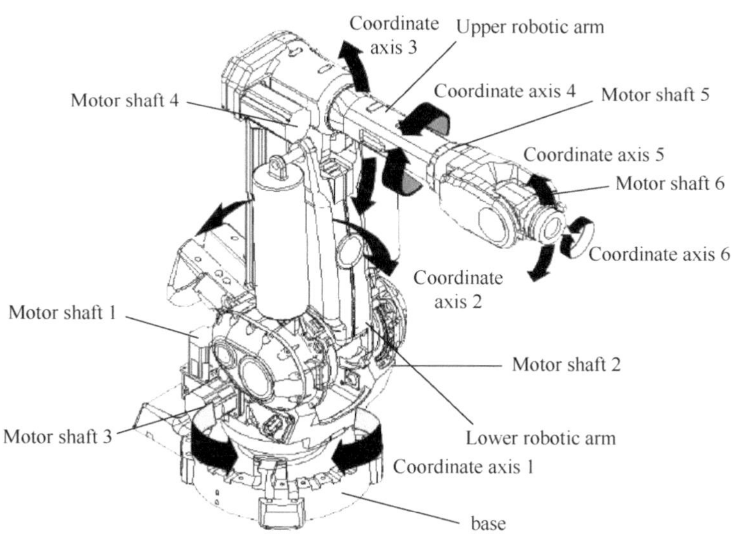

Fig. 4.17 Industrial robot

meter (single-axis torque sensor), finger sensor (ultra-small single-axis force sensor to detect the robot finger force) and six-axis force sensor.

According to the different detection methods of force sensors, they are divided into strain gauge sensors that detect strain or stress, piezoelectric element sensors that use piezoelectric effect, differential transformers that use displacement meters to measure the displacement generated by load, capacitive displacement meter sensors, etc. Strain gauge force sensing sensors are widely used in robots.

When selecting a force sensor, first pay special attention to the rating value, followed by the resolution, in the usual force control of the robot, the accuracy of the force is of little significance, the important thing is the resolution.

When the force sensor is actually installed and used on the robot, it is necessary to check the operating area in advance and remove obstacles. This is of great significance to ensure the personal safety of the experimenter and to ensure that the robot and peripheral equipment are not damaged.

(3) Distance sensor

Distance sensors can be used for robot navigation and avoidance of obstacles, as well as for positioning objects in robot space and determining their general shape characteristics. At present, there are two most common ranging methods: one is ultrasonic ranging method, and the other is laser ranging method. Ultrasonic wave is a mechanical vibration wave with a frequency of more than 20 kHz, and the distance is calculated by using the time interval of transmitting pulse and receiving pulse. The disadvantage of ultrasonic ranging method is that the beam is wide, its resolution is seriously limited, and it is mostly used for navigation and avoiding obstacles. The working principle of laser ranging method is: the helium–neon laser is fixed on the baseline, and the laser point is directed to the measured object by the reflector at one end of the baseline, the reflector is fixed on the motor shaft, and the motor is continuously rotated, so that the laser point is stably scanned to the measured object. The CCD(charge coupled device) camera receives the reflected light, uses the image processing method to detect the laser point image, and calculates the laser reflection Angle according to the position coordinates and the optical characteristics of the camera. The position of the reflection point can be calculated by using the principle of triangulation distance measurement.

(4) Other external sensors

In addition to the robot external sensors described above, auditory sensors, taste sensors and electromagnetic wave sensors can also be installed according to the special purpose of the robot, and these robots are mainly used for scientific research, Marine resources detection or food analysis, fire fighting and other special purposes. Most of these sensors are in the development stage and need to be further refined to enrich the specialized functions of the robot.

(5) Sensor fusion

Industrial robots are more and more complex, the type and number of sensors used in the system are more and more, each sensor has a certain use conditions and perceptual range, and can give the environment or the object part or the whole side of the information, in order to effectively use these sensor information, need to use some form of sensor information for comprehensive, fusion processing, different types of information One form of processing system is sensor fusion. The fusion technology of sensors involves new theories and methods in the fields of information, detection and control, such as neural network, knowledge engineering and fuzzy theory.

When the sensor detects the same environment or the same nature of the same object, the data provided by the sensor may be consistent or contradictory. If there is a contradiction, it needs systematic judgment to optimize. There are many methods of systematic decision, such as weighted average method, decision method and so on. In a navigation system, the position of the vehicle can be determined by the calculation of the positioning system (using recorded data such as speed, direction, etc.) or the observation of road signs (intersections, sidewalks, etc.). If the observation of the road sign is successful, the result of the observation of the road sign is used, and the value of the calculation method is corrected, otherwise the result of the calculation method is used.

The ideal target of multi-sensor information fusion should be the human sense, recognition, control system, but because there is no clear engineering description, so what kind of function of robot sensor fusion system is still a vague concept. With the improvement of robot intelligence level, the future multi-sensor information fusion theory and technology will gradually improve and develop.

In the future, more and more 3D vision and force sensors will be used on robots, and robots will become more and more intelligent. With the progress of sensing and identification systems, artificial intelligence and other technologies, robots have developed from being one-way controlled to their own storage and application of data, and gradually informatization. With the progress of multi-robot cooperation, control, communication and other technologies, robots are developing from independent individuals to the Internet and collaborative cooperation.

4.6 Development Trend of Intelligent Manufacturing Equipment Sensors

Sensors are an important symbol of the development of contemporary science and technology, with the rapid development of the Internet of Things, as the protagonist of intelligent perception, the development potential of sensors is growing. The continuous development of sensor technology will enable sensors to become more and more miniaturized and continue to become more intelligent, providing a new platform for future innovative products and services [13].

Today's smart sensors are emerging in an endless stream, and the research and development work of smart sensors is far from complete. It is constantly driven by the needs of lower cost, smaller size, smaller power consumption, higher performance and higher reliability, and the emergence of new sensing principles and new technologies continue to promote the development of smart sensors [14]. With the attention of national policies and the support of research and development, sensors are developing towards new principles, new materials and new technology sensors, micro-miniaturization, intelligence, multi-function and networking, as well as multi-sensor fusion and networking [15].

(1) New sensor research and development

New phenomena, new principles and new materials are the important basis for the development of sensor technology and the research of new sensors, and the discovery of each new principle and new material will give birth to new types of sensors. Further explore the sensitive functional materials with new effects, and develop a new type of physical property sensor with new principle. Physical sensor is also called solid state sensor, which includes semiconductor, electrolyte and strong magnetic body three categories. Among them, the highly sensitive threshold sensor developed by using the effects of quantum mechanics to detect weak signals is a new trend in the development of sensor technology. For example, the magnetic sensor using the absorption effect of nuclear magnetic resonance can extend the detection limit to 107 times the magnetic intensity of the earth; The Josephson effect thermal noise temperature sensor can measure the ultra-low temperature of 10–6. Extremely fast response infrared sensor using photon hysteresis effect. At present, the most advanced solid-state sensor can integrate differential pressure, static pressure and temperature three sensors on a chip at the same time, so that the differential pressure sensor has the function of temperature and pressure compensation.

(2) Miniaturization

The development of microelectronics technology and MEMS technology has promoted the development of sensors, and the volume of sensors is getting smaller and smaller, and the sensors, microprocessors, and actuators are integrated to form microelectromechanical systems. The size of microsensors is mostly millimeter, or even smaller. For example, pressure microsensors can be placed inside an injection needle and sent into a blood vessel to measure blood flow. In the future, with the reduction of sensor volume, micro-robots will play an increasing role in medical, military and other fields.

(3) Sensor integration and multi-functional

Semiconductor integrated circuit technology and its development ideas, micro-machining technology, thick film and thin film technology will be used in sensor manufacturing. The so-called integration is the use of semiconductor technology to produce sensitive components, information processing or conversion units and power supplies on the same chip; Multi-functional means that a sensor has the detection function of multiple parameters, such as semiconductor temperature and humidity sensor, multi-function gas sensor and so on. With the help of semiconductor evaporation technology, diffusion technology, lithography technology, precision machining and assembly technology, this development trend of sensors can be realized. For example, Honeywell's ST-3000 smart sensor is a semiconductor process that integrates a CPU, EPROM, and static pressure, pressure difference, and temperature three sensitive components on the same chip.

(4) The intelligence of the sensor

The combination of the sensor and the microcomputer forms an intelligent sensor with a microprocessor, which has the functions of detection and information processing and autonomous decision-making, and also has the functions of memory, storage, analysis, statistical processing and self-diagnosis, self-calibration and self-adaptation, and can carry out long-distance wireless communication. The smart sensor also has a configuration function and can be used flexibly. A variety of modular hardware and software can be set up in the smart sensor system, and the user can issue instructions through the microprocessor to change the combination of hardware and software modules of the smart sensor to complete different measurement functions.

(5) Biosensors and bionic sensors

Most of the sensors developed now are physical sensors, and chemical sensors and biological sensors should be actively developed in the future, especially the development of intelligent robot technology, and various sensors that simulate human sensory organs need to be developed, such as robot human sensory sensors, tactile sensors, taste sensors, etc. The creatures of nature are the models for human beings to learn from. Bionics is not only limited to the simulation of motion, but also the sensory organs of living things are worth studying and learning. For example, the smell of dogs (its sensitive threshold is 107 times that of humans), the vision of birds (its ability is tens of times that of humans), the hearing of bats, moths, and dolphins (active biological radar—ultrasonic sensors), and the proximity sense of snakes (equivalent to infrared temperature sensors with a resolution of 0.001 °C), all of which may be the breakthrough focus of future sensors.

(6) Image of the sensor

Modern sensors are no longer limited to the measurement of a point, but began to study one-dimensional, two-dimensional or even three-dimensional space measurement problems. Two-dimensional image sensors have been successfully developed, such as MOS type, CCD type, CID type all-solid camera devices.

(7) Wireless and networking

Wireless sensor network is a brand-new information acquisition and processing technology developed on the basis of sensor technology, computer technology and communication technology. It is an application-oriented network system close to the objective physical world, and has been widely used in the military field, precision agriculture, risk monitoring, environmental monitoring, intelligent transportation and many other fields [16]. With the development of sensor and network technology, the flexibility, dynamics and wireless communication of wireless sensor and wireless sensor network have broad development prospects.

(8) Low power consumption

Low-power sensors are one of the factors plaguing iot applications, and in addition to miniaturization, low-power sensors are also an important development direction. Taking environmental monitoring as an example, the sensors in the monitoring network are mostly in rivers or reservoirs, and are generally powered by solar energy or batteries, which is of great significance to reduce power consumption. For wireless sensors, the transmission power also needs to be reduced.

4.7 Summary of This Chapter

This chapter first introduces the concept, characteristics, classification and application of sensors, on this basis, focuses on the role and classification of intelligent manufacturing equipment sensors, intelligent sensor network and intelligent manufacturing equipment sensor selection and system design. Taking industrial robot as a typical representative of intelligent manufacturing equipment, the perception system of industrial robot is introduced. Finally, the development trend of intelligent manufacturing equipment sensor is briefly described.

References

1. Zhao Wei, Zhang Wenyin. Intelligent Logistics and Sensing Technology[M]. Beijing: Electronic Industry Press, 2016.
2. Gao Cheng. Sensors and Detection Technology[M]. Beijing: China Machine Press, 2015.
3. Wu Jianping, Peng Ying, Qin Zhangjian. Principle and Application of Sensors[M]. Beijing: China Machine Press, 2017.
4. Tero Karvinen, Kimmo Karvinen, Ville Valtokari. Comprehensive Guide to Sensor Practice[M]. Translated by Yu Xinlong, Li Ze. Beijing: Posts & Telecom Press, 2016.
5. Jacob Fraden. Handbook of Modern Sensors: Principles, Design and Applications[M]. Translated by Song Ping, Sui Li, Pan Zhiqiang, et al. Beijing: China Machine Press, 2019.
6. ON Semiconductor. Intelligent Passive Sensors Boost the Development of Intelligent Vehicles[J]. Automobile Technologist, 2016(11):69–71.
7. Michael J. McGrath, Cliodhna Ni Scanaill. Intelligent Sensors: Key Applications in Medical, Health and Environment[M]. Translated by Hu Ning, Wang Jun, Wang Ping. Beijing: China Machine Press, 2017.
8. Reza Ghodssi, Pinyen Lin. MEMS Materials and Processes Handbook[M]. Translated by Huang Anqing. Nanjing: Southeast University Press, 2014.
9. Wang Shuhua. Current Situation and Application of MEMS Sensors[J]. Micronanoelectronic Technology, 2011, 48(08):516–522.
10. Liu Yue, Bao Rongrong, Tao Juan, et al. Research Progress of Tactile Sensors and Their Applications in Intelligent Systems[J]. Science Bulletin, 2020, 65(01):70–88.

11. Jing Bo, Zhang Jie, Sun Yong. Intelligent Network Sensors and Wireless Sensor Networks[M]. Beijing: National Defense Industry Press, 2011.
12. Wang Xiaoqiang, Ouyang Jun, Huang Ninglin. Design and Implementation of ZigBee Wireless Sensor Networks[M]. Beijing: Chemical Industry Press, 2012.
13. Liu Jianhua, Sun Liansheng, Zhang Xu, et al. Connotation and Key Issues of 3D Digital Design and Manufacturing Technology[J]. Computer Integrated Manufacturing Systems, 2014, 20(03):494–504.
14. Gerard Meijer, Michiel Pertijs, Kofi Makinwa. New Technologies and Applications of Intelligent Sensor Systems[M]. Translated by Jing Xiangmeng, Ming Anjie, Liu Fengman, et al. Beijing: China Machine Press, 2018.
15. Fan Shangchun. Sensor Technology and Application[M]. Beijing: Beihang University Press, 2010.
16. Xu Kejun. Sensors and Detection Technology[M]. Beijing: Electronic Industry Press, 2016.

Chapter 5
Intelligent Manufacturing Equipment Control System Design

Intelligent control of intelligent manufacturing equipment is the development trend of control system. With the continuous development of network technology, various types of terminals with environmental awareness and computing modes based on network technology have promoted the Internet of Things to be applied more and more widely in the industrial field, and has been continuously integrated into all aspects of industrial production, elevating traditional industry to a new stage of intelligent industry. Among them, the most important applications are production process detection, real-time parameter acquisition, production equipment monitoring, material consumption monitoring, so as to achieve intelligent control of the production process. Whether the control system can reach the predetermined requirements is related to the success or failure of the system. The requirements of control systems for different purposes are different, which can generally be summarized as the stability, accuracy and rapid response of the system [1].

5.1 Overview of Intelligent Control

Intelligent control is a control mode with intelligent information processing, intelligent information feedback and intelligent control decision, which is the advanced stage of the development of control theory. It is mainly used to solve the control problems of complex systems that are difficult to be solved by traditional methods. The main characteristics of intelligent control research objects are uncertain mathematical models, high nonlinearity and complex task requirements. Intelligent control is based on control theory, computer science, artificial intelligence, operations research and other disciplines, and extends related theories and technologies, among which fuzzy logic, neural network, expert system, genetic algorithm and other theories are applied more, as well as adaptive control, self-organizing control and self-learning control and other technologies.

M. Xu et al., *Design and Case of Intelligent Electromechanical Equipment System*, Advanced and Intelligent Manufacturing in China,
https://doi.org/10.1007/978-981-95-3833-1_5

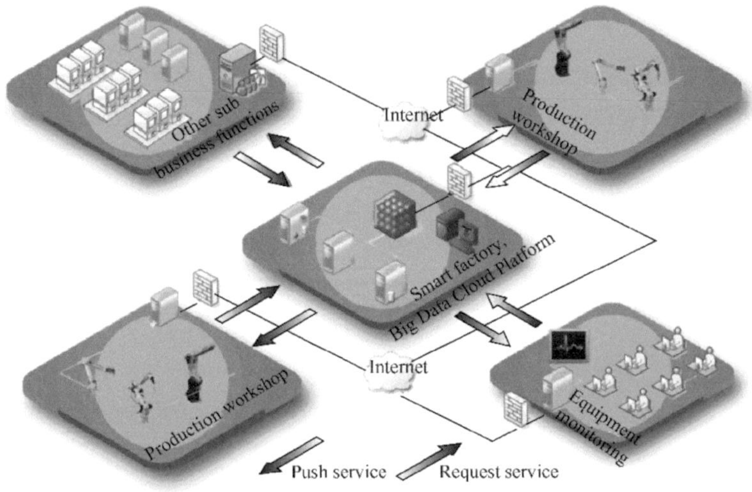

Fig. 5.1 Intelligent control of equipment

Intelligent control is widely used in machinery manufacturing industry. In modern advanced manufacturing systems, it is necessary to rely on those incomplete and inaccurate data to solve difficult or unpredictable situations, intelligent control provides some effective solutions to solve this problem. The intelligent control of the device is shown in Fig. 5.1.

The basic system of automatic control construction of intelligent manufacturing equipment in automobile manufacturing enterprises is composed of data service layer, IoT perception layer, platform service layer, etc. Through the IoT sensing layer, various intelligent devices such as robots, I/O devices and sensors can be accessed, and the data of intelligent manufacturing equipment can be analyzed into platform data and sent to the platform service layer. The platform service layer processes the equipment data and sends it to the data service layer. The data services layer is responsible for big data analysis and data storage. Platform services are the "brain" of the entire platform, responsible for all equipment management, data management, communication management, rights management, etc., and can release platform services in standard communication protocols to support the collaborative call of third-party systems. The iot awareness layer supports the access and control of all devices. The application of intelligent control in automobile manufacturing enterprises lays the foundation for the continuous expansion of intelligent manufacturing in automobile manufacturing enterprises in the future.

The purpose of intelligent control is to analyze and synthesize the system from the perspective of system function and overall optimization to achieve predetermined goals. The intelligent control system can realize the overall self-optimization and has the ability of self-adaptation, self-organization, self-learning and self-coordination. The core of intelligent control is high-level control, which can be represented by knowledge of non-mathematical generalized model and mathematical hybrid control process, using open closed-loop control and qualitative decision and quantitative control of the combination of multi-modal control, the complex system (such as nonlinear, fast time varying, complex multi-variable, environmental disturbance, etc.) effective global control, to achieve generalized problem solving, and has a strong The ability to tolerate fault. The intelligent control system has enough knowledge about the control strategy, the controlled object and the environment, and the ability to use these knowledge, and has the ability to compensate and self-repair and judge and make decisions.

The main difference between intelligent control and traditional control is that the traditional control method must rely on the model of the controlled object, while intelligent control can solve the control problem of the unmodeled system. The main methods of intelligent control technology include fuzzy control, knowledge-based expert control, neural network control and integrated intelligent control. The commonly used optimization algorithms are genetic algorithm, ant colony algorithm, immune algorithm and so on.

(1) Fuzzy Control

Fuzzy logic control, referred to as fuzzy control, is a computer digital control technology based on fuzzy set theory, fuzzy language variables and fuzzy logic reasoning. In 1974, E. H. Murdani of Britain composed the fuzzy controller for the first time according to the fuzzy control statement, and applied it to the control of boilers and steam engines, and achieved success. This pioneering work marks the birth of fuzzy cybernetics. Fuzzy control, which belongs to the category of intelligent control, is essentially a nonlinear control. Its characteristics are not only systematic theory, but also a lot of practical application background. Fuzzy control has been rapidly and widely applied in recent years.

Fuzzy control has made great progress both theoretically and technically and has become an important branch in the field of automatic control. Its typical application involves many aspects of production and life, for example, there are fuzzy washing machines and air conditioners in household electrical equipment; In the field of industrial control, there are water purification treatment, chemical reaction kettle, etc. In the special system and other aspects, there are subway parking, car driving, elevator, escalator, fuzzy control of robots, etc. [2].

(2) Expert Control

Expert system is a control system that uses expert knowledge to describe specialized or difficult problems. It is a program system with a large amount of expertise and experience. It applies artificial intelligence technology and computer technology to reason and judge according to the knowledge and experience provided by one or more experts in a certain field, simulates the decision-making process of human experts, and combines the theory and technology of expert system with that of other experts Control theory and technology are combined to emulate the experience of experts to achieve an intelligent control of system control in order to solve complex problems that require human experts to deal with well. In short, an expert system is a computer program system that simulates human experts to solve domain problems.

The expert control subject is composed of knowledge base and reasoning mechanism. By acquiring and organizing knowledge, the expert control object can be controlled by selecting appropriate rules according to certain strategy. Expert control can flexibly select the control rate, high flexibility; By adjusting the parameters of the controller, it can adapt to the changes of the object characteristics and the environment. Through expert rules, the system can work reliably in the case of nonlinear and large deviation, and has strong robustness. At present, expert system has been widely used in fault diagnosis, industrial design and process control, which provides a new method to solve industrial control problems and is an important technology to realize industrial process control [3].

(3) Neural Network Control

Neural network is a non-programmed, adaptive, brain-style information processing technology. It simulates the activity of human brain neurons, uses the connection and weight distribution between neurons to represent specific information, obtains a parallel distributed information processing function through network transformation and dynamic behavior, and imitates the information processing function of human brain nervous system in different degrees and levels, carries on self-learning by constantly modifying the weight of the connection, and advances on the basis of approximation theory Neural network modeling is carried out, and intelligent control is realized by direct self-correcting control, indirect self-correcting control and neural network predictive control. Due to its unique model structure and inherent nonlinear simulation capabilities, as well as highly adaptive and fault-tolerant characteristics, and with large-scale parallel, distributed storage and processing, self-organization, self-adaptation and self-learning capabilities, it is especially suitable for dealing with imprecise and fuzzy information processing problems that need to consider many factors and conditions at the same time, and has been widely used in control systems.

(4) Learning Control

Learning control refers to a technology that relies on the learning function of the control system itself to understand the characteristics of the control object and the external environment, and changes its own characteristics accordingly to improve the control performance of the system. It has certain abilities of recognition, judgment, memory and self-adjustment, mainly including genetic algorithm learning control and iterative learning control.

Genetic algorithm is a kind of search and optimization algorithm that simulates natural selection and genetic mechanism. It is designed and proposed according to the law of organism evolution in nature. Genetic algorithm is a computational model simulating the biological evolution process of natural selection and genetic mechanism of Darwinian biological evolution. It is a method to search for the optimal solution by simulating the natural evolution process, and can obtain better optimization results quickly when solving complex combinatorial optimization problems. As an optimization search algorithm, genetic algorithm has been widely used in combinatorial optimization, machine learning, signal processing, adaptive control and artificial life.

Iterative learning control is an important branch of learning control, which imitates the method of human learning, constantly repeats a control attempt of the same trajectory, revises the control law accordingly, and learns certain skills from experience through repeated training to obtain very good control effects. Iterative learning control can deal with dynamic systems with high uncertainty in a very simple way with strong adaptability and does not depend on the precise mathematical model of the dynamic system. It is an algorithm that iteratively generates optimized input signals and makes the system output approximate to the ideal value as much as possible. It is of great significance for the processing of nonlinear, complex, difficult to model and high-precision trajectory control problems.

5.2 Classification of Control Systems for Intelligent Manufacturing Equipment [4]

Intelligent control system is an automatic control technology that can drive intelligent machines to achieve control goals without human intervention, generally including hierarchical control system, fuzzy control system, neural network control system, expert control system and learning control system. It can be used separately according to the actual situation, or can be comprehensively applied to an actual intelligent control system or device to establish a hybrid or integrated intelligent control system. Figure 5.2 shows the general structure of the intelligent control system.

5.2.1 Hierarchical Control System

Hierarchical control system, proposed by Saridis et al., Purdue University, is an intelligent control theory based on adaptive control and self-organizing control system, which is mainly composed of three control levels. According to the level of intelligent control, it is divided into organization level, coordination level and execution level, and these three levels follow the principle of "accuracy increases with the reduction of intelligence". Hierarchical control system mainly includes two types of multi-level intelligent control theory based on knowledge/analysis and hierarchical hierarchical intelligent system theory whose accuracy decreases with the improvement of intelligence. The functional structure of hierarchical hierarchical control system is shown in Fig. 5.3.

(1) Organizational Level

The highest level of hierarchical intelligent control system is the organizational level, which is the brain of intelligent control, representing the dominant idea of the control system, with the ability to organize, learn and make decisions, the control function

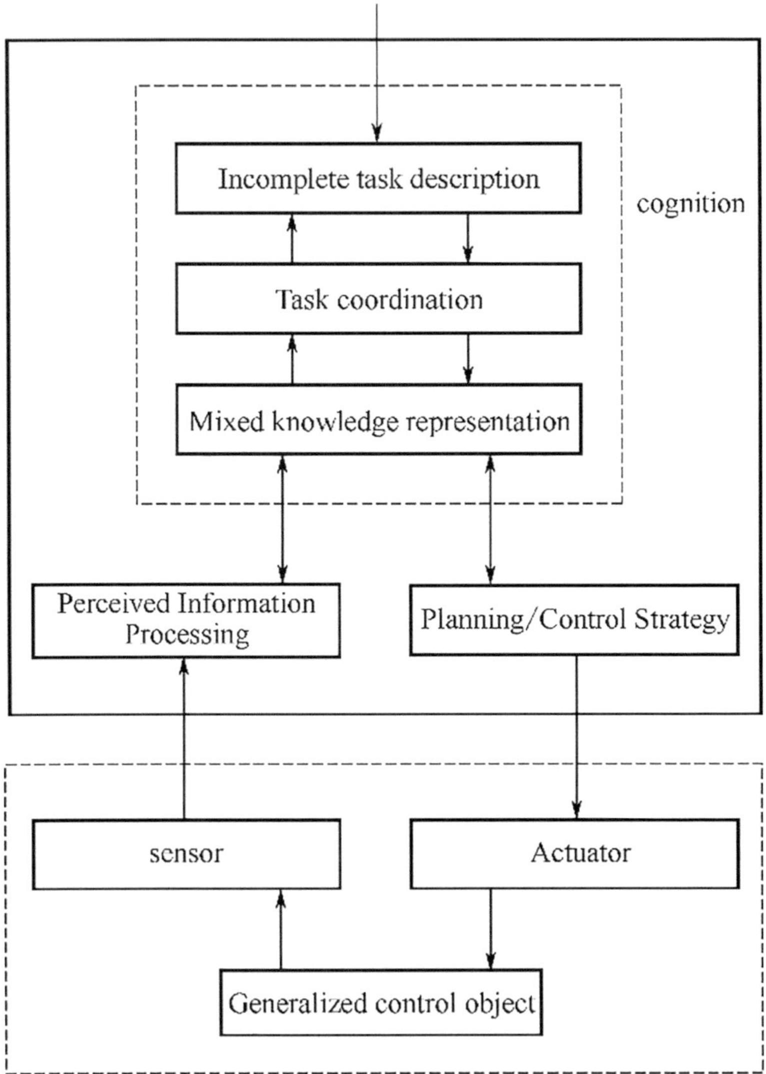

Fig. 5.2 General structure of intelligent control system

of executing the highest decision, and the real-time monitoring and guidance of all behaviors at the coordination level and the executive level through the human–machine interface. It can analyze input statements, identify control situations, organize tasks and propose appropriate task forms with a general understanding of task execution details. The organizational level requires low precision and high intelligent decision-making and learning ability.

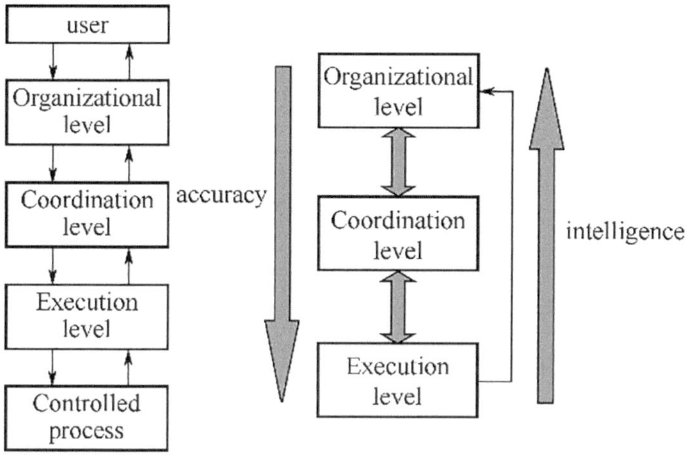

Fig. 5.3 Hierarchical intelligent control system structure

(2) Coordination Level

The coordination level is the interface between the organization level and the executive level, involving the decision-making mode and its representation, and using artificial intelligence and operations research methods to achieve control. It is the intermediate level of the control system, the main task is to coordinate the control function of each controller or coordinate the execution of each sub-task.

(3) Executive Level

The executive level is the lowest level of the intelligent control system, which executes the instructions of the organization level and the coordination level, and performs the control function of the relevant process. The executive level requires higher precision and lower intelligence.

5.2.2 Fuzzy Control System

The traditional control theory has good control ability for definite system, and has been widely used in production. For systems that are too complex or difficult to describe accurately, the traditional control methods often can not get good control effect. Fuzzy control system can realize fuzzy control, which is an automatic control system mainly composed of fuzzy controller, controlled object, detection module and feedback part. In 1965, Professor L. A. Zadeh of the University of California first proposed the fuzzy set theory, which used the concept of "membership function" to describe the fuzziness of things quantitatively, and put forward the concept of fuzzy

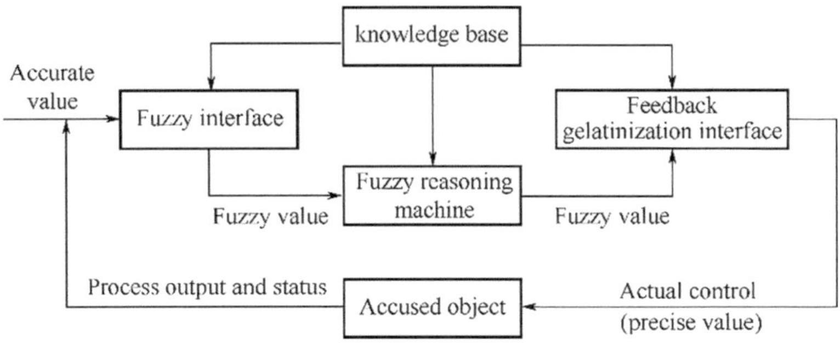

Fig. 5.4 Basic functional structure of fuzzy control unit

set. The idea of fuzzy set reflects the objective uncertainty in the real world and the subjective uncertainty in people's cognition.

Fuzzy control uses control laws to describe the relationship between system variables, provides a new mechanism for describing control laws based on natural language rules, and provides an alternative method to improve nonlinear controllers, which are generally used to control devices that contain uncertainties and are difficult to deal with with traditional nonlinear theory. Fuzzy control belongs to the category of intelligent control and is an important and effective form to realize intelligent control. It takes fuzzy sets, fuzzy language variables and fuzzy reasoning as its theoretical basis, and priori knowledge and expert experience as its control rules. Its basic idea is to use machines to simulate human control of the system, and use fuzzy controller approximate reasoning and other means on the basis of the fuzzy model of the controlled object Realize system control. Fuzzy control technology can be used in any situation where it is impossible or difficult to establish a mathematical model.

Fuzzy control simplifies the complexity of system design, does not depend on the precise mathematical model of the controlled object, and does not need to establish a complete mathematical model of the controlled object, especially for the control of nonlinear, time-varying, hysteretic and incomplete model systems. Fuzzy control uses linguistic fuzzy variables instead of numerical values to describe the system.

The fuzzy control unit consists of four functional modules: fuzzification, rule base, fuzzy reasoning and clarity. The basic functional structure is shown in Fig. 5.4. The fuzzy module realizes the fuzzy division of system variable domain and the fuzzy processing of clear input values. The rule base is used to store the control rules and system parameters based on language variables. Fuzzy inference is a nonlinear mapping relationship from input space to output space.

5.2.3 Neural Network Control System

Neural network refers to a network composed of a large number of artificial neurons that are similar to the nerve cells of the biological nervous system or a large number of processing units like biological neurons in parallel, with some intelligent and humanoid control functions, is a basically model-independent control method, suitable for uncertain or highly nonlinear control objects, and has strong self-adaptation And self-learning function. Neural network is an important branch field of intelligent control, which has made great progress in intelligent control, pattern recognition, computer vision, adaptive filtering and signal processing, nonlinear optimization, automatic target recognition, continuous speech recognition, sonar signal processing, knowledge processing, intelligent sensing technology, robotics, biomedical engineering and so on.

The basic structure of the biological nervous system is the neuron, also known as the nerve cell, which is the basic unit that processes the mutual information transmission between various parts of the human body. Neurons are composed of cell bodies, axons that connect to other neurons, and dendrites, other shorter branches that protrude outward (Fig. 5.5). The human brain typically has 1010 to 1011 neurons. The function of dendrites is to receive excitation from other neurons. All the signals received by the neuron cell body are simply processed and then output by the axon. The parts of the dendrites of a neuron that connect with the nerve endings of another neuron are called synapses.

A neural network consists of many interconnected processing units. These processing units are usually arranged in linear groups called layers. Each processing unit has many inputs, and each input has a corresponding weight associated with it. The processing unit will weigh the input sum, and through the transfer function to

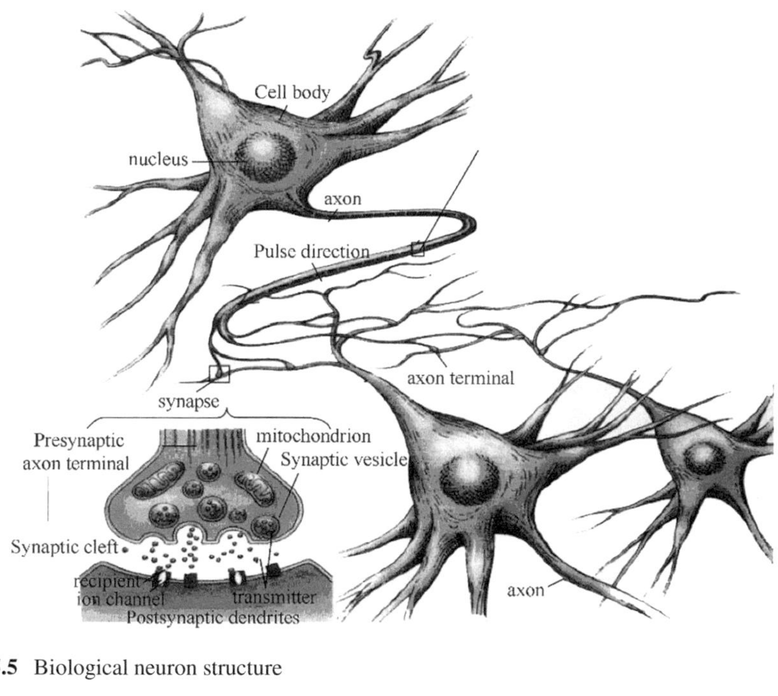

Fig. 5.5 Biological neuron structure

obtain the output, and then pass to the next layer of neurons. At present, many neuron models have been proposed, among which the earliest and most influential one is M-P model, which was first proposed by psychologist McCulloch and mathematician Pitts in 1943 on the basis of analyzing and summarizing the basic characteristics of neurons. It is the basis of most neural network models.

Neural network uses bionics to study advanced information processing in human brain and intelligent system. Control based on neural network can be regarded as a pattern recognition problem about controlled state, output or a performance evaluation function change signal. These signals are mapped into control signals by neural network. Even if the input information of neural network is insufficient, the pattern can be recognized quickly and the appropriate control signals can be generated. The control effect is reflected by the evaluation function of the system, which is a kind of variable signal input neural network, and serves as the learning algorithm or learning criterion of the neural network.

Techniques in the field of automatic control include system modeling and identification, parameter setting, pole assignment, internal model control, optimization, design, predictive control, optimal control, filtering and predictive fault-tolerant control. Applications for automatic control pattern recognition include handwritten characters, vehicle license plates, fingerprints, and voice recognition, as well as automatic target recognition, target tracking, robot sensor image recognition, and seismic signal identification. Applications in image processing include image edge monitoring, image segmentation, image compression and image recovery. Applications in the field of robotics include orbit control, eye-hand system control of operating robots, robot fault diagnosis and removal, intelligent adaptive mobile robot navigation, vision system control. Applications in the field of medical health include breast cancer cell analysis, optimization of transplant times, hospital quality improvement, etc.

Neural network control has been actively applied in the field of robotics. Robot is a nonlinear and uncertain system, the core is the robot control system, robot intelligent control is the forefront of the field of robot control research in recent years, has made a lot of achievements. The main purpose of the robot trajectory tracking control system is to make the robot's position, speed and other state variables track the given ideal trajectory by giving the driving torque of each joint. When the structure and mechanical parameters of the robot are determined, its dynamic characteristics are described by the mathematical model, and the design method provided by the automatic control theory can be used to design the robot controller based on the mathematical model.

5.2.4 *Expert Control System*

Experts are those who are very familiar with solving specialized problems, have a wealth of experience as well as detailed expertise in dealing with problems. A system designed based on the principle of expert control, which applies the concept and technology of expert system and simulates the control knowledge and experience of human experts, is called expert control system. Expert control system contains a large amount of expert level knowledge and experience in a certain field, and can use the knowledge of human experts and the experience of solving problems to deal with high-level problems in the field. The expert control system combines the theory and technology of the expert system with the control method and technology, and simulates the intelligence of the expert to realize the control of the system in the unknown environment. It has the characteristics of enlightening, transparency, flexibility, symbol operation, uncertainty reasoning, etc., and has been widely used in fault diagnosis and various industrial process control.

5.2.5 Humanoid Intelligent Control

The common point of intelligent control method research is to use the micro or macro structure function of human brain into the control system, so that the engineering control system has some kind of "human-like" or "human-like" intelligence, and transplant it into the engineering control system. The goal of the research on humanoid intelligent control is not the controlled object, but the controller itself, which directly detects, identifies, summarizes and summarizes the control experience, skills and various intuitive reasoning logic of the human, so that the structure and function of the controller can better simulate the functional behavior of the control expert from the macro level. The basic idea is to use the computer to simulate the human control behavior function in the control process, to identify and use the characteristic information provided by the dynamic process of the control system to the maximum extent, to inspire and intuitive reasoning, so as to realize the effective control of the object that lacks an accurate model. Intelligent control based on imitation of human intelligence makes up for the deficiency of traditional control system based on mathematical model, and shows strong vitality in industrial control.

Figure 5.6 shows the human-like hierarchical intelligent control system. The idea of hierarchical control (organization level, coordination level and execution level) is applied to the controller design, and hierarchical hierarchical control is carried out according to the degree of the controlled quantity deviation and the deviation change rate. Each hierarchical control strategy adopts human-like intelligent control scheme to realize the automatic, stable and optimal operation of the industrial process control system.

When the deviation of the controlled process system is large and the load changes widely, the control system adopts the expert control strategy of human-like operation. When the deviation e (t) and the deviation variation Δe (t) are slightly larger, fuzzy control is selected. When the deviation e (t) and Δe (t) are both small, self-tuning PID control and self-optimizing learning control are used.

The inner layer of the hierarchical intelligent control algorithm is self-tuning PID control, the second layer is fuzzy control (FC), and the outer layer is expert control

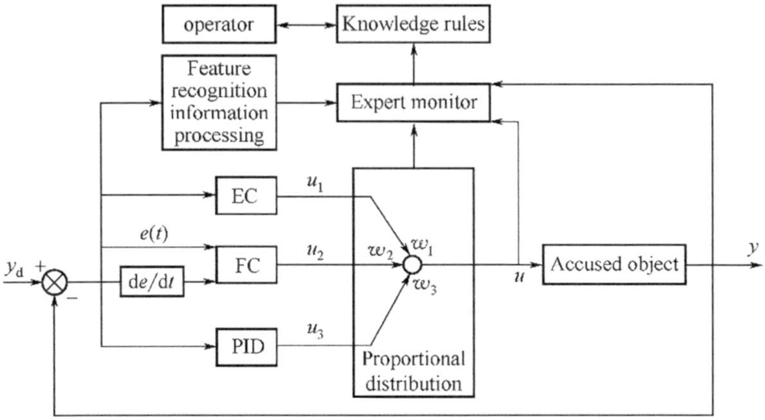

Fig. 5.6 Humanoid hierarchical intelligent control system

(EC). The working state of the three modules can be switched between each other under the monitoring of the expert control system.

5.2.6 Integrated Intelligent Control System

The intelligent control system which combines several intelligent control methods and mechanisms together is called integrated intelligent control system. The following are several examples to illustrate the integrated intelligent control system.

(1) Fusion of Fuzzy Logic Control and Neural Network

Fuzzy system and neural network control are two branches in the field of intelligent control. Both show strong fault tolerance in the process of information processing, and have their own basic characteristics and application ranges, as shown in Table 5.1. Fuzzy system is a kind of system designed after people's fuzzy logic thinking method. It is good at expressing empirical knowledge, can handle information with fuzzy elements, and allows imprecision of numerical quantities in the working process. However, its design parameters such as rule set and membership function are selected by experience, and it is difficult to design and adjust automatically, which is the main shortcoming of fuzzy system. The fault tolerance of neural network in the process of computing and processing information depends on the structural characteristics of the network itself. The fault-tolerant ability of human brain comes from the fuzziness of thinking method and the structural characteristics of brain itself.

Fuzzy logic control, neural network and expert control are three typical intelligent control methods. Usually, expert systems are built on the basis of expert experience, not on the operational data generated by industrial processes, and the inaccuracy and uncertainty of general complex systems are difficult to grasp even by domain

Table 5.1 Comparison between fuzzy systems and neural networks

Comparing objects	Fuzzy system	Neural network
Get knowledge	Expert experience	Algorithm example
Reasoning mechanism	Heuristic search	Parallel computing
Reasoning speed	Low	High
Fault tolerance	Low	High
Learning mechanism	Inductive formula	Automatically adjust weights
Natural language implementation	To make clear	Not obvious
Natural language flexibility	High	Low

experts, which makes the establishment of expert systems very difficult. As two typical intelligent control methods, fuzzy logic control and neural network control have their own advantages and disadvantages. The fusion of fuzzy logic and neural network—fuzzy neural network (Fuzzy Neural Network) is one of the hot spots in the research of intelligent control nowadays because it absorbs the advantages of fuzzy logic control and neural network control.

If we can use neural network to construct fuzzy system, we can use the learning method of neural network to automatically design and adjust the design parameters of fuzzy system according to the input and output samples, and realize the self-learning and self-adaptive function of fuzzy system. According to this idea, a fuzzy neural network system is developed. Fuzzy neural network is the product of the combination of fuzzy theory and neural network, which brings together the advantages of neural network and fuzzy theory, and integrates learning, association, recognition and information processing. A famous American scholar B. K Osko the ground-breaking work in this aspect, he published the book Neural Network and Fuzzy Systems systematically studies and summarizes the general principle of neural network and fuzzy system and method, research on neural network in the application of fuzzy system plays a big role. In recent years, fuzzy neural network has become one of the hotspots in the field of intelligent control and intelligent automation, and has achieved a lot of theoretical and application results, the more famous are fuzzy associative memory (FAM), fuzzy adaptive resonance theory (F-ART), fuzzy cognitive graph (FCM) and fuzzy multi-layer perceptron (FMLP).

There are roughly three ways of fusion of neural network and fuzzy technology:

① Neuro-fuzzy model. The model takes fuzzy logic control as the main body, uses neural network to realize the decision-making process of fuzzy logic control, and uses fuzzy logic control method as "sample" to train and learn the neural network off-line. The "sample" is the "teacher" of learning. After learning all the samples, the neural network is a fuzzy rule table, which has the function of self-learning and self-application Fig. 5.7 shows the structure of the neural-fuzzy model.

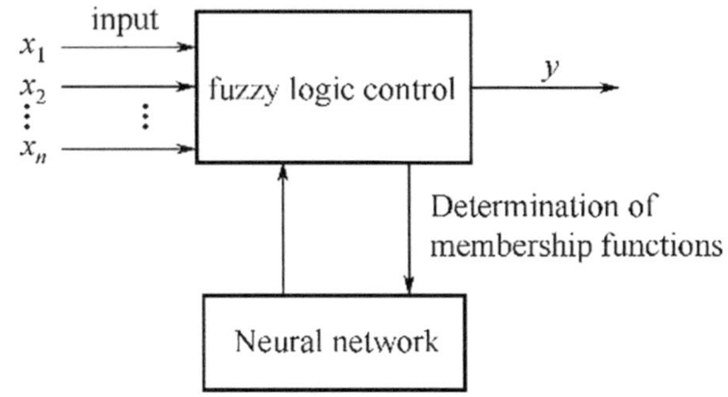

Fig. 5.7 Neural fuzzy model structure

② Fuzzy-neural model. The model takes the neural network as the main body, divides the input space into several different types of fuzzy inference combinations, makes fuzzy logic judgment on the system first, and takes the output of the fuzzy controller as the input of the neural network. The latter has the characteristics of self-learning intelligent control. Figure 5.8 shows the structure of the fuzzy-neural model.

③ Neuron-fuzzy model. In this model, neural network and fuzzy control directly process the input information according to the different properties of the input quantity, and directly act on the control object, which can give full play to their own control characteristics. The structure of the neuron-fuzzy model is shown in Fig. 5.9.

(2) Fuzzy Control System Based on Genetic Algorithm

Genetic algorithm (GA) is an innovative artificial optimization search algorithm based on natural selection and genetic mechanism, and according to the rules of

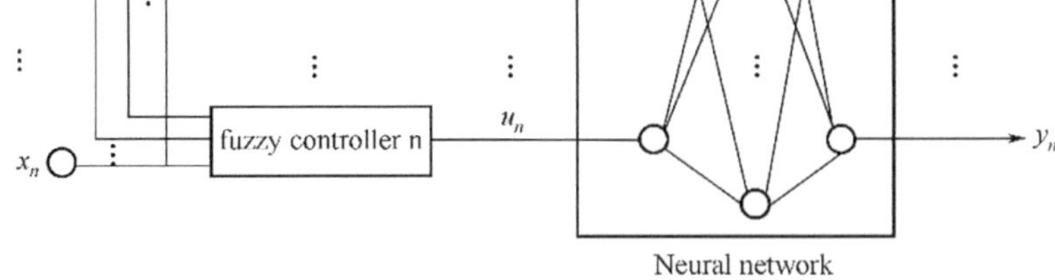

Fig. 5.8 Fuzzy neural model structure

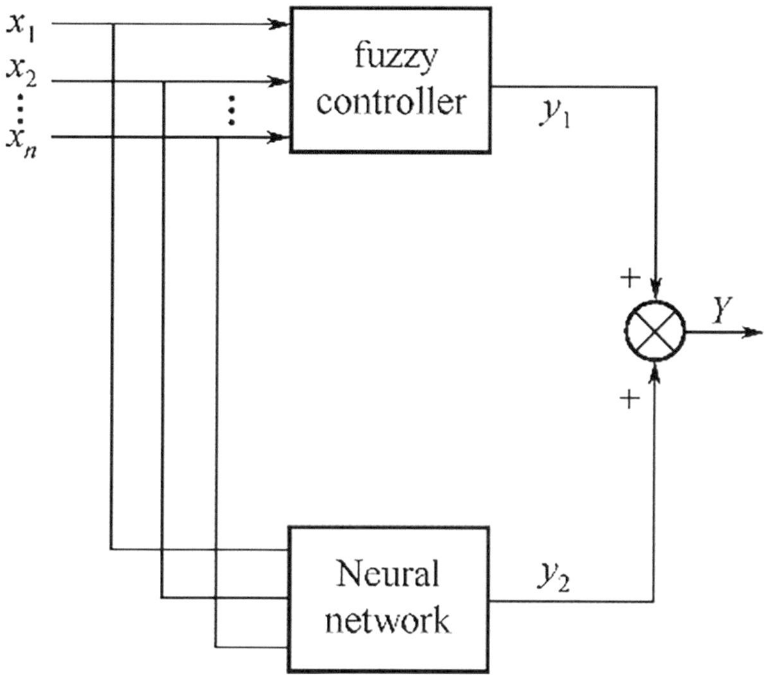

Fig. 5.9 Neuron fuzzy model structure

survival of the fittest and survival of the fittest. Robustness refers to the characteristics of survival ability through the coordination and balance between efficiency and function in different environments. The core of genetic algorithm is robustness.

Fuzzy control is an approximate reasoning method based on fuzzy set theory to simulate human brain activity. However, its control rules are invariable in the process of inference and can not adapt to the change of objects. Combining the optimization search technology of genetic algorithm with the fuzzy reasoning mechanism organically, the fuzzy reasoning rules can be changed according to the actual situation, thus giving the fuzzy controller the ability to acquire fuzzy reasoning knowledge automatically.

Figure 5.10 shows a fuzzy control system with genetic algorithm, where k is the fuzzy quantization factor.

(3) Expert Fuzzy System

Expert fuzzy control is a kind of intelligent control which combines expert system technology and fuzzy theory. It uses fuzzy logic, human experience knowledge and heuristic rules for solving control problems to construct control strategies. General expert fuzzy controller is a two-level coordination controller, which consists of basic control level and expert intelligent coordination level. The basic control level, as the

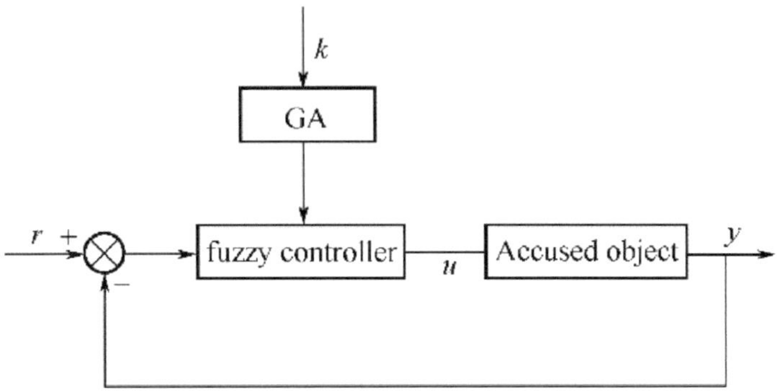

Fig. 5.10 Fuzzy control system with genetic algorithm

basic controller, forms a closed loop with the controlled process to complete real-time control; The expert intelligent coordination level monitors the performance of the control system online in real time, and adjusts the controller and parameters online according to the performance of the system, so as to effectively control the specific object. Learning system level online upgrade the knowledge base and database content of intelligent coordination level, so that the performance of the whole control system is gradually improved. Figure 5.11 shows the expert fuzzy control system.

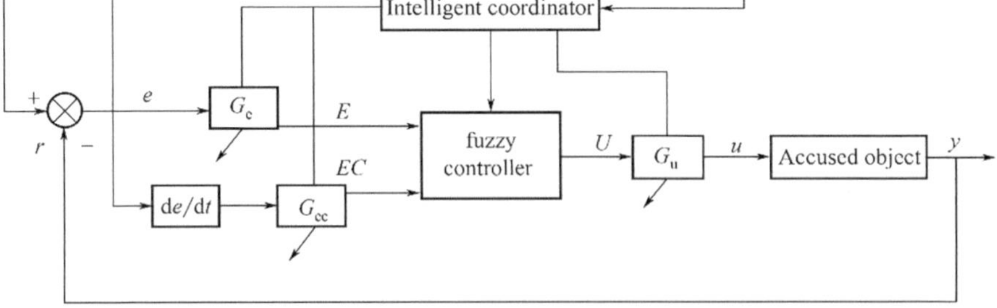

Fig. 5.11 Expert fuzzy control system

(4) Chaotic Fuzzy Control

Chaotic fuzzy control is to make the controlled chaotic system out of the chaotic state through the action of small control quantity, and achieve the expected periodic dynamic behavior, such as equilibrium state, periodic motion or quasi-periodic motion. The aim of chaos control is to eliminate bifurcation behavior and chaos in the description object. There is also a class of non-deterministic phenomena described by deterministic equations between deterministic phenomena and stochastic phenomena in nature, or called deterministic random phenomena—chaotic phenomena. The dynamic characteristics of neural networks in the human brain, brain neurons, deterministic nonlinear dynamic systems in industrial systems, etc., all show chaotic motion, which looks chaotic, but has a fine internal structure, which can attract and bind the system movement within a specific range. People's understanding of chaos control is that the chaotic system is controlled artificially and effectively through a certain method, so that it can develop to the actual state. Successfully suppress or eliminate chaos when it is harmful; When chaos is advantageous, chaos is used to produce the desired chaotic motion with certain characteristics, and even to produce some specific chaotic orbits. When the system is in a chaotic state, various outputs are produced by external control. The common characteristic of chaos control is to make use of the characteristics of chaos movement itself to achieve the purpose of control.

In industrial system, as a classical deterministic nonlinear dynamic system, there is chaos in general. How to suppress and eliminate chaos has aroused great attention of automation workers. Because chaos has randomness, ergodicity, regularity and non-periodicity hidden order, and because it is sensitive to initial conditions and parameter changes, people can build a neural network model of chaos dynamics from the chaos phenomenon in the human brain according to actual experience and rules. Such a model has intelligence itself, so it can carry out intelligent information processing and control better. Controlling and utilizing chaos has been applied in the fields of mechanics, communication, biology, medicine, chemical industry, machinery and ocean engineering. Figure 5.12 is a block diagram of a chaotic fuzzy control system, which outputs chaotic signals.

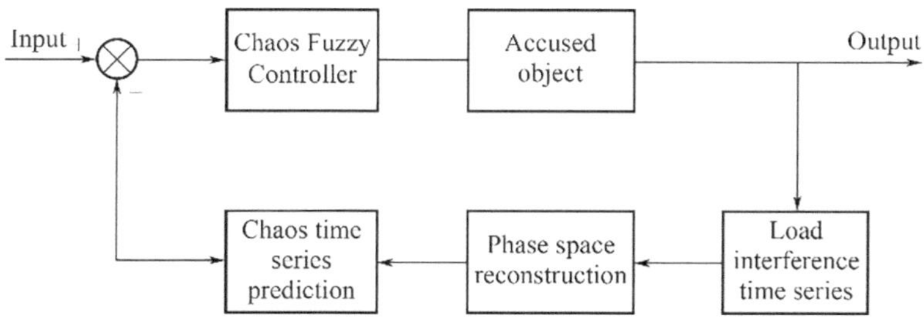

Fig. 5.12 Block diagram of chaotic fuzzy control system

5.3 Hardware Platform Design of Intelligent Manufacturing Equipment Control System

5.3.1 Overview

Intelligent control system includes two parts: hardware and software. In practical application, software and hardware should be closely combined to complete the work efficiently. The typical principle structure of intelligent control system consists of six parts, including actuators, sensors, perceptual information processing, planning and control, cognition, communication interface. The actuator is the output of the system and acts on external objects. An intelligent system can have many or even thousands of actuators, which must be coordinated in order to accomplish a given goal and task. The actuators are motors, positioners, valves, electromagnetic coils, transmitters, etc. Figure 5.13 shows the structure of the intelligent control system.

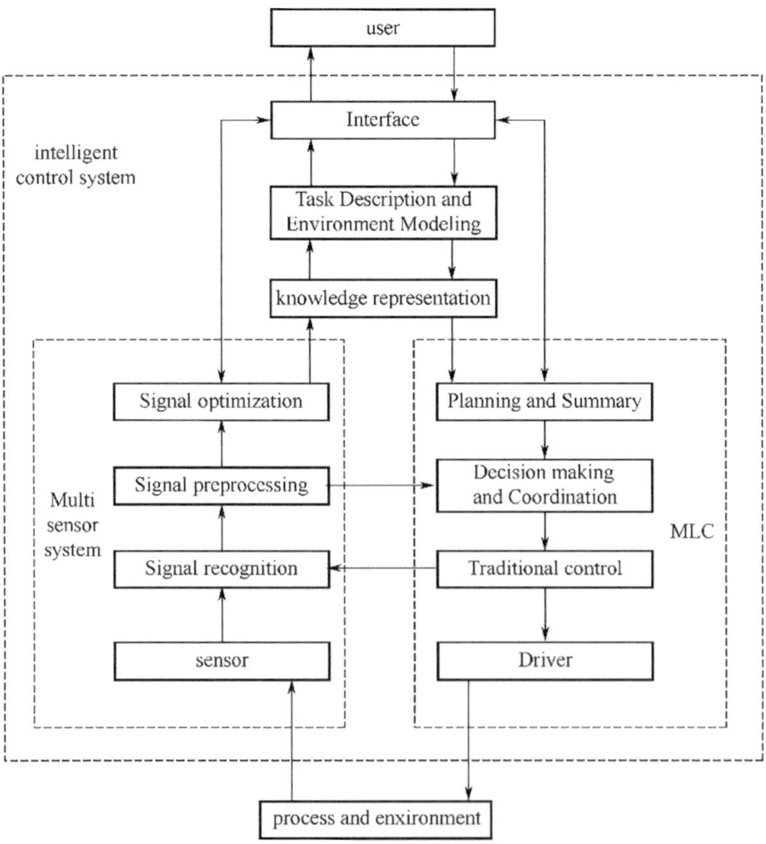

Fig. 5.13 Principle and structure of intelligent control system

The sensor is the sensory organ of intelligent manufacturing equipment, which is used to monitor the external environment and the state of the system itself, and provide input to the intelligent manufacturing equipment perception information processing unit. A wide variety of sensors, intelligent manufacturing equipment commonly used position sensors, force sensors, vision sensors, distance sensors, tactile sensors and so on.

The hardware structure of the small control system of intelligent manufacturing equipment is mainly composed of sensors, controllers and actuators. The sensor is used to collect information about the environment or its own working status, and pass this information to the controller. According to the control law written in advance, the controller processes the incoming information of the sensor and generates the control quantity to transmit to the actuator. The rich types and functions of sensors and the improvement of performance can provide comprehensive and accurate information to the controller, which greatly improves the working ability and efficiency of the entire control system. The controller is mainly a control circuit built around the micro control unit, such as single-chip microcomputer minimum system, FPGA digital circuit, etc. The actuator is mainly a variety of acousto-optic components, motors, steering gear and so on.

The medium and large control system mainly refers to the industrial control system, in the industrial control system, the hardware components and equipment of the small control system are difficult to meet the requirements of the complex industrial operation environment. Different from the sensors of small control systems, the information collected in the industry is mostly larger detection instruments, collecting information such as temperature, pressure, flow, level, composition and so on. Figure 5.14 shows the basic process control system structure of an industrial control system.

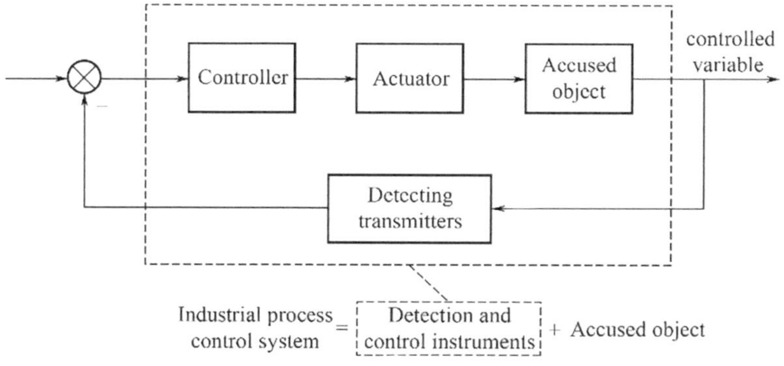

Fig. 5.14 Basic process control system structure

5.3.2 Common Control System Hardware

The core hardware of the control system is the main control board or controller. The main control board is generally composed of processing chips and peripherals, and is the core board used to process information and data and control system operation in embedded devices. The chips commonly responsible for processing information and data are microprocessors (MPU), microcontrollers (MCU), digital signal processors (DSP), etc., with storage units, external interfaces, external crystal oscillators, switching components, resistance capacitors, digital-to-analog/analog-to-digital converters, etc., which are used to exchange processing results with external devices. The occasions where the main control board is the core control hardware are mostly in embedded devices, and the commonly used embedded processors are:

① ARM processor. The main features are small size, low power consumption, low cost, high performance, 16/32 bit dual instruction set, and large market share;
② MIPS processor. The main features are high performance, wide positioning, 64 bit instruction set;
③ PowerPC processor. It is characterized by good scalability, high flexibility and wide application.
④ IntelAtom processor. It is characterized by low power consumption, small size and strong processing capacity.

Industrial controllers mainly include analog controllers, digital controllers and programmable logic controllers. Ddz-iii analog controller is a mainstream analog controller, which is composed of input circuit, PID operation circuit, output circuit, internal set circuit, indicating circuit, soft/hard hand operation circuit, etc., with higher accuracy, reliability and safety. There are many types of digital controller, taking the single loop programmable controller SLPC as an example, which is the representative model of YS-80 series, composed of CPU, ROM, RAM, D/A converter, input and output interface, communication interface, man–machine interface and so on. The biggest advantage of programmable logic controller (PLC) is programmable, which makes it more flexible, readable and modifiable, more reliable, its main structure is CPU, memory, input and output interface, etc., and the internal bus is used for data transmission.

Whether it is a small embedded control system or a medium or large industrial control system, sensors and detection instruments are indispensable to collect environmental or system status information such as temperature, pressure, flow, level, and composition. Temperature detection mainly includes thermocouple, thermal resistance, integrated sensor, etc. Pressure detection with elastic pressure gauge, electric pressure gauge, etc.; Flow detection has differential pressure type, rotor type, target type, elliptical gear type, eddy flow meter, etc. There are differential pressure type, capacitance type and ultrasonic type detection meter for level detection. The components are detected by combustible gas sensor, zirconia oxygen meter, gas chromatography analyzer, infrared gas analyzer, industrial acidity meter, etc. Other detection devices are photoresistor (light intensity detection), various cameras (image detection), sound sensitive components (sound detection).

In the embedded control system, the hardware acting as the execution link is mainly acousto-optic devices (such as LED lights, speakers, display lights), various small motors and steering gear; In the industrial control system, the hardware acting as the execution link is mainly control valves, large motors and so on.

5.3.3 Robot Control System

The robot control system is the brain of the robot, which controls all the actions of the robot, and is the key and core part of the robot system. A typical robot electrical control system is mainly composed of upper computer, motion controller, driver, actuator and feedback device.

The robot system completely adopts the full software form of PC. On the hardware platform of high-performance industrial PC and embedded PC, PLC and motion control functions can be realized through software programs to complete the logic control and motion control required by robot action. The real-time communication between PC and driver through high-speed industrial bus improves the production efficiency and flexibility of robot. This structure represents the development direction of robot control structure in the future [5, 6].

Pc-based control system configuration As shown in Fig. 5.15, the open motion control technology of the PC platform has the following features:

① Friendly man–machine interface. PC platform control system is more and more favored by users. Compared with the interface of single chip microcomputer and PLC scheme, PC (including display, keyboard, mouse, communication port, hard disk, floppy drive, etc.) has unparalleled input and output interaction ability. Of course, many embedded human–machine interfaces are also very friendly and portable, and have been widely used in the field of small and medium-sized control.

② Powerful. Because of the powerful computing power of PC and the advanced technology of motion control card, the motion control system based on PC can realize the advanced functions that single-chip microcomputer system and PLC system can not achieve. In particular, the performance of industrial computer is getting better and better, and it is suitable for various industrial environment control.

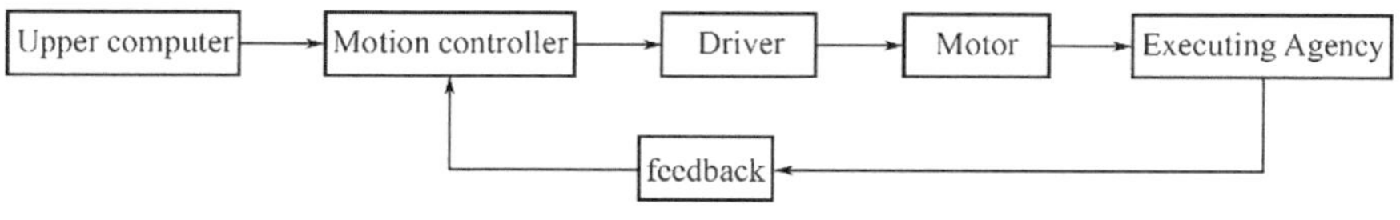

Fig. 5.15 Control system composition based on PC

③ Convenient development. Users can use VB, VC, C++ and other mainstream high-level programming languages to quickly develop human–machine interface, call mature and reliable rich motion function library, and quickly complete the development of large-scale control software. Developed software is easily portable to similar machines.

④ High cost performance. Due to the decline in the cost of PC, the cost of motion control card is also declining, with a high cost performance, so that the control system based on PC and motion control has a good comprehensive cost advantage in most motion control occasions.

The popularity of PC and the development of motion control card have met the requirements of standardization, flexibility and openness of new numerical control systems, which have been widely used in the development and transformation of automatic control systems of various industrial equipment, national defense equipment, intelligent medical devices and other equipment.

Motion control cards (Figs. 5.16 and 5.17) are widely used in various fields of equipment automation in the manufacturing industry. Motion control card is based on the computer bus, the use of high-performance microprocessor (such as DSP) and large-scale programmable devices to achieve multiple servo motor multi-axis coordinated control of A high-performance stepper/servo motor motion control card, including pulse output, pulse counting, digital input, digital output, D/A output and other functions, it can issue a continuous, high frequency pulse train By changing the frequency of the pulse emitted to control the speed of the motor, by changing the number of pulses emitted to control the position of the motor, its pulse output mode includes pulse/direction, pulse/pulse mode. The pulse count can be used for position feedback of the encoder to provide accurate position of the machine and correct errors generated during transmission. Digital input/output points can be used for limits, origin switches, etc. The library functions include S-shape, T-shape acceleration and deceleration, linear interpolation and arc interpolation, multi-axis linkage function, etc. The motion control card is generally a master–slave control structure with the PC mechanism: the PC is responsible for the management of the human–computer interaction interface and the real-time monitoring of the control system (for example, the management of the keyboard and mouse, the display of the system status, the planning of the motion trajectory, the transmission of control instructions, the monitoring of external signals, etc.); The control card completes all the details of motion control (including pulse and direction signal output, automatic speed up and down processing, origin and limit signal detection, etc.). Motion control cards are equipped with open function library for users to develop and construct the required control system under DOS or Windows system platform.

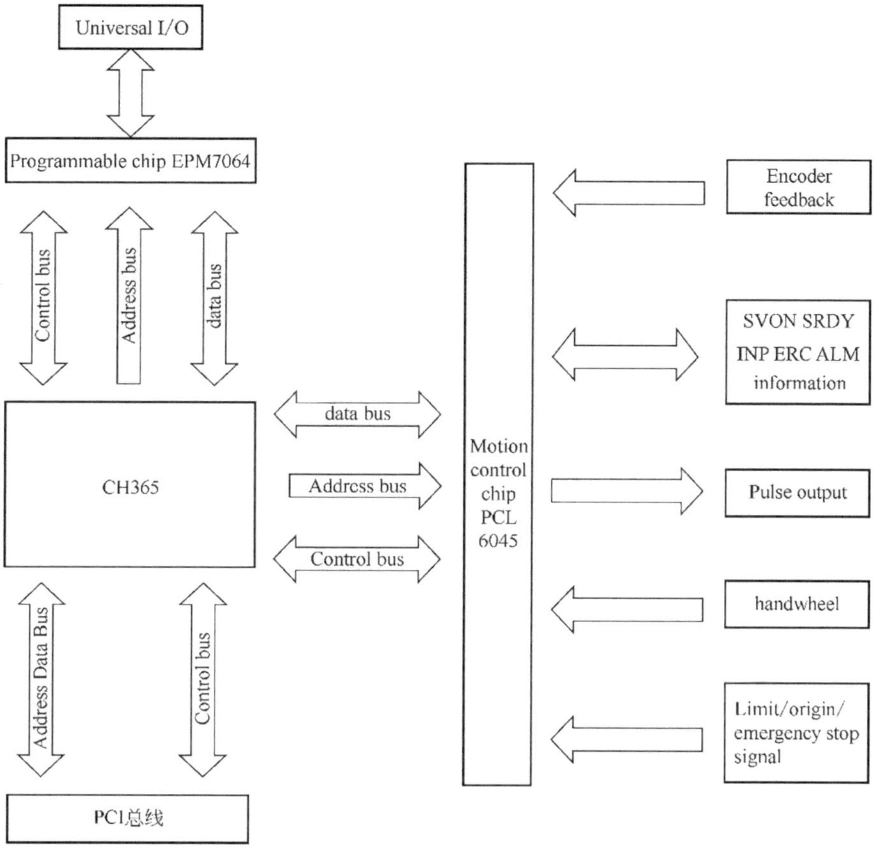

Fig. 5.16 System block diagram of motion control card

5.4 Software Design of Intelligent Manufacturing Equipment Control System

5.4.1 Overview

The controlled quantity of the control system is the physical quantity required to be controlled. The controlled quantity may be required to maintain a constant value during the production operation, such as temperature, pressure, liquid level, voltage, etc. It may also be required to operate according to a certain established law in production, such as flight trajectories, recording curves, etc. The control device is a mechanism that exerts control effect on the controlled object, and it can use different principles and methods to control the controlled object. The feedback control based on the feedback control principle is the most basic way. The real-time control information of the controlled object by the control device comes from the feedback information

Fig. 5.17 A certain model of multi-axis motion control card

of the controlled quantity, which is used to constantly correct the deviation between the controlled quantity and the input quantity to realize the controlled objectObject control.

The system is composed of various components with different structures. From the function of automatic control, a system must contain two parts of the controlled object and the control device, and the control device is composed of basic working components with certain functions. In different systems, components of completely different structures can have the same function. Therefore, according to the different controlled objects and components used, the automatic control system has various forms, but generally, it is composed of the given link, the measurement link, the comparison link, the operation and amplification link, and the execution link, as shown in Fig. 5.18.

(1) Given Link

The given link is the link that gives the input signal, which is used to determine the "target value" of the controlled object, and the given link can send a signal through various forms (power, non-power, digital, analog, etc.). For example, the input device of the CNC machine tool feed system is the given link.

(2) Measurement Process

The measurement step is used to measure the controlled object and convert the controlled object into another physical quantity (generally electricity) that is easy

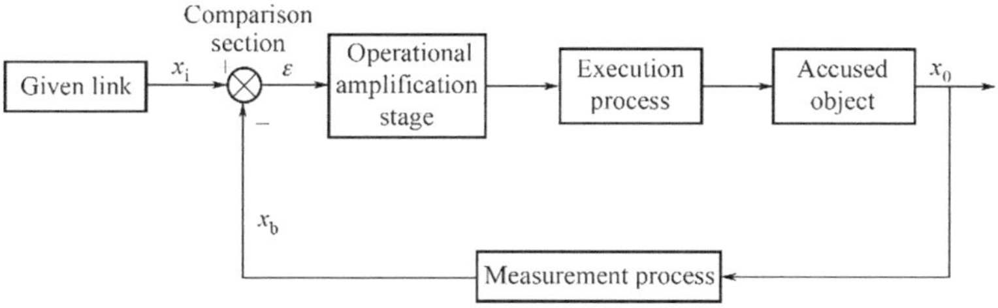

Fig. 5.18 Composition of automatic control system

to transmit. For example, a potentiometer converts a mechanical Angle into a voltage signal, a tachometer motor converts a speed into a voltage signal, and a grating measuring device converts a linear displacement into a digital signal. The aforementioned thermal element also belongs to this category.

(3) Comparison Stage

In this link, the input signal xi is compared with the feedback xb of the controlled variable xo issued by the measurement link, such as amplitude comparison, phase comparison, displacement comparison, etc., to obtain a low-power deviation signal $\varepsilon(\varepsilon = \text{xi} - \text{xb})$, and the deviation signal is the output of the comparison link.

(4) Operation and Amplification Process

In order to achieve control, it is necessary to correct the deviation signal, and then carry out power amplification in order to promote the execution link, commonly used amplifiers are current amplification, electrical amplification, hydraulic amplification, etc.

(5) Execution Phase

The execution link receives the control signal sent by the amplification link and drives the controlled object to operate according to the expected law. The execution link is generally an active power amplifier device, and energy conversion is carried out during the work. For example, the electric energy is converted into mechanical energy through a DC motor to drive the controlled object to do mechanical motion.

The given link, the measurement link, the comparison link, the operation and the amplification link together with the execution link, constitute the control part of the control system, the purpose is to control the controlled object.

In addition to the hardware configuration with good performance, the control system should also have fully functional software to realize the functions of real-time monitoring, numerical calculation, data processing and various control algorithms. The software should be real-time and respond quickly to system-specific inputs. The control software should also have the ability to concurrently process information, support multi-task parallel operation, be able to share resources and be able to effectively communicate in real time. In addition, it also requires a good human–machine interface, able to respond to accidental events in a timely manner, and make correct judgments and processing.

5.4.2 Common Software Design Methods for Intelligent Control Systems

Software design methods commonly used in control systems include structured programming, top-down programming and modular programming. These three design methods are often integrated together to ensure the fast realization of software through top-down, step-by-step refinement, modular design and structured coding.

(1) Structured Program Design

In the process of program design, the program flow chart is generally prepared first according to the function of the program, and then according to the program flow chart with VC++ and other high-level languages to write the program. When the program is large in scale and complex in structure, it is not easy to draw the flowchart of the program. Structured program design is the basic principle of detailed design which mainly focuses on module function and process design. Structured programming represents the relationship between program text and its corresponding process in a clear and understandable way, thus organizing the design and coding of programs. Structured programming is a subset of procedural programming that uses logical structures for written programs to make understanding and modification more efficient and easier.

The structured program design idea is clear, emphasizing the structure of the program, the software system is divided into several functional modules, each module is programmed separately according to the requirements, and then the modules are connected and combined to form the corresponding software system. The core idea of structured programming method is that "a module only has one entrance and one exit". Each module is connected through a sequential, selective, and cyclic control structure, and only one entry is allowed to be called by other modules. Because the modules are independent of each other, when designing one module, it will not be implicated by other modules, so the original more complex problem can be simplified into a series of simple module design. The independence of modules also brings a lot of convenience for the expansion of existing systems and the establishment of new systems, because existing modules can be made full use of building blocks for expansion. Any algorithm function can be realized by the combination of three basic program structures composed of program modules: sequence structure, selection structure and loop structure.

(2) Top-Down Programming

When designing more complex programs, a top-down approach is generally adopted. From the top down is from the whole to the part and then to the details, that is, first consider the overall goal, clarify the overall task, and then divide the problem into several parts, and each part is refined until it is decomposed into a problem that is easier to solve. The top-down programming method refers to that the main control program is first started, and then each function is segmented into smaller function modules according to the interface relationship until the lowest level module is designed. Top-down is an orderly step-by-step hierarchical decomposition and refinement of the program design method, the level is clear, easy to write, easy to debug, is widely used by program design engineers design method.

(3) Modular Programming

After defining the overall task of software design, it is necessary to enter the design of the overall structure of software. At this time, the top-down approach is generally adopted, and the total task is gradually subdivided from the top to the bottom, until the basic unit can be dealt with concretely. If this basic program unit is clearly defined and can be designed, debugged, error-corrected, and ported independently, it is called a module. Modular design, simply put, is that the writing of the program is not the beginning of the entry of computer statements and instructions, but first with the main program, subroutine, subprocess and other frameworks to describe the main structure and process of the software, and define and debug the input and output link relationship between the various frameworks, and gradually refine, to obtain a series of functional block as a unit of algorithm description. The method of program design based on function blocks is called modularity. The purpose of modularity is to reduce the complexity of the program and simplify the operation of program design, debugging and maintenance. Each module is developed independently, tested, and finally assembled into the entire software. The modular overall structure has the advantages of clear structure concept, flexible combination and easy debugging, connection and error correction, which makes the program design more simple and intuitive, thus improving the legibility and maintainability of the program. When dealing with faults or changing functions, it is often only designed local modules without affecting the whole, so it is an ideal structure often used. The key of modular approach is how to decompose the system into modules and how to design modules, and what rules to follow in module design. The system is decomposed into modules to obtain the highest module cohesion, that is, there is the maximum correlation within a module, and modules that only achieve a single function have high cohesion. Keep coupling to a minimum, that is, the relationship between different modules is as weak as possible. The depth of the chain between modules should not be too much, that is, the level of the module should not be too high, and it should generally be controlled at about 7 layers, with clear interfaces, good information concealment, and moderate module size.

5.5 Automatic Control Technology of Modern Industrial Equipment

5.5.1 *Overview*

Automatic control refers to the use of additional equipment or devices in the absence of direct human participation, so that the machine, equipment and other controlled objects or a certain working state or parameter of the production process automatically in accordance with the predetermined law. Based on mathematical theory, automatic control uses the feedback principle to automatically affect the dynamic system, so that the output value is close to or reaches the predetermined value.

Modern industrial automation control system are programmable logic controller (PLC), distributed control system (Distributed Control System, DCS) and fieldbus control system (FCS), etc. FCS combined with DCS, industrial Ethernet, advanced control and other new technologies will have a strong vitality. Figure 5.19 shows the industrial networking structure.

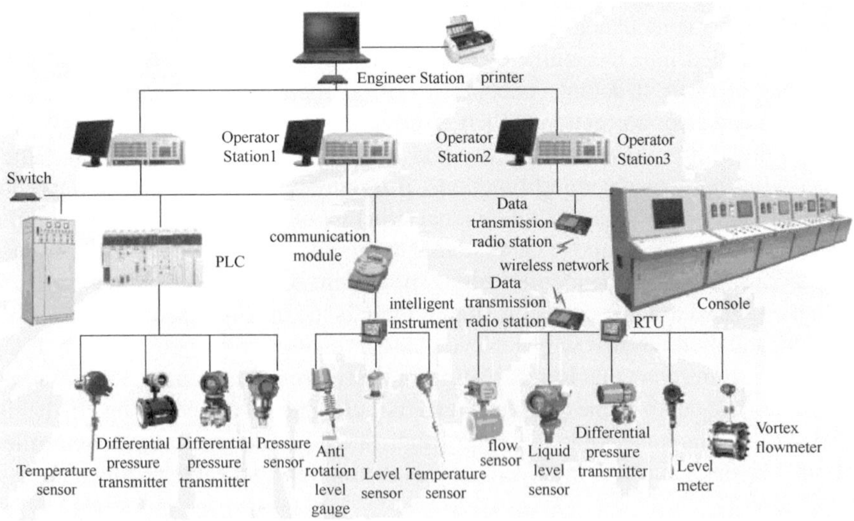

Fig. 5.19 Industrial network structure

5.5.2 *Programmable Logic Controllers*

PLC (Programmable Logic Controller) control system is an electronic system with digital operation, which is specially designed for industrial environment application. It is a digital operation controller with microprocessor for automatic control and a new industrial control device designed to replace relay contactor control system. It uses programmable memory to store, execute logical operations, sequence control, timing, counting and arithmetic operations, and controls various types of machinery or production processes through digital and analog inputs and outputs. The PLC control system consists of CPU, instruction and data memory, input/output interface, power supply, digital-analog conversion and other functional units. It uses a kind of programmable memory to store programs, execute user-oriented instructions such as logical operation, sequence control, timing, counting and arithmetic operation, and control various types of machinery or production processes through digital or analog input/output. It has the advantages of strong versatility, high reliability, simple instruction system, convenient programming and small size, and has become the core part of industrial control. Widely used in mechanical manufacturing, metallurgy, electric power, textile, environmental protection and other industries, especially in mechanical processing and machine tool control, it has become the most ideal first choice controller for reforming and developing electromechanical integration products such as machine tools [7, 8].

(1) Development Process of PLC

Since the United States launched PLC to replace the traditional relay control device in the 1960s, PLC has been rapidly developed, the function of PLC has been continuously improved, and it has been widely used in the world industrial control field. With the continuous development of computer technology, signal processing technology, sensor technology, control technology, network technology and the continuous improvement of user needs, PLC adds analog processing and motion control functions on the basis of switching quantity processing, in addition to powerful logic control functions, in motion control, process control and other fields also play a very important role.

Before the appearance of PLC, mechanical control and industrial production control were realized with industrial relays. In a complex control system, hundreds of various relays may be used, and the wiring and installation work is very large. If the control process and requirements change, the components and wiring in the control cabinet also need to make corresponding changes, which is high cost and long construction period.

In 1968, General Motors (GM), in order to adapt to the needs of the continuous update of the model, proposed to combine the advantages of the computer and the advantages of the relay control system to make a universal control device, and the computer programming method to simplify the program input, with a process-oriented, problem-oriented "natural language" programming. The American Digital Equipment Corporation (DEC) developed the world's first PLC in 1969. Because PLC is powerful and easy to operate, it is quickly applied and promoted in the field of industrial control.

At present, in order to meet the different needs of large and small enterprises, and further expand the scope of application of PLC in the field of industrial automation, small PLC to reduce the size, enhance the function, speed up, low price direction, so that it can more widely replace relay control; Large and medium-sized PLC to large capacity, high reliability, high speed, multi-function, networking direction, so that it can carry out comprehensive automatic control of large-scale, complex systems. Table 5.2 shows the development process of a PLC.

(2) The Composition of PLC

PLC adopts a typical computer structure, the hardware is mainly composed of CPU module, I/O interface module, RAM, ROM, power module. The central processing unit (CPU) is the core of the PLC, it is the operation, control center, the work receives, stores the user program, data and input signals, diagnoses the working state, reads the user program, interprets and executes, and completes the various operations specified in the user program. Memory is divided into system program memory and user program memory. The function of I/O interface module is to connect the industrial field device with the CPU module, including the switching I/O interface module, the analog I/O interface module, the intelligent I/O interface module and the peripheral communication interface module. The power module provides power for the working process of the PLC. Figure 5.20 shows the PLC hardware diagram.

(3) The Working Principle of PLC

The working process of PLC controller is generally divided into three stages: input sampling, user program execution and output refresh, which is called a scan cycle. During operation, the CPU of the PLC controller repeats the above three stages at a certain scanning speed. In the input sampling phase, the PLC controller reads all the input status and data successively in a scanning manner and stores them in the corresponding cell in the I/O image area. After the end of the input sampling, the user program execution and output refresh phase. In both phases, even if the input state and data change, the state and data of the corresponding cell in the I/O image area will not change. Therefore, if the input is a pulse signal, the width of the pulse signal must be greater than one scan cycle, so that the input can be read in any case. In the execution phase of the user program, the PLC controller always scans the user program in order from top to bottom. When scanning each ladder diagram, it always scans the control line composed of contacts on the left side of the ladder diagram first, and performs logical operations on the control line composed of contacts in the order of first left and then right, first up and then down, and then refreshes the status of the corresponding bit of the logic coil in the system RAM storage area according to the results of logical operations, or refreshes the output coil in the I/O

Table 5.2 The development process of PLC

Generation times	Time	Main features
First generation	The first one was born in the early 1970s	The CPU is composed of small and medium-sized integrated circuits, and the memory is magnetic core memory; Simple in function, mainly capable of controlling conditions, timing, and counting; The model is single and has not formed a series; The reliability is slightly higher than that of relay contactor systems; There is no established programming language
Second generation	From the early 1970s to the late 1970s	The CPU uses a microprocessor and the memory uses EPROM, which has greatly developed the technology of PLC. PLC has functions such as logical operation, timing, counting, numerical calculation, data processing, computer interface, and analog control. The software develops self diagnostic programs, further improving reliability; The system is beginning to develop towards standardization and systematization; There is a distinction between integral and modular structures in terms of structure, and the overall functionality is transitioning from specialized to universal
Third generation	From the late 1970s to the mid-1980s	The emergence of single-chip computers, the entry of semiconductor memory into industrial production, and the use of large-scale integrated circuits have promoted the further development of PLCs, transforming them into specialized industrial computers. Its characteristics are: the CPU uses 8-bit and 16 bit microprocessors, greatly enhancing the functionality and processing speed of the PLC; Capable of remote I/O with communication function; Added various special functions; The self diagnostic function and fault-tolerant technology are developing rapidly; In terms of software, a process oriented ladder diagram language and its disguised statement table (also known as logical symbols) have been developed; The volume of PLC has been further reduced, the reliability has been greatly improved, and the cost has become larger and lower

(continued)

Table 5.2 (continued)

Generation times	Time	Main features
The fourth generation	From the mid-1980s to the mid-1990s	With the rapid development of computer technology, it promotes the complete computerization of PLC. PLC fully utilizes 8-bit and 16 bit microprocessor chips, with a processing speed of up to 1 microsecond per step. Functionally, it has high-speed counting, interrupt, A/D, D/A, PID, etc., which can meet the requirements of process control and enhance networking capabilities
The fifth generation	From the mid-1990s to present	RISC (Instruction Set CPU) chips are widely used in the computer industry, and surface mount technology and processes have matured, greatly reducing the size of PLC machines. PLCs use 16 bit and 32-bit microprocessor chips. CPU chips are also developing towards specialization. Capable of powerful numerical operations, function operations, and large-scale data processing; Various intelligent modules have been developed; The human–machine intelligent interface with LCD micro reality is widely used, and advanced ones have developed to touch screens; In addition to handheld programmers, laptops and powerful programming software are extensively used

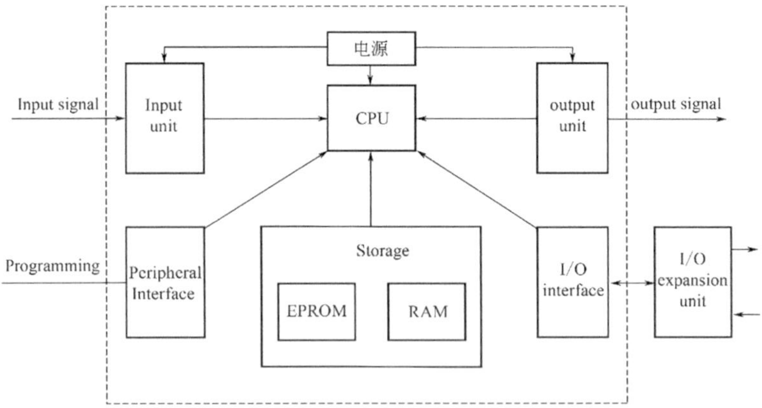

Fig. 5.20 PLC hardware block diagram

image area. The state of the corresponding bit, or determine whether to perform the special function instructions specified in the ladder diagram. When the scanning user program is finished, the PLC controller enters the output refresh stage. During this period, the CPU refreshes all output latches according to the corresponding state and data in the I/O image area, and then drives the corresponding peripherals through the output circuit.

The result of running the scan user program is different from the result of running the hard logic of the relay control device in parallel. In general, the PLC samples all the input signals in order and reads them into the input image register for storage, which is used when the PLC executes the program. The data in the current input and output image register is processed by arithmetic, and then the results are written into the output image register for storage. When the PLC refreshes the output lock, it is used to drive the user device Cycle, generally within 100 ms, PLC cycle scan working process is shown in Fig. 5.21. The scan cycle of PLC controller includes self-diagnosis, communication, etc., and a scan cycle is equal to the sum of all the time of self-diagnosis, communication, input sampling, user program execution, output refresh, etc.

PLC program readability, easy modification, reliability, versatility, easy expansion, easy maintenance can be comparable to computer programs, coupled with its small size, light weight, convenient installation and debugging, so that its design and processing cycle is greatly shortened, maintenance is also convenient.

(4) Characteristics of PLC

① High reliability. PLC mostly adopts single chip microcomputer, high integration, coupled with the corresponding protection circuit and self-diagnosis function, improves the reliability of the system, and has been widely used in industrial control.
② Programming is easy. PLC programming uses relay control ladder diagram and command statements, the image is intuitive, the programming workload is much less than the microcomputer instructions, easy to master, convenient to use, and even does not need too much computer expertise, you can program.
③ Convenient configuration. PLC adopts the building block structure, the user only needs to drag the mouse through the configuration software to simply combine, it can flexibly change the function and scale of the control system, therefore, suitable for any control system.
④ Complete I/O modules. PLC for different field signals have the corresponding template can be directly connected with the device in the industrial field, and through the bus and the CPU motherboard.
⑤ Easy to install. Compared with computer systems, the installation of PLC does not require a dedicated computer room, nor does it require strict shielding measures. When used, only the detection device and the I/O interface terminals of the actuator and PLC are correctly connected to work normally.
⑥ Running speed is fast. Since PLC is executed by program control, both reliability and running speed are incomparable to relay logic control, and the difference between PLC and microcomputer control systems has become smaller and smaller in recent years.
⑦ Good environmental adaptability. PLC in the realization of various amounts of I/O control at the same time, but also has the ability to output analog voltage and digital pulse, so that it can control a variety of can receive these

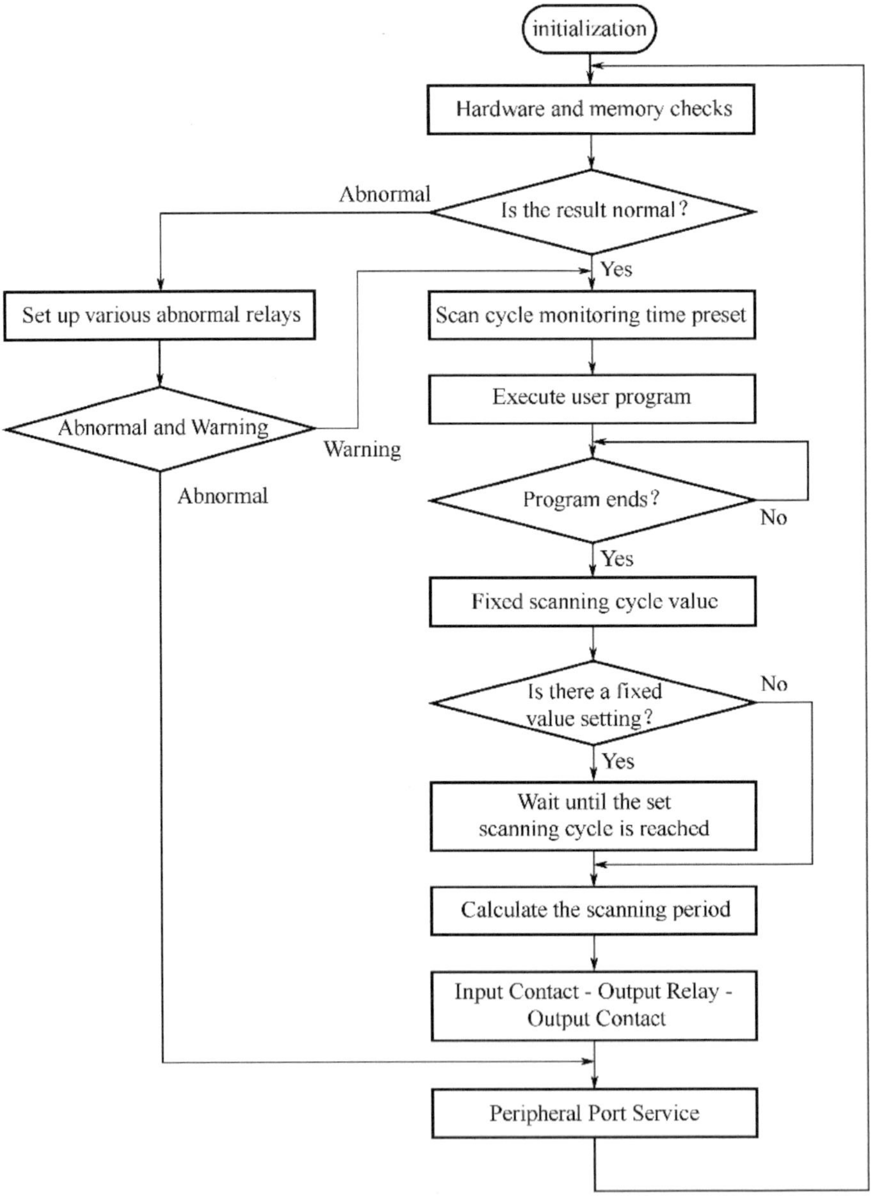

Fig. 5.21 The cyclic scanning process of PLC

signals of servo motors, stepper motors, variable frequency motors, etc., and anti-interference ability is strong, can be applied to various fields of industrial control, the application range is very wide.

(5) Classification of PLC

① Count by I/O. The more I/O points the PLC control system handles, the more complex the control relationship is, the larger the program memory capacity required by the user, the more PLC instructions and other functions are required, and the faster the process of instruction execution. According to the number of PLC I/O points can be divided into minicomputer, medium-sized machine, mainframe three categories (Table 5.3).

Table 5.3 Characteristics of small, medium, and large PLCs

PLC model	Characteristic	Typical representative
Small-scale	Mainly based on switch control, small PLC input and output points are generally below 256 points, and the user program memory capacity is around 4K words. The current high-performance compact PI C also has certain communication capabilities and a small amount of analog processing ability. The characteristics of this type of PLC are: low price and compact size. Suitable for controlling a single device and developing small-scale electromechanical integrated products	Integrated PLC products such as SIEMENS S7-200 series, OMRON CPM2A series, MITUBISH FX series, and AB SIC500 series
Medium-scale	Input, the total output points are between 256 and 2048 points, and the user program memory capacity reaches about 8K words. Medium sized PLC not only has control functions for both analog and digital quantities, but also has strong digital computing capabilities. Its communication and analog processing functions are powerful, and medium-sized machines are more abundant than small-sized machines. Medium sized machines are suitable for more complex logic control systems and process control systems for continuous production lines	SIEMENS' S7-300 series, OMRON's C200H series, AB's SLC500 series, etc
Large-scale	The total number of points is above 2048, and the user program storage capacity reaches over 16K words. Large PI The performance of C is comparable to that of large PIC input and output industrial control computers, with the ability to calculate, control, and regulate. It also has a powerful network structure and communication networking capabilities. It can also be interconnected with other types of controllers and connected to the upper computer to form a centralized and decentralized production process and product quality control system. Suitable for equipment automation control, process automation control, and process monitoring systems	SIEMENS' S7-400 series, OMRON's CVM1 and CS1 series, AB's SLC5/05 series, etc.

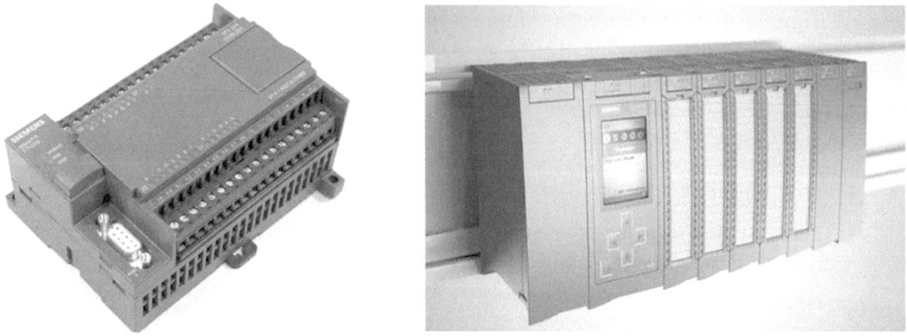

(a) S7-200 (b) S7-1500

Fig. 5.22 SIEMENS S7 series PLC

② According to the structural form. According to the different structure of PLC, PLC can be divided into two categories: integral and modular. The integral structure is characterized by the basic components of the PLC CPU board, input board, output board, power board and so on, which are compact installed in a standard housing to form a whole, constituting a basic unit or expansion unit of the PLC. The basic unit is equipped with an expansion port, which is connected to the expansion unit through an expansion cable, and is equipped with analog input/output module, thermocouple, thermal resistance module, communication module and many other special function modules to form different configurations of PLC. Integral structure PLC small size, low cost, easy to install. Micro and small PLCS are generally integrated structures, such as the S7-200 series of Siemens (Fig. 5.22a).

The modular structure of the PLC is composed of some module units, these standard modules such as CPU module, input module, output module, power module and various functional modules, etc., these modules can be inserted on the frame and the substrate. Each module function is independent, the overall size is unified, and can be flexibly configured according to needs. Large and medium-sized PLCS use this mode, for example, the S7-300, S7-400, and S7-1500 series of Siemens, as shown in Fig. 5.22b.

The average price of each I/O point of the integrated PLC is cheaper than the modular type, and the integrated structure is generally used in small control systems. However, the modular PLC hardware configuration is convenient and flexible, the number of I/O points, the proportion of input points and output points, the use of I/O modules and other aspects of the choice is larger than the integral PLC, and can be flexibly selected according to the actual situation, maintenance replacement module, determine the fault range is also very convenient, so the more complex, higher requirements of the system generally choose modular PLC.

S7-1500, a typical representative of large and medium-sized PLC, has greatly improved the indicators of S7-300/400. The maximum number of programming blocks for CPU1516-3PN is 6000. The maximum number of data blocks is 5 MB.

The maximum number of FB, FC, and OB blocks is 512 KB. 5 MB of working memory for programs and 1 MB of working memory for data. SIMATIC memory card Max 2 GB; There are 2048 S7 timers and counters respectively, and the number of IEC timers and counters is not limited; Bit memory (M)16 KB; Up to 8192 I/O modules, up to 32 process image partitions, process image input and output of 32 KB respectively; Up to 32 modules per rack; The motion control function supports up to 20 speed control shafts, positioning shafts and external encoders with high-speed counting and measurement functions.

③ Classified by function. According to the different functions of PLC PLC can be divided into low-grade, mid-grade, high-grade three categories. Low-grade PLC has logical operation, timing, counting, shifting and self-diagnosis, monitoring and other basic functions, but also a small amount of analog input/output, arithmetic operations, data transmission and comparison, communication and other functions, mainly used for logic control, sequence control or a small amount of analog control of stand-alone control system.

In addition to the functions of low-grade PLC, mid-range PLC also has strong analog input/output, arithmetic operations, data transmission and comparison, number system conversion, remote I/O, subroutines, communication networking and other functions. Some can also add interrupt control, PID control and other functions, suitable for complex control systems.

In addition to the functions of the high-end PLC, it also adds symbolic arithmetic operations, matrix operations, bit logic operations, square root operations and other special functional functions, tabulation and table transmission functions. High-grade PLC machines have stronger communication networking functions, which can be used for large-scale process control or constitute a distributed network control system to achieve factory automation.

Of course, the price of high-grade PLC has also increased. System design, can not pursue high, according to the actual needs, reasonable selection of PLC, so as not to cause waste of resources and cost waste. In small I/O control situations, low-grade or integral PLC can generally meet the requirements.

(6) Application of PLC [9]

PLC controller has been widely used in steel, petroleum, chemical, electric power, building materials, machinery manufacturing, automobile, textile, transportation, environmental protection and cultural entertainment and other industries at home and abroad, the use of the situation can be roughly summarized as the following categories.

① Logic control of switching quantity. This is the most basic and most extensive application field of PLC controller, which replaces the traditional relay circuit to achieve logic control and sequence control, which can be used for the control of a single device, and can be used for multi-machine group control and automated assembly line. Such as injection molding machines, printing presses, stapler machines, combined machine tools, grinding machines, packaging production lines, electroplating lines, etc.

② Analog quantity control. In the industrial production process, there are many continuously changing quantities, such as temperature, pressure, flow, liquid level and speed and other analog quantities. In order for A programmable controller to handle analog, it is necessary to realize A/D conversion and D/A conversion between analog and digital quantities. PLC manufacturers have produced supporting A/D and D/A conversion modules, so that programmable controllers are used for analog control. Not only large and medium-sized machines have this function, but also some minicomputers have this function.

③ Motor control. Motion control is the most used control mode in the industrial field, PLC controller can be used for circular motion or linear motion control. In terms of the configuration of the control mechanism, the position sensor and the actuator were directly connected with the switching I/O module in the early days, and now a dedicated motion control module is generally used. For example, a single or multi-axis position control module that can drive a stepper motor or a servo motor. Almost all the products of the world's major PLC controller manufacturers have motion control functions, which are widely used in various machinery, machine tools, robots, elevators, etc.

④ Process control. Process control refers to the closed-loop control of analog quantities such as temperature, pressure and flow. As an industrial control computer, PLC controller can compile a variety of control algorithm programs to complete closed-loop control. PID regulation is a common method used in closed-loop control system. Large and medium-sized PLC have PID module, many small PLC controllers also have this function module. PID processing is generally run a special PID subroutine. Process control is widely used in metallurgy, chemical industry, heat treatment, boiler control and so on.

⑤ Data processing. Modern PLC controller has mathematical operation (including matrix operation, function operation, logic operation), data transmission, data conversion, sorting, table lookup, bit operation and other functions, can complete the data collection, analysis and processing. These data can be compared with the reference values stored in the memory to complete certain control operations, and can also be transmitted to other intelligent devices by means of communication functions, or they can be printed and tabulated. Data processing is generally used in large control systems, such as unmanned flexible manufacturing systems; It can also be used in process control systems, such as some large control systems in the paper, metallurgy, and food industries.

⑥ Data collection and monitoring. Since PLC is mainly used for field control, it is very necessary to collect field data. On this basis, PLC is connected with the upper computer or touch screen, which can observe the current value of these data and carry out statistical analysis in time. Some PLCS also have a data recording unit, which can insert the memory card of a general personal computer into the unit to save the collected data.

⑦ Communication and networking. PLC controller communication includes communication between PLC controllers and communication between PLC controllers and other intelligent manufacturing equipment. With the development of computer control, the factory automation network has developed rapidly, and the PLC controller manufacturers attach great importance to the communication function of the PLC controller, and have launched their own network systems. PLC controllers produced now have communication interfaces.

5.5.3 DCS Control System

Distributed control system (Distributed Control System, DCS) is developed in the mid—1970s microprocessor based distributed computer control system. It is a combination of control technology, computer technology, communication technology, graphic display technology and network technology. The device is a new distributed computer control system which uses computer technology to monitor, operate, manage and control the production process centrally.

DCS has the characteristics of decentralization and centrality, autonomy and coordination, flexibility and expansibility, advanced nature and inheritance, reliability and adaptability, friendliness and novelty.

① Dispersion. DCS is not only decentralized control, but also the meaning of geographical dispersion, equipment dispersion, function dispersion and risk dispersion. The purpose of decentralization is to disperse the risk and improve the reliability and safety of the system.

② Concentration. Centralization refers to centralized monitoring, centralized operation and centralized management. DCS communication network and distributed database is the concrete embodiment of centralization, using the communication network to connect the physically dispersed equipment into a unified whole, using the distributed database to achieve the integration of the whole system information, to achieve centralized monitoring, centralized operation and centralized management.

③ Autonomy. The autonomy of DCS means that each computer in the system works independently. Process control station for signal input, operation, control and output; Operator station monitoring, operation and management; The function of the engineer station is configuration.

④ Coordination. Coordination refers to the interconnection and coordination of the computers in the system with the communication network to achieve the overall function of the system.

⑤ Flexibility and extensibility. DCS adopts modular structure to provide all kinds of functional modules, which can be flexibly configured to form simple and complex control systems, and can also modify the control scheme at any time according to the change of production process and flow.

⑥ Man–machine interface (MMI) friendly. The operator station uses color LED large screen display, interactive graphic screen (general appearance, group, point, trend, alarm, operation guidance flow chart screen), etc., the application of multimedia technology, visual image, simple operation.

⑦ Adaptability. DCS uses high-performance electronic components, advanced production technology and various anti-interference technologies, which can make DCS adapt to the harsh working environment.

In addition, DCS not only uses a series of redundant technologies to reduce the occurrence of failures, such as the control station host, I/O board, communication network and power supply dual, but also uses a hot backup work mode to automatically check the fault, once the fault automatically switched. At the same time, the hardware and software faults of the system are checked in real time by the fault diagnosis and maintenance software, and the fault shielding technology is adopted to make the fault impact as small as possible.

(1) DCS Development Process

DCS was first proposed by Honeywell in the United States, and officially launched the world's first distributed control system—TDC-2000 to the market in December 1975, after which its development went through three major stages (Table 5.4). The function of the system gradually expands from the bottom layer to the top layer. The control function of the system is gradually developed from a single loop control to a combination of logic control, sequence control, program control, batch control and formula control.

The technical characteristics of the new generation of DCS include full digital, information and integration, the field analog instrument to the field digital instrument, and the field bus interconnection, the software function modules in the control station dispersed in each field digital instrument, and can be unified configuration to form a control loop, to achieve thorough decentralized control.

DCS as a new generation of industrial automation process control equipment in the world is widely used in petroleum, chemical, metallurgical, textile, electric power, food and other industries, China in petroleum, metallurgy, chemical and electric power and other industries has also been widely promoted and applied.

(2) Composition of DCS

A basic DCS should include at least one field control station, an operator station, an engineer station, and a system network four parts. In addition, there should be corresponding DCS software. Figure 5.23 shows the DCS structure.

① Field control station. Field control station is the core of DCS, complete the control function of the system, system performance, reliability and other important indicators are also dependent on the field control station. The hardware generally adopts industrial computer systems, including field measurement units, input/output devices of execution units, and process I/O. The part of the internal main CPU and memory used for data processing, calculation and storage is called the logical part, while the field I/O is called the field part.
② Operator station. The operator station mainly completes the functions of the human–machine interface, generally using desktop general-purpose computers such as PCS, due to real-time display and monitoring, it generally requires a

Table 5.4 Development of DSC system

Development stage	Main features	Representative
1st generation	The basic architecture of DCS is composed of five parts: process control unit, data acquisition unit, CRT operation station, upper management computer, and high-speed data channel connecting various units and computers	Honeywell's TDC-2000 is representative, as well as Yokogawa's Yahpark system, Foxboro's Spectrum system, Bailey's Netwook90 system, and others
2st generation	The introduction of a local area network has organized process control stations, central operation stations, system management stations, and gateways according to the concept of network nodes, further increasing the scale and capacity of the system, providing greater room for expansion, and gradually improving its functionality. In addition to loop control, sequence control, logic control, and other functions have been added, strengthening the functionality of the system management station and enabling optimization control and production management functions	Honeywell's TDGC-3000, Fisher's PROVOX, Taylor's MOD300, Westinghouse's WDPF, etc.
3st generation	Further expansion has been achieved in terms of functionality, adding an upper layer network and incorporating production management functions into the system, forming a three-layer functional structure of direct control, supervisory control and coordinated optimization, and upper layer management; In terms of networking, various manufacturers have widely adopted standard networking products; The five configuration languages defined by IEC61131-3 are adopted by most DCS manufacturers; In terms of the products that make up the system, besides the on-site control station, which is basically a proprietary product of various DCS manufacturers, the hardware and basic software of human–machine interface workstations, servers, and various functional stations have become the mainstream of control systems	Foxboro's I/A Series, Honeywell's TDC-3000/UCN, Yokogawa's Centum XL/μ XL, Bailey's INFI-90, Westinghouse's WDPF II, etc.

large-size liquid crystal display, or expand the multi-screen display to broaden the operator's observation range. In order to improve the display speed of the picture, the operator station is generally configured with large memory and excellent performance of the graphics card.

③ Engineer station. The engineer station mainly performs the application configuration of DCS. Application configuration can achieve a variety of applications, only the completion of the correct configuration, a universal DCS can become a specific control application of the operational system. The tasks to be completed in configuration are: define a specific system, determine the control function,

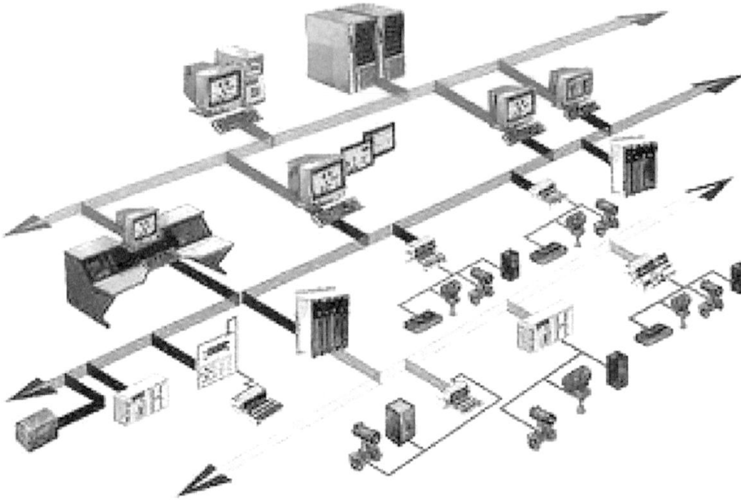

Fig. 5.23 Structure of DCS

control input/output, control loop algorithm, parameters selected in the control calculation, man–machine interface set in the system to achieve human management and monitoring of the system, as well as alarm, report and historical data records. In addition, the system may be equipped with special function stations. It is a computer that performs a specific function, such as a history station dedicated to recording historical data; Advanced computing station for advanced control operation functions; Management station for production management, etc. The main function of the server is to complete the work of the supervisory control layer.

④ DCS network. DCS network includes system network, fieldbus network and high-level management network. The system network is the bridge of effective data transmission between DCS stations with different functions to achieve the overall function of the system. The real-time, reliability and data communication ability of the system network are related to the performance of the whole system. The fieldbus network digitizes the field detection and transmission unit and the control and execution unit, and connects the system with the field through the fieldbus. The high-level management network enables DCS to develop from a simple low-level control function to a higher level of data acquisition, supervision and control, production management and other plant-wide control and management system, making DCS a computer management and control system to achieve factory automation.

⑤ DCS software. Corresponding to DCS workstation, DCS software includes field control station software, operator station software and engineer station software.

The main function of the site control station software is to complete the direct control of the site, including loop control, logic control, sequence control and hybrid control. Field I/O drive, complete process input/output; The input process quantity is preprocessed; Field data are collected in real time and stored in the local database in the field control station. Control calculation is based on the control algorithm and detection data, related parameters to calculate the amount of real-time control. Through the field I/O driver, the control quantity is output to the field.

The main function of operator station software is man–machine interface and HMI processing, including graphic display, operation command interpretation and execution, field data and status monitoring and abnormal alarm, historical data archiving and report processing. The operator station software mainly includes: a. Graphics processing, the graphics file generated by the configuration software for static screen display and dynamic data display and periodic data update; b. Operation command processing; c. Trend curve display of historical data and real-time data; d. Display of alarm information, display, record and processing of event information; e. Historical data recording and storage, dumping and archiving software; f. Report recording and printing; g. Formation, display, printing and storage of system run logs.

Part of the software of the engineer station is running online, mainly to complete the diagnosis and monitoring of the DCS system itself, alarm when abnormal is found, and give detailed abnormal information through the display screen of the engineer station. The other part of the main function is configuration.

DCS development process is mainly the use of system configuration software, according to the actual needs of the control system, the generation of various application software process. Configuration software includes basic configuration and application software configuration. The basic configuration is to configure information for the system, such as the number of various stations in the system, their index flags, the maximum number of points per control station, the shortest execution cycle and memory capacity. The configuration of the application software mainly includes the following aspects:

① The configuration of the control loop. The configuration of the control loop is to make use of the basic functional modules provided by the system to form a variety of practical control systems. Figure 5.24 shows the real-time DCS control of a production unit. At present, the configuration methods provided by various DCS mainly include the specified operation module connection mode, the decision table mode, and the step record mode.

The specified operation module connection method is to call a variety of independent standard operation modules, connect a variety of control loops with lines, and finally automatically generate control software, which is a configuration method of information flow and control functions are very intuitive. The decision table is a pure form, as long as according to the requirements of the configuration table, fill in the content one by one or answer the question, this way is very conducive to the user's configuration operation. Step recording method is a writing method based on language instructions, which has large programming freedom and various complex functions can be realized by some techniques, but the configuration efficiency is low. In addition, because this configuration

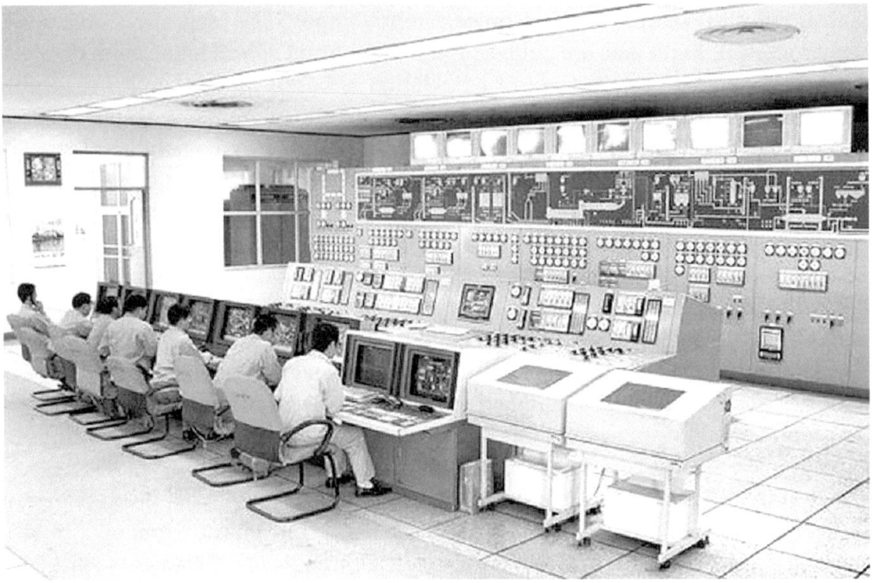

Fig. 5.24 Real time control example of DCS in a certain production unit

method is not intuitive enough, it often has high requirements for configuration engineers in terms of technical level and configuration experience.

② Real-time database generation. Real-time database is the most basic information resource of DCS. These real-time data are stored and managed by the real-time database. In DCS, the common way to establish and modify real-time database records is to generate database files with general database tool software, and the system directly uses this data format for management or uses some method to convert the generated data files to the format required by DCS, which is very convenient and efficient.

③ The generation of industrial process picture. DCS has rich control system and detection system screen display functions. It is a basic requirement of DCS configuration to express various measurement and control parameters and system status in the form of characters, rod diagrams, curves, etc. combined with the general appearance, grouping, control loop, flow chart, alarm and other screens. In addition, it can also display various variable catalog screen, operation guidance screen, fault diagnosis screen, engineer maintenance screen and system configuration screen according to needs.

④ Generation of historical database. All DCS support historical data storage and trend display functions, history database usually by the user in the condition of no programming, through the screen editing and compilation technology to generate a data file, the file defines the structure and scope of each historical data record. The data in the historical database is generally divided into groups, and the data type and sampling time in each group are the same. Information about each data point is defined at generation time.

⑤ Report generation. The report printing function of the operator station of DCS is configured through the report generation part of the configuration software. There are great differences in the report printing function of different DCS. In general, DCS supports the following two types of report printing functions: one is periodic report printing, and the other is triggered report printing, and users generate different report forms according to their needs and preferences.

5.5.4 Fieldbus

With the development of control technology, computer technology, communication technology, network technology, etc., the field of information exchange is rapidly covering all levels from field equipment to control and management, from the workshop, workshop, factory, enterprise and even the market around the world, in this background gradually formed an enterprise information system based on network integrated automation system, fieldbus is adapted to this situation New technology is on display. Fieldbus control system (FCS) Fieldbus Control System, is centralized digital control system, distributed control system (DCS) of a new generation of control system, known as the cross-century new technology of automatic control. The fieldbus takes the measurement and control equipment as the network node and the twisted pair and other transmission media as the link to connect the measurement, control and execution equipment located in the production site with digital computing and digital communication capabilities into a network system, and follows the standard communication protocol to achieve data transmission and communication between multiple measurement and control equipment and between the field equipment and the remote monitoring computer Information exchange, forming an automatic control system to meet the needs of various applications. In essence, fieldbus is a local area network. The software, hardware and protocol follow the standard specification and are used as communication methods in the control system. FCS ADAPTS to the development of industrial control systems in the direction of digitalization, decentralization, networking and intelligence, and makes industrial automation products once again updated.

Fieldbus plays a very important role in the development of industry, and is widely used in petroleum, chemical industry, electric power, medicine, metallurgy, processing and manufacturing, transportation, national defense, aerospace, agriculture and buildings.

Similar to other types of local area networks, the model structure of fieldbus networks is also based on the 7-layer Open System Interconnection Reference Model (OSIRM). In order to reduce the extra time cost caused by inter-layer operation and information format conversion, various fieldbus network models have simplified OSIRM to varying degrees, generally adopting only the physical layer, data link layer and application layer of the OSI reference model. Some buses also add a user layer on top of the application layer [10].

(1) Development of FCS

Fieldbus was born in the mid-1980s. As an interconnected communication network between field intelligent manufacturing equipment in process automation, manufacturing automation, transportation and other fields, and a comprehensive technology with intelligent sensors, control, computers, digital communications, and networks as the main content, fieldbus communicates the connection between the production process field control equipment and the higher control management network, in order to completely break the information isolation of the automation system. The island created the conditions. Fieldbus control system is not only an open communication network, but also a fully distributed control system. As a link of intelligent manufacturing equipment, it connects the intelligent manufacturing equipment connected on the bus and as a network node to a network system, and further constitutes an automation system to realize the comprehensive automation function of basic control, compensation calculation, parameter modification, alarm, display, monitoring, optimization and control integration.

The fieldbus control system breaks through the defects caused by the closed system of the private network in the DCS system, and turns the closed, dedicated solution into a solution based on openness and standardization, which can connect the automation equipment of different manufacturers that comply with the same protocol into a system through the fieldbus network, and at the same time combine the DCS centralized and distributed The system structure has become a fully distributed structure, the control function is completely decentralized to the site, and the basic control function can be realized by relying on the on-site intelligent manufacturing equipment itself.

In 1984, ISA/SP50, part of the Standards and Implementation Working Group of the American Instrumentation Association (ISA), began to develop fieldbus standards. In 1985, the International Electrotechnical Commission decided that the Proway Working Group was responsible for the research and development of fieldbus architecture and standards; In 1986, Germany began to develop the process fieldbus (Process Fieldbus) standard, referred to as PROFIBUS, which opened the field bus standard development and product development prelude.

In 1992, 80 companies including SIEMENS and ABB jointly established ISP (Interoperable System Protocol), which began to develop fieldbus standards based on PROFIBUS. In 1993, led by Honeywell and other companies, the WorldFIP (Factory Instrumentation Protocol) organization was established, with more than 120 companies joining the organization and developing fieldbus standards based on the French standard FIP.

In 1994, ISP and WorldFIP North America merged and established the Field bus Foundation (FF), which promoted the formulation of field bus standards and product development, and promulgated the low-speed bus H1 standard in the first quarter of 1996, interconnecting different manufacturers' FF compliant instruments The control system and communication network make H1 low speed bus enter the practical stage. At the same time, some influential bus standards have been derived in different industries. Most of them are gradually formed on the basis of corporate standards and are supported by other companies, manufacturers, users and even international organizations [11].

(2) Characteristics of FCS

① Openness. The traditional control system is a self-enclosed system, which can only communicate through the serial port or parallel port of the workstation. Fieldbus is committed to the establishment of a unified factory underlying network of open systems, users can according to their own needs and objects, products from different suppliers into the size of the system, can be connected to other devices or systems anywhere in the world that comply with the same standards. Communication protocols are consistent and open, and information can be exchanged between devices from different manufacturers. Fieldbus developers are to users according to their own needs and considerations, the products from different suppliers to form a system of random size, through the fieldbus to build an open interconnection system in the field of automation.

② Interoperability and reliability. Fieldbus in the selection of the same communication protocol, as long as the selection of the appropriate bus network card, jack and adapter can achieve the interconnection between devices, the system of information transmission and communication, greatly reduce the wiring and line checking workload, effectively improve the reliability of control. Interoperability refers to the realization of information transmission and communication between interconnected devices and systems, and devices with similar performance from different manufacturers can be replaced with each other.

③ Intelligence and functional autonomy of field equipment. The signal transmission of traditional equipment is a one-way transmission of analog signal, and the error generated in the transmission process is large, and the system is difficult to judge the fault quickly but runs with the fault. Fieldbus adopts two-way digital communication, the sensor measurement, compensation calculation, engineering quantity processing and control and other functions are dispersed to the field equipment to complete, can diagnose the operating status of the equipment at any time, and only rely on the field equipment to complete the basic functions of automatic control.

④ The system structure is highly decentralized. The fieldbus has formed a new architecture of fully distributed control system, which fundamentally changes the existing DCS which combines centralized and distributed control system, simplifies the system structure and improves the reliability.

⑤ Good environmental adaptability. Working at the front end of the production site, as the field bus at the bottom of the factory network, is designed to adapt to the field environment, can support twisted pair, coaxial cable, optical cable, radio frequency, infrared and power lines, etc., with strong anti-interference ability, can use the two-wire system to achieve power transmission and communication, and can meet the safety and explosion-proof requirements.

⑥ Flexibility. Various controllers, actuators and sensors are connected through the field bus, fewer cables, easy to lay, low cost, and the system design is more flexible, high reliability of signal transmission and strong anti-interference ability. The control system based on fieldbus will gradually replace the original control system, and the complex wiring harness will be replaced by fieldbus.

⑦ Less hardware. Because the intelligent manufacturing equipment scattered in the field bus system can directly perform a variety of sensing, control, alarm and calculation functions, the number of transmitters can be reduced, and the functional units such as signal conditioning, conversion, isolation and complex wiring are no longer required, and industrial PC can also be used as an operating station, thereby saving hardware investment and the area of the control room.

⑧ Low installation cost. The fieldbus system is simple, and multiple devices can be connected to one bus cable, reducing the amount of cables, terminals, slot boxes, and Bridges. When adding a field control device, connect it to the existing cable nearby.

⑨ Low maintenance cost. The on-site control equipment has the ability of self-diagnosis and simple fault handling, and sends the relevant diagnosis and maintenance information to the control room through digital communication, so that users can query the operation, diagnosis and maintenance information of all equipment, so as to analyze the cause of the fault early and quickly eliminate it, shorten the maintenance downtime, simple system, and reduce the maintenance workload.

⑩ User system integration initiative. Users can freely choose the equipment provided by different manufacturers to integrate the system, and the initiative in the process of system integration is in the hands of users.

⑪ High accuracy and reliability. Compared with analog signals, fieldbus equipment is intelligent and digital, which fundamentally improves the accuracy of measurement and control and reduces the transmission error. At the same time, because the structure of the system is simplified, the equipment and the connection are reduced, the internal function of the field instrument is strengthened, the signal transmission is reduced, and the reliability of the system is improved.

(3) Mainstream Fieldbus Types

At present, there are more than 40 kinds of fieldbuses in the world, but no fieldbus can cover all the should, according to its ability to transmit data can be divided into sensor bus, device bus, field bus three categories.

① FF field bus. The FF Fieldbus Foundation is an international organization jointly established by WORLDFIPNA and ISP Foundation in June 1994, with the goal of establishing a single, open, interoperable international standard for fieldbuses. FF fieldbus is based on ISO/OSI open system interconnection model, and takes its physical layer, data link layer and application layer as the corresponding layers of FF communication model, and adds a user layer on the application layer.

The Foundation fieldbus has two communication rates: low speed H1 and high speed H2. The transmission rate of H1 is 3125Kbps, the communication distance can be up to 1900 m (can be extended by repeater), and the bus power supply is supported. H2 has two transmission rates of 1 and 2.5 Mbps, and the communication distance is 750 and 500 m. The physical transmission medium can support stranded wire, optical cable and wireless transmission, and the protocol conforms to IEC1158-2 standards. Figure 5.25 shows the structure of a FF fieldbus.

② LonWorks fieldbus. The LonWorks fieldbus (Fig. 5.26) is a local operating network launched in 1992 by Echelon Corporation in the United States, initially used primarily for building automation, but soon expanded to industrial field networks. LonWorks technology has a complete platform for

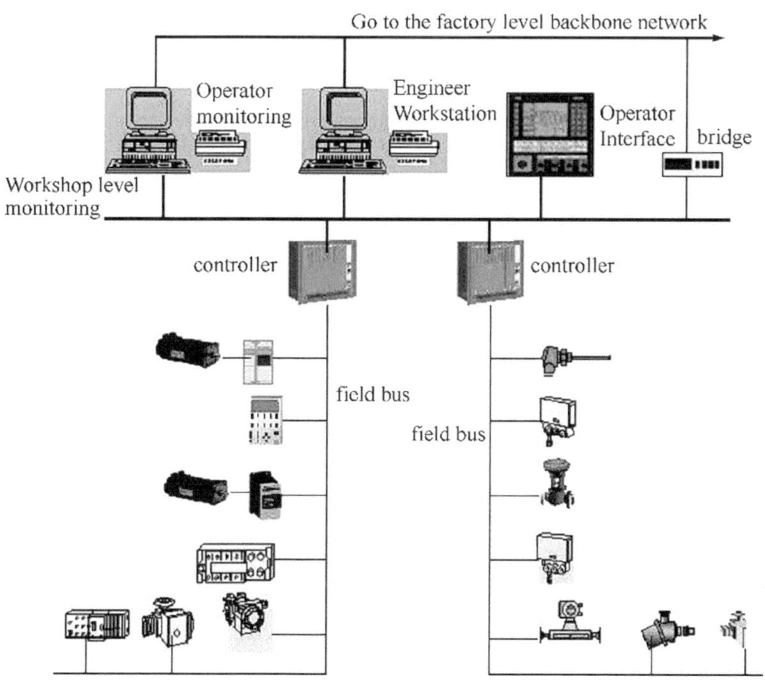

Fig. 5.25 Field-bus structure

developing control network systems, including all the hardware and software needed to design, configure, install, and maintain control networks. The basic unit of LonWorks network is the node, a network node includes the neuron chip (NeuronChip), the power supply, a transceiver and the I/O circuit with the interface of the monitoring device, the core is the neuron chip. Provides complete system resources with three integrated cpus, one of which is used to execute user-written applications and the other two to complete network tasks. The 11 I/O pins on the neuron chip can be programmed to provide 34 different I/O object interfaces, supporting a variety of signal modes such as level, pulse, frequency, coding, etc. Its two 16-bit timers/counters are available for frequency and timing; It provides a communication port that allows operation in single-ended, differential, and dedicated modes, with transmission rates up to 1.25 Mbps [12] (Fig. 5.27).

LonWorks fieldbus integrates many functions of today's fieldbus, while having some characteristics of local area network, and is widely used in aerospace, agriculture, computer/peripheral equipment, diagnosis/monitoring, test equipment, medical and health, military/defense, office equipment systems, robots, security guards, transportation equipment and other fields. Its versatility shows that it is not a bus for a particular domain,

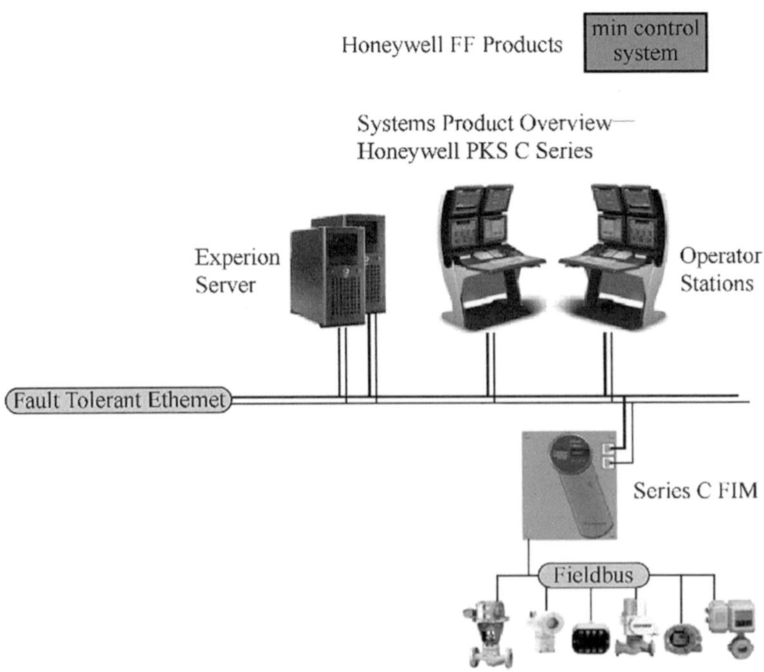

Fig. 5.26 A certain field bus structure

Fig. 5.27 LonWorks field bus

but a network technology that can integrate control systems in different domains into a more complex system based on LonWorks [13].

③ PROFIBUS Field bus. PROFIBUS is a field bus standard for automation technology (Fig. 5.28), which is jointly developed by 14 companies such as Siemens and five research institutions in Germany. It is a program bus network, with ProFIbus-DP and PROFIBUS-PA.

Profibus-dp (distributed around, Decentralized Peripherals) used in factory automation, many sensors and actuators can be controlled by a central controller, also can choose diagnostic machine using standard or know the status of each module.

FIBUS-PA (Process Automation) is used in process automation systems, controlled by process control system monitoring and measurement equipment, is an intrinsically safe communication protocol, which can be applied to explosion-proof areas (Ex-zone0 and Ex-zone1 in the classification of industrial explosion-proof danger areas). The physical layer matches IEC61158-2, allowing the communication cable to provide power to the field device, even if there is a fault, the power flow can be limited, avoiding the situation that may cause explosion. When using network power, a PROFIBUS-PA network is limited in the number of devices that can be connected. PROFIBUS-PA has a communication rate of 31.25Kbps. PROFIBUS-PA uses the same communication protocol as PROFIBUS-DP. As long as there is a conversion device, it can be connected to the PROFIBUS-DP network, and the PROFIBUS-DP with a faster rate is used as the network backbone to transmit signals to the controller. In applications

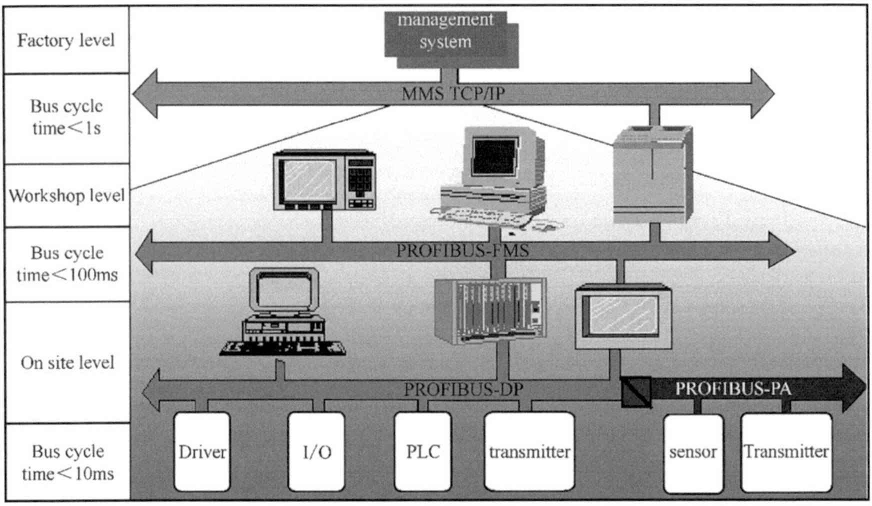

Fig. 5.28 PROFIBUS field bus

where automation and process control need to be handled simultaneously, PROFIBUS-DP and PROFIBUS-PA can be used simultaneously.

The transmission rate of the PROFIBUS fieldbus is 9.6Kbps to 12 MBPS, the maximum transmission distance is 1 km at 12 MBPS, and the repeater can be extended to 10 km. The transmission medium can be twisted pair or optical cable, and up to 127 sites can be connected. PORFIBUS supports master–slave system, pure master station system, multi-master multi-slave mixed system and other transmission modes. The master station can actively send information and has the control of the bus. In multi-master system, information is passed between master stations by token, and the station that gets the token has the bus control in the specified time. When the master station has control, it can send or request information to the slave station in the master–slave mode.

④ CAN fieldbus. CAN (Fig. 5.29) is the controller of the local area network (CAN) Controller Area Network, referred to as, was originally developed by Germany BOSCH company perform component used in automobile internal measurement and data communication between, now become an international one of the most widely used field bus, the bus specification has been ISO international standard Weaving is an international standard, and the CAN protocol is also based on the open Systems interconnection model of the International Standards Organization. The signal transmission medium is twisted pair, the communication rate is up to 1 Mbps/40 m, the direct transmission distance is up to 10 km/Kbps, and up to 110 devices can be connected. Because of its high performance, high reliability, real-time and other advantages, it has been widely used in industrial automation, transportation, medical instruments and construction industries.

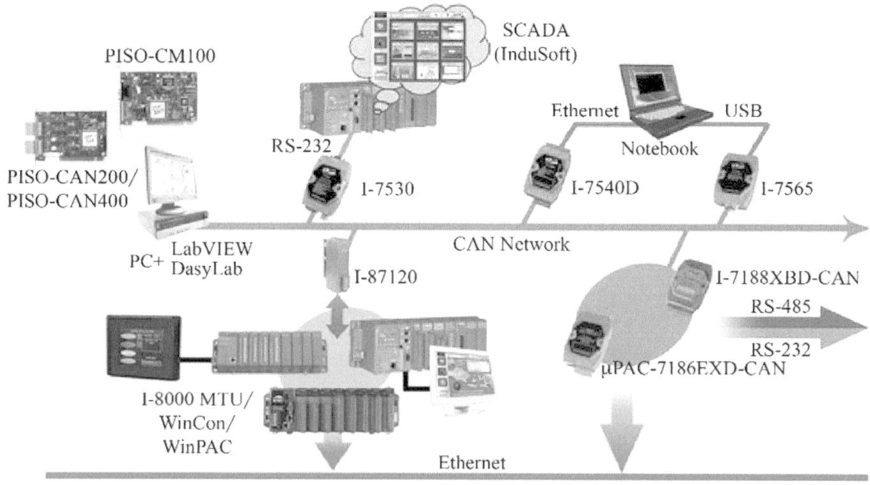

Fig. 5.29 CAN bus

The application of CAN bus can reduce the body wiring, the signal transmission between modules only needs two signal lines, and all the other lines across the body are no longer needed except the bus, which greatly saves costs. The data of CAN bus system is stable and reliable, and the interline interference is small and the anti-interference ability is strong. A certain model has multiple control networks such as body, comfort and multimedia, among which the body control uses the CAN network to control the engine, gearbox, ABS and other body safety modules, and shares the speed, speed, oil temperature and so on to the whole car to achieve intelligent car control, such as automatically locking the door at high speed, automatically opening the door when the airbag pops up. Fig. 5.30 shows the CAN bus of a vehicle.

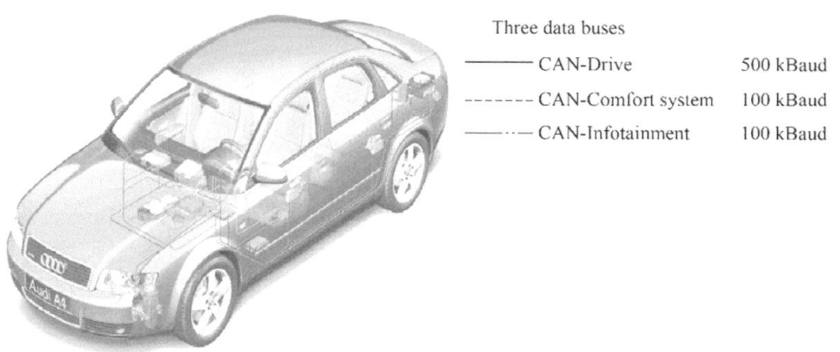

Fig. 5.30 A certain vehicle CAN bus

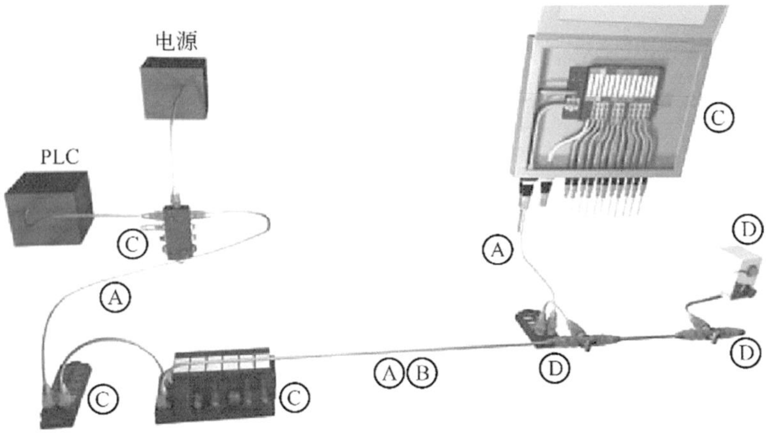

Fig. 5.31 Devicenet bus

⑤ Devicenet fieldbus. Devicenet (Fig. 5.31) is a low-cost communication bus that uses the Controller Local Area Network (CAN) as its underlying communication protocol, and its application layer has defined lines for different devices, including information exchange, security devices, and large control systems. It connects industrial equipment to the network, eliminating expensive hard wiring costs and improving communication between devices.

Devicenet defines the physical layer, data link layer and application layer in the seven-layer architecture of the OSI model. Flat cables can be used in the network to transmit signals and power small devices. Three bit rates are allowed: 125, 250 and 500Kbps, the length of the trunk is inversely proportional to the bit rate, and a single network can have up to 64 nodes, with the function of duplicate node address detection. Features that allow multiple masters in a single network can be used in high-noise environments.

The Devicenet specifications and protocols are free and open. The main features of Devicenet are: short frame transmission, the maximum data of each frame is 8 bytes; Non-destructive bit-by-bit arbitration technology; The network can connect up to 64 nodes; Data transmission baud rate of 125Kbps, 250Kbps, 500Kbps; Point-to-point, multi-master or master/slave communication mode; Physical and data link layer protocols of CAN are adopted.

⑥ HART Fieldbus. HART communication protocol (Highway Addressable Remote Transducer), is introduced in 1985 the United States ROSEMOUNT company a for field intelligent instrument and the communication protocol between the control room equipment, has a relatively low bandwidth, moderate response time of communication, has become a global intelligence Industry standard for instrumentation.

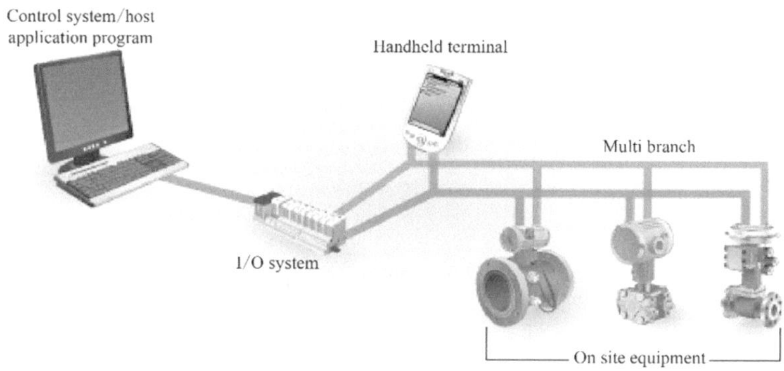

Fig. 5.32 HART field bus

The HART fieldbus (Fig. 5.32) uses the unified device description language DDL to describe device features. HART can use the bus power supply to meet the essential safety and explosion-proof requirements, the protocol can two-way transmission of digital information, breaking through the traditional instrument can only receive control information from the host, the transmission of large amount of information, each HART device includes such as equipment status, diagnostic alarm, process variables, units, loop current, manufacturers and other up to 40 standard information items. In digital communication mode, a pair of cables can handle multiple variables, and in field instrumentation, the HART protocol supports 256 process variables, and HART-compatible products and host systems from different vendors can all work together.

⑦ CC-Link fieldbus. CC-Link (Control & Communication Link) is an open fieldbus, as shown in Fig. 5.33, its data capacity is large, the communication speed of multi-level optional, is a composite, open, adaptable network system, the transmission rate of 10 Mbps, excellent performance, simple to use, widely used, not only solved Industrial field wiring complex problems, while providing excellent noise resistance and compatibility. CC-Link is a network based on the device layer, but can also cover the higher level of control layer and the lower level of sensing layer.

CC-Link is an open fieldbus with advanced technology, excellent performance, wide application, simple use and low cost. The network control system developed with CC-Link has the advantages of real-time, open, complete protection functions, fast communication rate, advanced network, convenient wiring, etc., which is conducive to centralized monitoring of decentralized systems and improvement of system automation level To reduce the labor intensity of workers.

⑧ INTERBUS Field bus. INTERBUS is an earlier field bus launched by the German company Phoenix (Fig. 5.34), which became the international standard IEC61158 in February 2000. As an open bus system, INTERBUS is

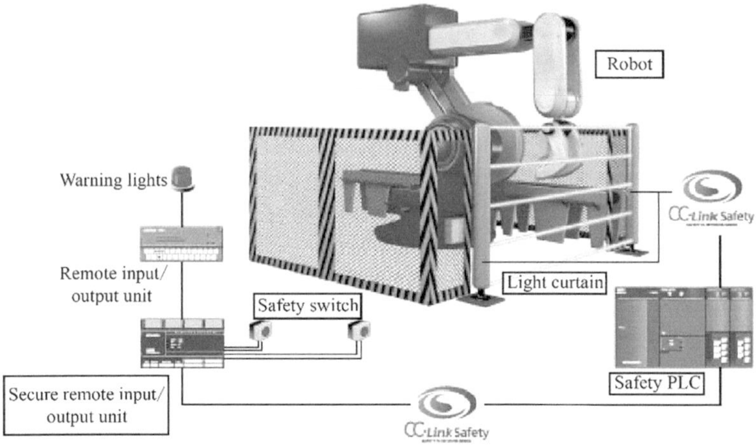

Fig. 5.33 CC-Link field bus

compatible with any PC platform and can be used in 80% of the world's PLCS. INTERBUS implements the control mode of open control, which is composed of open bus system INTERBUS, Microsoft open program, WindowsNT structure and industrial PC. The goal is to realize the common and unified information flow in office and production site based on the same platform. Through all levels and not stay, achieve full vertical integration (Complete Vertical Integration, CVI). Using open control, individual components of different hardware and software manufacturers can be integrated into an automated system, truly realizing the interoperability of different manufacturers, without any interface interface, widely used in automotive, tobacco, warehousing, paper, packaging, food and other industries, becoming an international fieldbus leader.

(4) Prospect and Development Trend of Fieldbus Technology

Fieldbus technology has been widely used in the field of industrial control, and the development of fieldbus technology has become a focus topic in the field of industrial automation. With the development of network technology, the field of low-speed fieldbus will continue to develop and improve, and high-speed fieldbus technology will also flourish. At present, fieldbus products are mainly low-speed bus products, which are used in the field of low running speed, and the performance requirements of the network are not very high. No matter what kind of fieldbus, it can better realize the process control with slow speed requirements. One of the key technologies of fieldbus is interoperability, so it is the desire of all users to achieve standardization and unification of fieldbus technology. The field bus technology with development prospect includes: hardware and software technology of intelligent instrument and network equipment development; Configuration resistance, including network topology, network

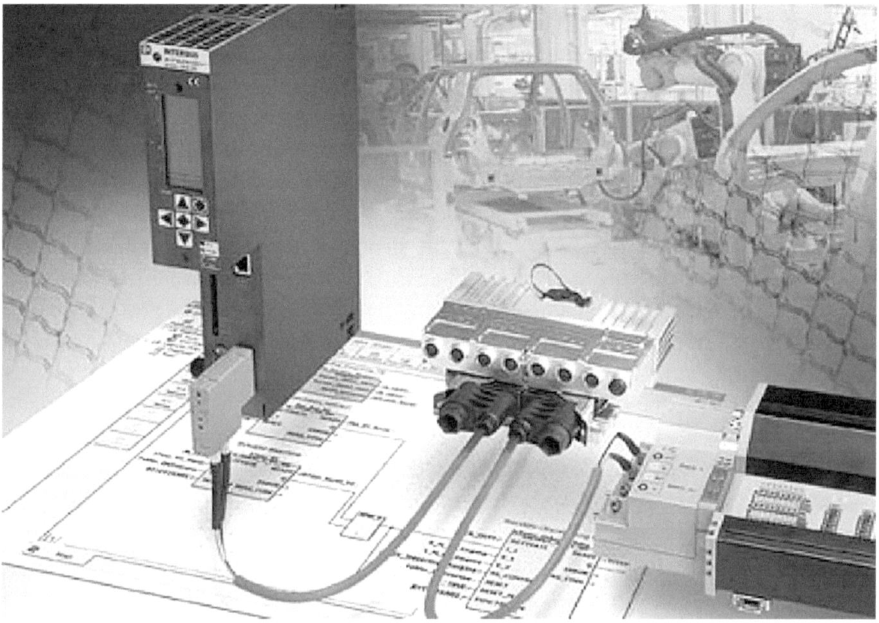

Fig. 5.34 INTERBUS field-bus structure

equipment, network segment interconnection, etc. Network management technology, including network management software, network data operation and transmission; Man–machine interface, software technology; Fieldbus system integration technology.

The high-speed fieldbus is mainly used for the interconnection of the control network, connecting the control computer, PLC and other intelligent devices with high processing speed, and realizing the connection between the low-speed fieldbus network Bridges, which is the necessary technology to fully realize the fully decentralized control structure of the system. The future competition for fieldbuses may be the design and development of high-speed fieldbuses [14, 15].

5.5.5 PC Numerical Control

(1) Composition of PC System

The composition structure of PC system [16] is quite different in different application fields. Commonly used for ordinary PC, industrial PC and PC104-based embedded PC system.

In the ordinary PC system, the CPU, memory, I/O controller, interface circuit, bus slot and other hardware modules are configured on a large motherboard, and it is fixed in the chassis with a regulated power supply, hard disk, optical drive, etc., thus constituting the host of the PC system. The host is then connected to standard external devices such as display, keyboard, and mouse, which constitutes a basic PC system. According to the needs of the application, on the basis of the basic system, the printer, scanner, pen pad, plotter, projector, microphone, sound device and other peripheral equipment can be further connected to form a PC system for specific applications.

The industrial PC system mostly uses the structure of passive backboard + independent plug-in board. The passive baseplate is fixed in the high-strength chassis. The hardware circuits (such as the CPU board, input and output circuits, and application modules) of the host are made into independent boards, which are plugged into the passive baseplate and secured by the fastening devices in the chassis. The chassis of industrial PC and its power supply system are designed with high anti-interference, special dust prevention measures are also taken in the inlet and outlet, and some even use a fully sealed structure to cool through the air conditioning system. In addition, in order to ensure the reliable operation of hard disk, optical drive and other devices in the industrial field, special seismic measures are also taken in the chassis.

The PC104 system designs the unit circuit of the computer into independent modules of the same size, such as CPU module, electronic disk module, network module, signal acquisition module, etc., and then connects these modules through the stack to form a complete system. The PC104 system generally does not design an independent computer chassis, but the PC system composed of a stack is embedded into the application system as a component, and is integrated with it to form a complete device.

Although the PC system in different application fields is not the same in external form and composition, and its physical structure is also different, the hardware structure and software system of the core part of the PC system are the same, and the same or compatible specifications and standards are adopted at the information level. For example, ordinary PC and PC104 embedded computer, although in the external form, components and physical structure are completely different, but the two types of PC system used by the system software (operating system, etc.) can be exactly the same, the core of the system and the information exchange between the external device based on the specifications and standards can also be exactly the same. This allows applications that can run on ordinary PCS to run directly on PC104 systems; External devices that can be connected to an ordinary PC can also be directly connected to the PC104 system. For specific tasks, there is not much difference between the two, except that the former is less expensive and the latter is more reliable.

(2) The Working Principle of PC System

The core hardware of PC system includes CPU, memory, etc., and peripheral hardware includes I/O controller and interface circuit (such as hard disk controller, floppy controller, USB controller, network interface, keyboard interface, mouse interface, parallel interface, serial interface, etc.). Application software is a software system configured to accomplish certain tasks. An operating system is a software platform that supports the running of application software.

As can be seen from Fig. 5.35, the working process of a PC system can be regarded as a process of information processing and transformation. For a specific application (such as design, calculation, analysis, etc.), the original information required to complete the task is first entered into the PC through the input devices of the PC system, such as floppy drives, flash drives, scanners, networking devices, etc.; Secondly, through the hardware system and operating system in the PC host, the input information is transmitted to the application software to complete the task. The application software can cooperate with the operating system and the PC core hardware to complete the information transformation and processing task and produce the required output information. Then through the operating system and hardware system, the output information is sent to the PC outside; Finally, the output information is displayed and saved to a suitable carrier through the output devices of the PC system, such as printers, plotters, projectors, and external storage devices.

It should be pointed out that the process of PC system to complete the task is generally a man–machine coordination, repeated interaction process. In this process, the operator needs to communicate with the computer repeatedly through interactive devices such as keyboard, mouse, and display to complete the task. For example, the use of PC system engineering design, in the design process, designers need to interact with the computer repeatedly according to their own knowledge and experience in order to produce an ideal design.

Figure 5.36 shows the basic structure of a PC numerical control system. The controller of PC numerical control system (PC numerical control device) is a digital controller with PC as the core, and its basic structure is composed of two parts: hardware system and software system.

The hardware system consists of PC hardware platform and additional CNC hardware module. Whether it is an ordinary PC or an industrial PC, its hardware platform is universal, without the need for CNC system manufacturers to produce themselves.

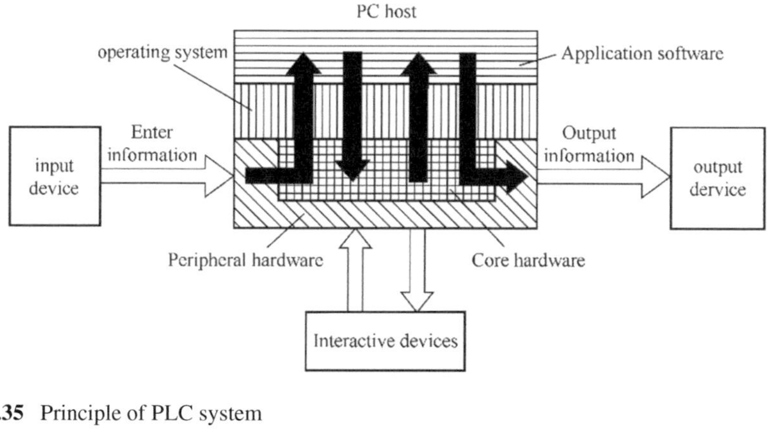

Fig. 5.35 Principle of PLC system

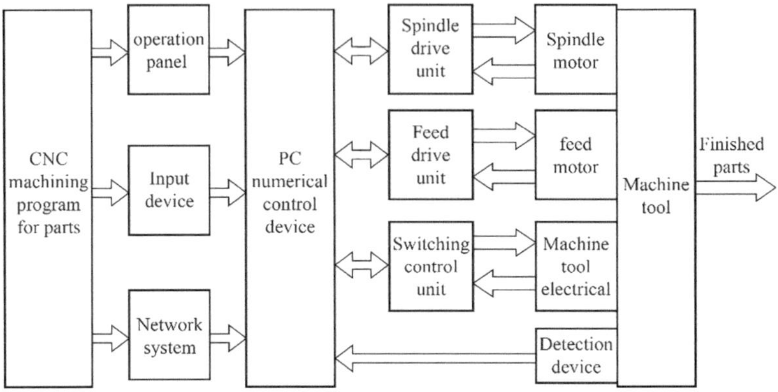

Fig. 5.36 Basic structure PC numerical control system

CNC hardware module is a function module attached to the PC hardware platform to complete specific CNC tasks, and these modules generally need to be designed and manufactured by the CNC system manufacturer.

The software system includes PC operating system and CNC application software. PC operating system belongs to the system software, is universal, can be purchased from the market. CNC application software is composed of various information processing software modules and control software modules to complete CNC tasks, which has a strong pertinence and needs to be developed by CNC system manufacturers themselves. The function and performance of PC numerical control system are mainly determined by the numerical control application software, so the development of numerical control application software is the main task of the development of PC numerical control system.

5.5.6 Advanced Control Technology Methods

(1) Neural Network Learning Intelligent Control

Neural network refers to a network composed of a large number of artificial neurons that are similar to the nerve cells of the biological nervous system, or a large number of processing units like biological neurons in parallel. This neural network has some intelligent and humanoid control functions. Learning algorithm is the main feature and research topic of neural network. Neural networks have the function of learning similar to human, learning refers to the effective self-regulation of the body in the complex and changeable environment. The learning process of neural network is the process of modifying the weighted coefficient of the input, and finally making its output reach the expected value. Commonly used learning algorithms include Hebb learning algorithm, WidrowHoff learning algorithm, backpropagation learning algorithm – BP learning algorithm, Hopfield feedback neural network learning algorithm and so on.

Neural network uses a large number of neurons, according to a certain topology and learning adjustment method, to realize parallel computing, distributed storage, variable structure, highly fault-tolerant, nonlinear operation, self-organization, learning or self-learning, etc., and has unique capabilities in the control of parameters, structures or environments of intelligent control, self-organization, self-learning and so on.

(2) Adaptive Control

Adaptive control, like conventional feedback control and optimal control, is also a control method based on mathematical model. The research object of adaptive control is to control the system with a certain degree of uncertainty. Uncertainty means that the mathematical model describing the controlled object and its environment is not completely deterministic and contains some unknown factors and random factors. The adaptive control is based on less prior knowledge about the model and the disturbance, so it needs to constantly extract the information about the model during the operation of the system, so as to gradually improve the model. According to the input and output data of the object, the model parameters are constantly identified, which is called online identification of the system. As the production process continues and online identification is used, the model will become more and more accurate and closer to reality. The model changes with the environment, so the control function based on this model will be constantly adjusted and improved, and the control system has a certain adaptability.

(3) Robust Control

Robustness refers to the characteristics of the control system to maintain certain performance under the perturbation of parameters of a certain structure and size. The definition of performance is different, which can be divided into stable robustness and performance robustness. A fixed controller designed with the robustness of a closed-loop system as its goal is called a robust controller. The main robust control theories include Kharitonov interval theory, $H\infty$ control theory, structure singular value theory (μ theory) and so on. The research of robust control began in the 1950s and is one of the hot spots in the international self-control field.

Robust control combines the advantages of classical frequency domain design theory and modern control theory, systematically gives the technology and means of loop shaping in frequency domain, and fully considers the influence of system uncertainty, which can not only ensure the robust stability of the control system, but also optimize some performance indexes. The state-space method has the advantages of accurate calculation and optimization in time domain, and many kinds of control problems can be transformed into $H\infty$ robust control theory standard problems, which are general and suitable for practical engineering applications.

5.6 PLC Control System Design

In the design of PLC control system, first of all, we should be familiar with the controlled object and calculate the input/output device, then select PLC and determine the hardware configuration, on this basis, design the electrical schematic diagram and design the console (cabinet), and finally compile the control program, debug the program and prepare technical documents, as shown in Fig. 5.37.

First of all, we must be familiar with the controlled object and design the process layout, which is the basis of system design. Understand in detail the process of the controlled object and its requirements for the control system, the relationship between various mechanical, hydraulic, pneumatic, instrumentation, and electrical systems, the working mode of the system (such as automatic, semi-automatic, manual, etc.), the relationship between PLC and other intelligent devices in the system, the type of human–machine interface, the way of communication networking, the type and scope of alarm, power outage and emergency Handling of conditions, etc. In addition, the user input device (such as buttons, operating switches, limit switches, sensors, etc.), the output device (such as relays, contactors, signal indicators, etc.), and the controlled object driven by the output device (such as motors, solenoid valves, etc.) should be selected.

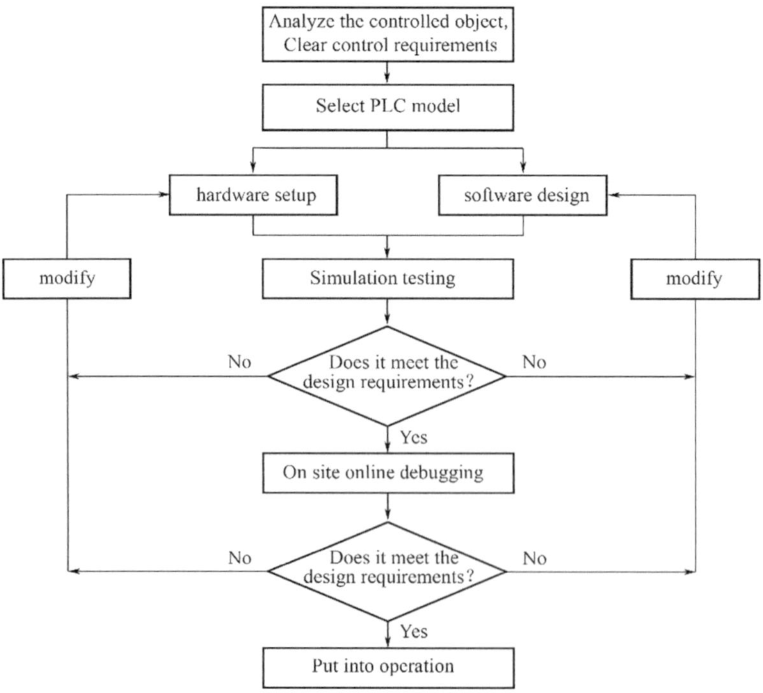

Fig. 5.37 Design of PLC control system

Determine which signals need to be input to the PLC, which load is driven by the PLC, and classify the nature and quantity of each input and output – whether it is a digital quantity or an analog quantity, whether it is a direct flow or an AC quantity, and the size of the voltage level, to provide a basis for the selection and hardware configuration of the PLC. Finally, the control object and control function are classified, which can be divided according to the signal use or according to the control area, to determine the physical location of the detection equipment and control equipment, and to analyze the form, function, scale and relationship between each detection signal and control signal. After the signal point is determined, the process layout or signal diagram is designed.

5.6.1 Hardware Design of PLC Control System

With the promotion and popularization of PLC in industrial control field, the types and quantities of PLC products are increasing. Foreign PLC products, domestic products have dozens of series, hundreds of models. There are many varieties of PLC, its structure form, performance, capacity, instruction system, configuration method, programming method, price and so on are different, and the use of occasions are also different. When designing the control system, PLC should be reasonably selected according to the actual situation.

(1) Selection of PLC Models

The selection of PLC models should be based on the premise of meeting the control requirements to ensure reliability, easy maintenance and use, and the best cost performance. According to the field application requirements, usually can be selected according to the control function or input and output points. The I/O points of the overall PLC are fixed, so the user has less room for choice and is used for small control systems; Modular PLC provides a variety of I/O cards or cards, so users can choose and configure the I/O points of the control system more reasonably, and the function expansion is convenient and flexible, generally used for large and medium-sized control systems.

The selection of input and output modules should be considered in accordance with the application requirements. For the input module, the application requirements such as signal level, signal transmission distance, signal isolation and signal power supply mode should be considered. For the output module, the type of output module should be considered. Generally, the relay output module has the characteristics of low price, wide voltage range, short life and long response time. Thyristor output module is suitable for frequent switching and low inductive power factor load, but the price is expensive and the overload capacity is poor. The output module also has DC output, AC output and analog output, which should be consistent with the application requirements.

The power module is introduced into the PLC at the same time in the introduced equipment, but it should be designed and selected according to the requirements of the product manual, and the general PLC power supply should be designed and selected 220VAC power supply, consistent with the domestic grid voltage. In important occasions, uninterruptible power supply or regulated power supply should also be used. In order to prevent voltage fluctuations from damaging the PLC, it is necessary to isolate the input and output signals, sometimes using simple diodes or fuse tubes.

In addition, when choosing PLC, the cost performance should be considered. When considering economy, factors such as application scalability, operability and input–output ratio should be considered at the same time to compare and take into account, and finally select a more satisfactory product. The increase of points has an impact on the selection of CPU, memory capacity, control function range, etc., and should be fully considered in the estimation and selection to improve the cost performance of the control system.

① Selection of operation control function. The operation function of PLC includes logic operation, timing and counting, data shift, comparison function, algebraic operation, data transmission, PID operation and other advanced operation functions, and the current PLC has communication function. Design and selection should be based on the requirements of practical applications, reasonable selection of the required computing functions.

For small single equipment that only needs digital quantity control, general small PLC (such as Siemens S7-200 series, OMRON CPM1/CPM2 series, etc.) can meet the requirements. For the application system based on digital quantity control and with A small amount of analog quantity control, such as the continuous quantity control of temperature, pressure, flow, etc., which is often encountered in industrial production, the analog input and output module with A/D conversion and D/A conversion should be selected, and the corresponding sensor, transmitter and drive device should be connected, and the small PLC with strong operation and data processing function (such as West) should be selectedS7-200 or S7-300 series from Siemens).

For engineering projects with more complex control and higher control function requirements, such as PID operation, closed-loop control, communication networking and other functions, the scale and complexity of visible control are selected for mid-range or high-end machines (such as Siemens S7-300 or S7-400 series, etc.).

The control function includes PID control operation, feedforward compensation control operation, ratio control operation, etc. It should be determined according to the control requirements whether PID control unit, high-speed counter, analog unit with speed compensation, ASC code conversion unit, etc., need to consider the actual situation to determine.

② Programming function. PLC control system has five standardized programming languages: sequence function diagram (SFC), ladder diagram (LD), function module diagram (FBD) three graphical languages and statement table (IL), structure text (ST) two text languages. The selected programming language should comply with its standards (IEC6113123), at the same time, it should also support a variety of language programming forms, such as C, Basic, etc., to meet the control requirements of special control occasions.

There are two PLC programming methods: offline programming and online programming. Offline programming means that PLC and the programmer share a CPU, the programmer in programming mode, the CPU only provides services for the programmer, does not control the field equipment, after the completion of programming, the programmer switches to the running mode, the CPU controls the field equipment, can not be programmed. The off-line programming method can reduce the system cost, but it is not convenient to use and debug. Online programming means that the CPU and the programmer have their own CPU, the host CPU is responsible for field control, and in a scan cycle with the programmer for data exchange, the programmer to the online program or data sent to the host, the next scan cycle, the host according to the newly received program to run. This method has high cost, but the system debugging and operation are convenient, and it is often used in large and medium-sized PLCS.

Portable simple programmer is mainly used in small PLC, its control scale is small, the program is simple, the simple programmer can be used. CRT programmer is suitable for large and medium-sized PLC, in addition to preparing and entering programs, but also can edit and print program text. Since IBM-PC has been popularized, IBM-PC and its compatible machine programming packages are a good programming tool for PLC.

③ Scanning speed. PLC uses scanning mode to work, the processing speed is related to the length of the user program, CPU processing speed, software quality and so on. At present, the PLC contact has fast response and high speed, and the execution time of each binary instruction is 0.2–0.4 ns, so it can adapt to the application needs of high control requirements and corresponding requirements. Scan cycle (processor scan cycle) should meet: small PLC scan time is not more than 0.5 ms/K; The scanning time of large and medium-sized PLC is not more than 0.2 ms/K. The appropriate model should be selected according to the situation in the design.

④ Networking communication function. PLC as the main control device of factory automation, most products have communication networking capabilities. Select the communication mode as required. Large and medium-sized PLC systems should support a variety of fieldbuses and standard communication protocols (such as TCP/IP), and should be able to connect with the factory management network (TCP/IP) when needed. Communication protocols should conform to ISO/IEEE communication standards. The communication interface of PLC system should include serial and parallel communication interface (RS-232C/422A/423/485), RIO communication interface, industrial Ethernet, common DCS interface, etc. Large and medium-sized PLC communication bus (including interface equipment and cables) should be 1:1 redundant configuration, the communication bus should meet international standards, and the communication distance should meet the actual requirements of the device.

⑤ Other requirements. Considering the special requirements of the controlled object for closed-loop control of the analog, high-speed counting, motion control and human–machine interface (HMI), PLC with corresponding special I/O modules can be selected. In some application scenarios with high reliability requirements, redundant control systems or backup systems should be used.

In system design, in addition to hardware, PLC programming is also very important. In addition to a detailed understanding of the hardware, the user should have an understanding of the software functions of the selected PLC product. For the network control structure or the control system that needs to be managed by the upper computer, the communication software package is the main basis for choosing PLC. Communication software packages are often used with communication hardware, such as modems.

PLC is usually directly used for industrial control, the general industrial site can work reliably, in the selection of environmental conditions should be given full consideration. General PLC and its external circuit (including I/O module, auxiliary power supply, etc.) can work reliably.

(2) PLC Capacity Estimation

The capacity of PLC includes the total number of input and output points and the storage capacity of the user memory. In the selection of PLC models should be based on the actual situation, do not blindly pursue too high performance indicators, to meet the actual requirements. In general, in addition to meeting the requirements of the control system in terms of input and output points and memory capacity, it is recommended to leave an appropriate margin to be used for bad point backup or equipment expansion functions and system expansion.

① Determination of I/O points. The number of PLC I/O points is determined based on the actual input and output points of the system. When determining the number of I/O points, an appropriate margin should be considered for maintenance and expansion. Usually I/O points can be considered as 10% to 15% of the actual need; If there are a large number of I/O modules, spare modules are reserved according to the preceding ratio. When actually ordering, it is also necessary to round out the input and output points according to the product characteristics of the manufacturer's PLC.

② Determination of memory capacity. The memory capacity is the size of the hardware storage unit that can be provided by the programmable controller itself, and the program capacity is the size of the storage unit used by the user application project in the memory, so the program capacity is less than the memory capacity. The amount of memory capacity occupied by the user program is related to many factors, such as I/O points, control requirements, computing processing capacity, program structure, etc. Since the user application has not yet been written, it can only be roughly estimated before the program is written. There is no fixed formula for the estimation of memory memory capacity, and different formulas are given in many literatures, which are generally 10 to 15 times the number of digital I/O points, plus 100 times the analog I/O points, which is the total number of words in memory (16 bits for a word), and then consider the margin according to 25% of this number.

(3) I/O Module Selection

In the PLC control system, the various measurement parameters of the object need to be input into the PLC, after CPU calculation and processing, and then the result is output in the form of a digital quantity, and the input/output (I/O) module is required between the PLC and the production process. Production equipment or control system of switches, buttons, relay contacts, etc., only on or off two state information, such signals need to be picked up through the digital input module to achieve. The most common input module is 24 V DC input, as well as DC 5 V, 12 V, 48 V, AC 115 V, 220 V and so on. The control of the indicator light on and off, the start and stop of the motor, the on and off of the thyristor, the opening and closing of the valve, etc., only needs to use "1" and "0" binary logic, and is driven by the digital output module.

Parameters such as temperature, pressure, liquid level, flow rate, etc., can be converted into the corresponding analog signal by different detection devices, and then input it into the PLC through the analog input module, and converted into a digital quantity. Many actuators of production equipment or processes often require analog signals to be controlled, and the control signal output by PLC is a digital quantity, which requires a corresponding module to convert it into analog quantity. This module is the analog output module.

(4) Assign Input/Output Points

After the selection of PLC models, the number of input/output points is an important factor in determining the price of the control system and the rationality of the design, so in the case of completing the same control function, the input/output points can be simplified through reasonable design. After PLC model and input/output (I/O) module selection, the overall configuration diagram of PLC system is designed. Then according to the process layout, referring to the specific PLC related instructions or manuals, the input signal and the input point, the output control signal and the output point correspond one by one, and draw the I/O wiring diagram, that is, the PLC input/output electrical schematic diagram.

(5) Safety Circuit Design

The safety loop is a chain control loop that protects the load or control object and prevents operation errors or control failures, generally considering short circuit protection, interlock, and voltage shutdown. In the PLC external output of each load circuit to install fuses for short circuit protection, fuse specifications should be reasonable choice according to the load parameters. The control software should ensure the interlock of the circuit, in addition, the hardware interlock measures should be taken in the PLC external wiring to ensure the safe and reliable operation of the system. The power supply line of the PLC external load should have voltage loss protection measures, when the temporary power failure and then restore the power supply, you must press the "start" button of the PLC external load to start by itself. In an emergency, press the "emergency stop" button to cut off the load power supply, interrupt the program, and protect personal property and equipment safety. In some exceeding the limit may produce dangerous occasions, such as elevators, tower cranes, etc., should also set the limit protection, when the limit protection action, directly cut off the load power supply, while the signal input PLC.

5.6.2 Software Design of PLC Control System

Software design is the core of PLC control system design. The application software design of PLC refers to the process of compiling and forming the corresponding files for the user control program according to the hardware structure and process requirements of the control system, using the corresponding configuration software and programming language. To design the PLC application software, it is necessary to fully understand the production process, technical characteristics and control requirements of the controlled object. The control functions of the system are completed through the application software of PLC. Software design includes system initialization program, main program, subroutine, interrupt program, fault emergency measures and auxiliary program design, small switching control system generally only main program. According to the general requirements and the specific situation of the control system, determine the basic structure of the program, draw the control flow chart or function flow chart, simple system can be designed by empirical method, complex system is generally designed by sequential control design method.

(1) PLC Application Software Design Content

The main content of PLC application software design includes: determine the program structure, define the input/output, intermediate mark, timer, counter and data area and other parameter tables, prepare the program, write the program specification. The PLC application software design also includes the configuration of human machine interface (HMI) devices such as text display or touch screen and other special function modules.

(2) Familiar with the Controlled Object to Develop Equipment Operation Plan

PLC hardware design includes PLC and peripheral circuit design, electrical circuit design and anti-interference measures design.

After selecting the PLC model and distributing I/O points, the main content of hardware design is the design of the schematic diagram of the electrical control system, the selection of electrical control components and the design of the control cabinet. The schematic diagram of the electrical control system includes the main circuit and the control circuit. The selection of electrical components is mainly based on the control requirements to select buttons, switches, sensors, protective appliances, contactors, indicators, solenoid valves, etc.

After the completion of these works, based on this, according to the requirements of the production process, the logical relationship between each input/output and various operations is analyzed, the detection amount and control method are determined, and the operation content and operation sequence of each equipment in the system are designed. For more complex systems, the system can be controlled according to physical location or control function. Figure 5.38 shows the PLC control process of the key equipment of the large insulating glass automatic production line developed by the author's team. More complex systems generally need to draw a system control flow chart to clearly indicate the order and conditions of actions. Since the PLC program execution is a cyclic scanning working mode, the PLC program block diagram starts the input scanning again after the output refresh and the cycle execution. Tables 5.5, 5.6, 5.7, 5.8 and 5.9 lists the PLC control parameters of the chip combining machine.

(3) Familiar with Programming Languages and Software

PLC design needs to master the programming language and programming software, according to the relevant manual to understand the programming software and its operating system in detail, select one or several suitable programming language forms, familiar with its instruction system and parameter classification and prepare some test procedures on the computer operation training, on the simulation platform for trial run.

(4) Define Parameter Table

Parameter table definitions include definitions for input/output, intermediate flags, timers, counters, and data areas. There is no uniform standard for the definition format and content of the parameter table, which varies according to the situation of the system and personal preferences, but the content contained is basically the same. The general design principle of parameter tables is to be as detailed as possible. The input/output signal table should be defined according to the PLC input/output electrical schematic before the programming begins. Each kind of PLC input point number and output point number have their own clear provisions, after determining the PLC model and configuration, to allocate the input/output signal PLC input/output number (address), and compiled into a table.

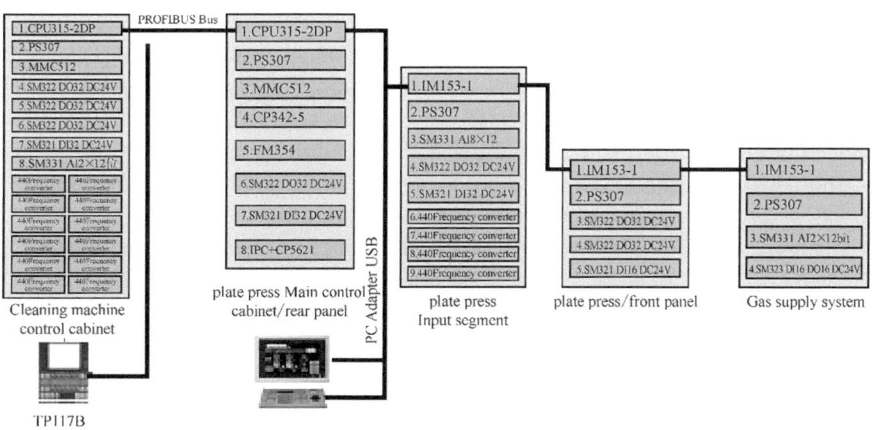

Fig. 5.38 PLC control process of laminating machine

Table 5.5 Corresponding I/O variable table for each control cabinet

Control cabinet	I/O module	Address
Control cabinet I (main control cabinet)	SM321 DI32	0.0–3.7
	SM322 DO32	0.0–3.7
	FM354-1	304–319
	FM354-2	320–335
Control cabinet II (input + detection)	SM321 DI32	4.0–7.7
	SM322 DO32	4.0–7.7
	SM331 AI8 × 12	256–271
Control cabinet III (front panel of the main unit)	SM321 DI16	8.0–9.7
	SM322 DO32	8.0–11.7
	SM322 DO32	12.0–15.7
Control cabinet IV (inert gas)	SM323 DI 16 DO 16	I2.0–15.7
		I0.0–11.7
	SM331 AI2 × 12	272–275
	SM331 AI2 × 12	276–279

The input/output signal table should clearly indicate the position of the template, the input/output address number, the signal name and the signal type, etc. The input/output definition table comments should be as detailed as possible. The addresses should be arranged in the order from small to large. Points that are not defined or standby need to be numbered to facilitate the search and use during programming, debugging, and modification of the program, and to facilitate the expansion of functions in the future. Intermediate flags, timers, counters and data areas are generally defined when used in the programming process, and are consolidated during or after the programming process along with input/output signal tables.

(5) Programming

Simple PLC program, can be written by translation method. The components in the original relay-contactor control circuit schematic diagram are replaced by the functional equivalent components in the PLC of the selected model, and the relay-contactor control circuit is translated into the PLC ladder program diagram. For the program design of sequence control mode or step control mode, function diagram or state flow diagram can be used, which is clear and intuitive.

For more complex logic control, when programming, Boolean logic algebra is taken as the theoretical basis, logic variable "0" or "1" is taken as the research object, and three basic logic operations of "and" or "not" are taken as the analysis basis, and the "on" and "off" state of the contact is expressed by logic variable "0" or "1"The expression of the multivariable "and" logic relationship can be directly converted into a ladder diagram of contact series. An expression with a multivariable "or" logical relationship can be directly converted into a ladder diagram with parallel

Table 5.6 Variable table of control cabinet I

Signal description	Symbol	PLC I/O address	Upper address (WinCC)	Data type
Front limit of the shaft	Spindle-L-1	I0.0		
Rear limit of the shaft	Spindle-L-2	I0.1		
Front limit of activity film	Mov Piece-L-1	I0.2		
Rear limit of activity film	Mov Piece-L-1	I0.3		
		I0.4		
		I0.5		
		I0.6		
Maintenance limit	Examine limit	I0.7		
Conveyor belt left cylinder + position	Bel-L-Cylinde+	I1.0		
Conveyor belt left cylinder position	Bel-L-Cylinde−	I1.1		
Conveyor belt cylinder + position	Bel-M-Cylinde+	I1.2		
Conveyor belt cylinder position	Bel-M-Cylinde+	I1.3		
Conveyor belt right cylinder + position	Bel-R-Cylinde+	I1.4		
Conveyor belt right cylinder position	Bel-R-Cylinde+	I1.5		
		I1.6		
		I1.7		
Activity film right cylinder 1 + position	M-L-Cldr-1+	I2.0		
Activity film right cylinder 1 − position	M-L-Cldr-1−	I2.1		
Activity film right cylinder 2 + position	M-L-Cldr-2+	I2.2		
Activity film right cylinder 2 − positionI	M-L-Cldr-2−	I2.3		
Activity film left cylinder 1 + position	M-L-Cldr-1+	I2.4		

(continued)

Table 5.6 (continued)

Signal description	Symbol	PLC I/O address	Upper address (WinCC)	Data type
Activity film left cylinder position 1	M-L-Cldr-1–	I2.5		
Activity film left cylinder 2 + position	M-L-Cldr-2+	I2.6		
Activity film left cylinder 2 – position	M-L-Cldr-2–	I2.7		
		I3.0		
		I3.1		
		I3.2		
		I3.3		
		I3.4		
		I3.5		
		I3.6		
		I3.7		
Servo I power on	Serve-I-On	Q0.0		
Servo II powered on	Serve-II-On	Q0.1		
Fan on	Fan-On	Q0.2		
Vacuum pump on	Vacuum-On	Q0.3		
Conveyor belt left cylinder + position	Belt-L-Cylinder+	Q0.4		
Conveyor belt left cylinder position	Belt-L-Cylinder–	Q0.5		
Conveyor belt cylinder + position	Belt-M-Cylinder+	Q0.6		
Conveyor belt cylinder position	Belt-M-Cylinder+	Q0.7		
Conveyor belt right cylinder + position	Belt-R-Cylinder+	Q1.0		
Conveyor belt right cylinder position	Belt-R-Cylinder–	Q1.1		
On the left door cylinder+	Dr-L-Clde-U+	Q1.2		
On the left door cylinder–	Dr-L-Clde-U–	Q1.3		
Left door cylinder mounted	Dr-L-Clde-U Static	Q1.4		
Under the left door cylinder+	Dr-L-Clde-D+	Q1.5		

(continued)

Table 5.6 (continued)

Signal description	Symbol	PLC I/O address	Upper address (WinCC)	Data type
Under the left door cylinder−	Dr-L-Clde-D−	Q1.6		
Left door cylinder downward positioning	Dr-L-Clde-U Static	Q1.7		
On the right door cylinder+	Dr-L-Clde-U+	Q2.0		
On the right door cylinder−	Dr-L-Clde-U−	Q2.1		
Under the right door cylinder+	Dr-L-Clde-D+	Q2.2		
Under the right door cylinder−	Dr-R-Clde-D−	Q2.3		
Repair the safety cylinder+	Examine-Cldr+	Q2.4		
Repair the safety cylinder−	Examine-Cldr−	Q2.5		
Overhaul the large cylinder+	Examine-BigCldr+	Q2.6		
Overhaul the large cylinder−	Examine-BigCldr−	Q2.7		
Activity film left cylinder 1 + position	M-R-Cldr-1+	Q3.0		
Activity film left cylinder position 1	M-R-Cldr-1−	Q3.1		
Activity film left cylinder 2 + position	M-R-Cldr-2+	Q3.2		
Activity film left cylinder 2 − position	M-R-Cldr-2−	Q3.3		
Activity film right cylinder 1 + position	M-R-Cldr-1+	Q3.4		
Activity film right cylinder 1 − position	M-R-Cldr-1−	Q3.5		
Activity film right cylinder 2 + position	M-R-Cldr-2+	Q3.6		
Activity film right cylinder 2 − position	M-R-Cldr-2−	Q3.7		

Table 5.7 Variable table of control cabinet II

Signal description	Symbol	PLC I/O address	Upper address (WinCC)	Data type
Input segment sensor1	Input Sensor-1	I4.0		
Input segment sensor2	Input Sensor-2	I4.1		
Input segment sensor3	Input Sensor-3	I4.2		
Input segment sensor4	Input Sensor-4	I4.3		
Input segment sensor5	Input Sensor-5	I4.4		
Input segment sensor6	Input Sensor-6	I4.5		
Output segment sensor1	Outnput Sensor-1	I4.6		
Output segment sensor2	Outnput Sensor-2	I4.7		
Foot switch	Foot switch	I5.0		
		I5.1		
		I5.2		
		I5.3		
		I5.4		
		I5.5		
		I5.6		
		I5.7		
Height measuring sensor1	H-M-Sensor 1	I6.0		
Height measuring sensor2	H-M-Sensor 2	I6.1		
Height measuring sensor3	H-M-Sensor 3	I6.2		
Height measuring sensor4	H-M-Sensor 4	I6.3		
Height measuring sensor5	H-M-Sensor 5	I6.4		
Height measuring sensor6	H-M-Sensor 6	I6.5		
Height measuring sensor7	H-M-Sensor 7	I6.6		
Height measuring sensor8	H-M-Sensor 8	I6.7		
Height measuring sensor9	H-M-Sensor 9	I7.0		
Height measuring sensor10	H-M-Sensor 10	I7.1		
Height measuring sensor11	H-M-Sensor 11	I7.2		
		I7.3		
		I7.4		
		I7.5		
		I7.6		
		I7.7		
Positioning in the X direction on the upper frame	Frame-X orientation	I256.0–257.7		
Positioning in the Y direction of the upper frame-1	Frame-Y orientation-1	I258.0–259.7		

(continued)

Table 5.7 (continued)

Signal description	Symbol	PLC I/O address	Upper address (WinCC)	Data type
Positioning in the Y direction of the upper frame-2	Frame-Y orientation-2	I260.0–261.7		
Height detection	H-Measurt	I262.0–263.7		
Glass thickness detection	Thickness-Measure	I264.0–265.7		
Aluminum frame width detection	Aluminum Measure	I266.0–267.7		
		I268.0–269.7		
		I270.0–271.7		
Power on the frequency converter	Transducer On	Q4.0		
		Q4.1		
		Q4.2		
		Q4.3		
		Q4.4		
		Q4.5		
		Q4.6		
		Q4.7		
Height measuring cylinder+	H-Measure Cldr+	Q5.0		
Height measuring cylinder−	H-Measure Cldr−	Q5.1		
Glass thickness measuring cylinder+	T-Measure Cldr+	Q 5.2		
Glass thickness measuring cylinder−	T-Measure Cldr−	Q5.3		
		Q5.4		
		Q5.5		
		Q5.6		
		Q5.7		
On the upper frame cylinder+	Frame-Cldr-U+	Q6.0		
On the upper frame cylinder−	Frame-Cldr-U−	Q6.1		
Under the upper frame cylinder+	Frame-Cldr-D+	Q6.2		
Under the upper frame cylinder−	Frame-Cldr-D−	Q6.3		
		Q6.4		
		Q6.5		
		Q6.6		
		Q6.7		
DC I forward rotation	Dc Motor-I+	Q7.0		
DC I reversal	Dc Motor-I−	Q7.1		
DC II forward rotation	Dc Motor-II+	Q7.2		

(continued)

Table 5.7 (continued)

Signal description	Symbol	PLC I/O address	Upper address (WinCC)	Data type
DC II inversion	Dc Motor-II−	Q7.3		
DC III forward rotation	Dc Motor-III+	Q7.4		
DC III inversion	Dc Motor-III−	Q7.5		
		Q7.6		
		Q7.7		

contacts. An expression with a multivariable "and or" or "or" logical relationship can be directly converted into a ladder diagram of series and parallel contacts.

If there is operating system support, try to use high-level programming language forms, such as ladder diagram language. In the process of writing, according to the actual needs, the intermediate flag signal table and the storage unit table are defined one by one, and enough common temporary area should be set aside to save the use of memory.

Many small PLCS use simple programmers that can only enter instruction codes. After the ladder diagram is designed, it is also necessary to make the ladder diagram a code program according to the instruction statement and list the program list. After being familiar with the selected PLC instruction system, it is easy to write the statement table program according to the ladder diagram.

Similar to other high-level language programming, PLC programming process to make timely comments on the programmed program, so as not to forget the relationship between them, increase the readability and error correction of the program. Comments should include the function of the program segment, logical relations, design ideas, signal source and destination, so as to facilitate the reading and debugging of the program, improve the efficiency of programming.

(6) Testing of Procedures

After the PLC program is written, it is necessary to test the program. Program testing is an important part of the whole program design work, which can check the actual running effect of the program. Program testing and program writing are often carried out alternately, testing can find some problems in the program, and then make programming modifications, so repeated, improve the reliability of the program. When testing, start with each functional unit module, set the input signal, observe the effect of the change of the input signal on the system, and use the instrument if necessary. After the completion of the test of each functional unit module, the whole program is connected to test the interface of each part until it is satisfied.

Program testing can be done in the laboratory or in the field. If the program test is carried out on site, it is necessary to isolate the PLC from the field signal to avoid safety accidents due to imperfect procedures.

Table 5.8 Variable table of control cabinet III

Signal description	Symbol	PLC I/O address	Upper address (WinCC)	Data type
Suction cup micro switch1	Sucker etrl switch-1	I8.0		
Suction cup micro switch2	Sucker etrl switch-2	I8.1		
Suction cup micro switch3	Sucker etrl switch-3	I8.2		
Suction cup micro switch4	Sucker etrl switch-4	I8.3		
		I8.4		
		I8.5		
		I8.6		
		I8.7		
		I9.0		
		I9.1		
		I9.2		
		I9.3		
		I9.4		
		I9.5		
		I9.6		
Sucker1	Sucker1	Q8.0		
Sucker2	Sucker2	Q8.1		
Sucker3	Sucker3	Q8.2		
Sucker4	Sucker4	Q8.3		
Sucker5	Sucker5	Q8.4		
Sucker6	Sucker6	Q8.5		
Sucker7	Sucker7	Q8.6		
Sucker8	Sucker8	Q8.7		
Sucker9	Sucker9	Q9.0		
Sucker10	Sucker10	Q9.1		
Sucker11	Sucker11	Q9.2		
Sucker12	Sucker12	Q9.3		
Sucker13	Sucker13	Q9.4		
Sucker14	Sucker14	Q9.5		
Sucker15	Sucker15	Q9.6		
Sucker16	Sucker16	Q9.7		
Sucker17	Sucker17	Q10.0		
Sucker18	Sucker18	Q10.1		

(continued)

Table 5.8 (continued)

Signal description	Symbol	PLC I/O address	Upper address (WinCC)	Data type
Sucker19	Sucker19	Q10.2		
Sucker20	Sucker20	Q10.3		
Sucker21	Sucker21	Q10.4		
Sucker22	Sucker22	Q10.5		
Sucker23	Sucker23	Q10.6		
Sucker24	Sucker24	Q10.7		
Sucker25	Sucker25	Q11.0		
Sucker26	Sucker26	Q11.1		
Sucker27	Sucker27	Q11.2		
Sucker28	Sucker28	Q11.3		
Sucker29	Sucker29	Q11.4		
Sucker30	Sucker30	Q11.5		
Sucker31	Sucker31	Q11.6		
Sucker32	Sucker32	Q11.7		
Sucker33	Sucker33	Q12.0		
Sucker34	Sucker34	Q12.1		
Sucker35	Sucker35	Q12.2		
Sucker36	Sucker36	Q12.3		
Sucker37	Sucker37	Q12.4		
Sucker38	Sucker38	Q12.5		
Sucker39	Sucker39	Q12.6		
Sucker40	Sucker40	Q12.7		
Sucker41	Sucker41	Q13.0		
Sucker42	Sucker42	Q13.1		
Sucker43	Sucker43	Q13.2		
Sucker44	Sucker44	Q13.3		
Sucker45	Sucker45	Q13.4		
Sucker46	Sucker46	Q13.5		
Sucker47	Sucker47	Q13.6		
		Q13.7		
Cylinder on door+	Baffle-I-Cldr-U+	Q14.0		
Cylinder on door−	Baffle-I-Cldr-U−	Q14.1		
Cylinder in door+	Baffle-I-Cldr-M+	Q14.2		
Cylinder in door−	Baffle-I-Cldr-M−	Q14.3		
Lower cylinder of door+	Baffle-I-Cldr-D+	Q14.4		

(continued)

Table 5.8 (continued)

Signal description	Symbol	PLC I/O address	Upper address (WinCC)	Data type
Lower cylinder of door–	Baffle-I-Cldr-D–	Q14.5		
Upper cylinder of door II+	Baffle-I-Cldr-U+	Q14.6		
Upper cylinder of door II–	Baffle-I-Cldr-U–	Q14.7		
Cylinder in door II+	Baffle-I-Cldr-M+	Q15.0		
Cylinder in door II–	Baffle-I-Cldr-M–	Q15.1		
Lower cylinder of door II+	Baffle-I-Cldr-D+	Q15.2		
Lower cylinder of door II–	Baffle-I-Cldr-D–	Q15.3		
		Q15.4		
Guide wheel cylinder	Guide whed Cldr	Q15.5		
		Q15.6		
		Q15.7		

(7) Procedure Specification

The program specification is a technical document for the design and comprehensive description of the entire program content. The purpose is to let the user of the program understand the basic structure of the program and the treatment of some problems, as well as the reading method and the matters that should be paid attention to in the use of the program. The program specification generally includes the basis of program design, the basic structure of the program, the analysis of each functional unit, the formula and principle used, the source of each parameter and the operation process, the test situation of the program, etc.

Each of the steps in the above process is an indispensable part of the application design. To design an excellent PLC application program, we must do every link of work. However, the core of application design is the writing of the program, and the other steps serve it.

(8) Common Programming Methods

The programming methods of PLC mainly include experience design method and logic design method. Logic design method is based on logic algebra theory, write the input and output logic expression, and then convert into ladder diagram. Because the general logic design process is more complicated and the cycle is longer, most of them use the empirical design method. If the control system is more complex, you can use a flowchart. Empirical design method is based on some typical applications, according

Table 5.9 Variable table of control cabinet IV

Signal description	Symbol	PLC I/O address	Upper address (WinCC)	Data type
		I10.0		
		I10.1		
		I10.2		
		I10.3		
		I10.4		
		I10.5		
		I10.6		
		I10.7		
		I11.0		
		I11.1		
		I11.2		
		I11.3		
		I11.4		
		I11.5		
		I11.6		
		I11.7		
Gas flow detection	Gas flux examine	I272.0–273.7		
		I274.0–275.7		
Gas supply pressure switch 1	Gas Pressure Switch 1	I276.0–277.7		
Gas supply pressure switch 2	Gas Pressure Switch 2	I278.0–279.7		
Gas solenoid valve 1	Gas I valve	Q16.0		
Gas solenoid valve 2	Gas II valve	Q16.1		
		Q16.2		
		Q16.3		
		Q16.4		
		Q16.5		
		Q16.6		
		Q16.7		
Partition I solenoid valve	Subarea I valve	Q17.0		
Partition II solenoid valve	Subarea II valve	Q17.1		
Partition III solenoid valve	Subarea III valve	Q17.2		
		Q17.3		

Fig. 5.39 Double cavity insulated glass

to the specific requirements of the controlled object to the control system, select some basic links, appropriate combination, modification, improvement, so that it becomes a program that meets the control requirements. Many of the basic links mentioned here are converted from the relay contactor control line, which is very similar to the relay contactor line drawing method, and the signal input, output mode and control function are roughly the same. For engineers and technicians who are familiar with the design principle of the relay contactor control system, they can quickly master the ladder diagram language design. The quality and efficiency of the program design have a lot to do with the experience of the programmer. Empirical design method has no universal law to follow, we must accumulate and enrich ourselves in actual combat, and gradually form our own design style.

Taking the large mechatronic intelligent manufacturing equipment developed by the research team of the author—the key equipment of insulating glass production line chip machine as an example, the PLC process is introduced.

Figure 5.40 shows the process for assembling pieces of inert gas dielectric insulating glass (Fig. 5.39), such as double cavities (three layers of glass and two aluminum frames).

Glass I process:

When the closing machine is in standby mode, the two pressing plates are open (default width), and there is no glass signal at the output end of the cleaning machine, the closing machine is always in standby state;

When there is glass at the output end of the cleaning machine, the transmission wheel of the input section of the closing machine begins to act, the glass I enters the input section, when the sensor detects the glass end enters the input section, the transmission wheel stops acting, the operator visually checks the cleanliness of the glass, and labels (labels can also be attached to the outside of the glass II), step on the foot switch, the transmission wheel action, the glass continues to move forward;

When the sensor of the detection section detects that there is glass entering, the transmission wheel of the detection section stops, and at the same time, the glass thickness, height, width and aluminum frame thickness are automatically detected (the width detection is carried out during the glass entering process);

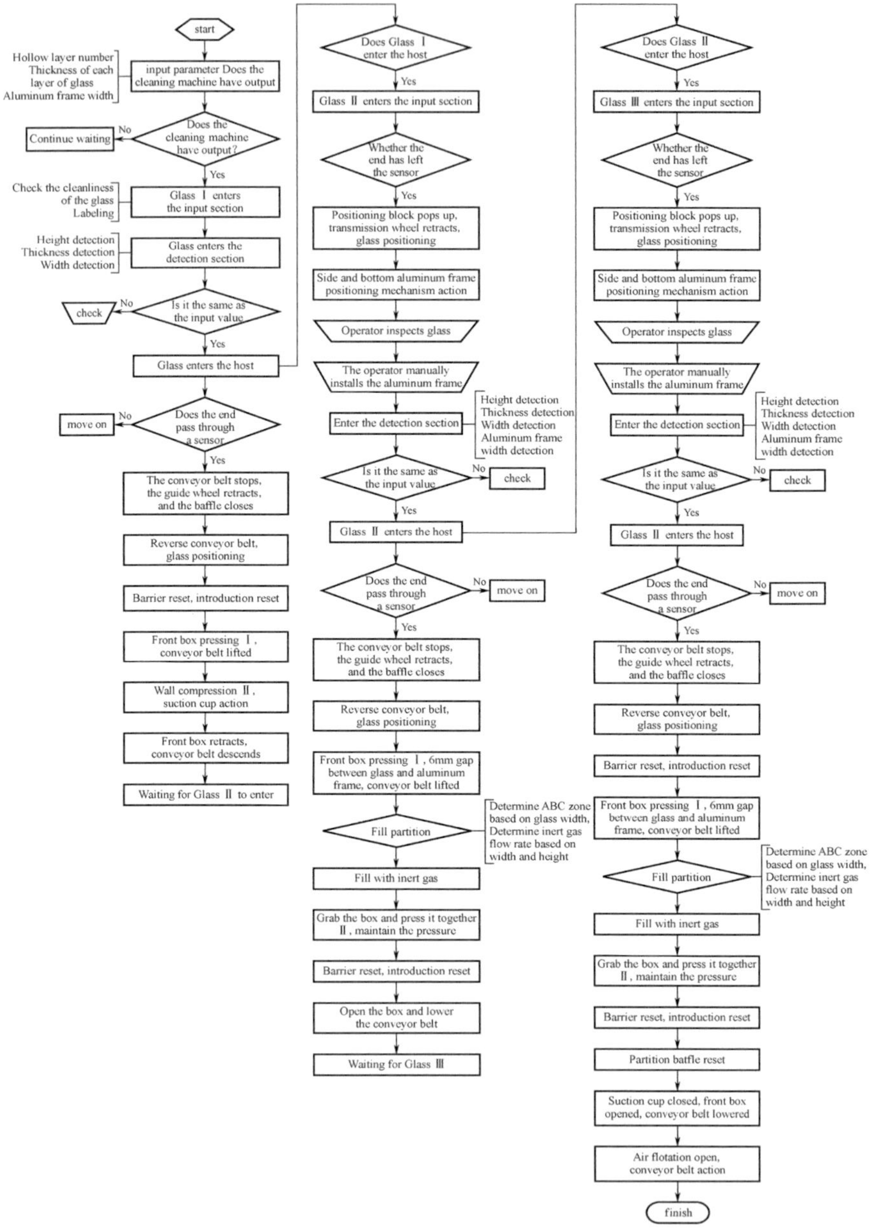

Fig. 5.40 Double cavity hollow glass laminating process (inert gas medium)

After the detection is completed, it is confirmed that when the main machine is idle, the glass I enters the main machine (main machine drive belt), and when the end passes a certain distance Δx from the sensor on the right side of the main machine, the conveyor belt stops, the small guide wheel backs back, the right baffle closes, the conveyor belt moves in reverse, and the glass backs up against the baffle. The baffle opens and the small guide wheel resets;

The front box of the host is closed (stroke I), the conveyor belt is lifted, the front box is pressed (stroke II is in place), the suction cup moves to absorb the glass I, the front box is returned, the box is opened, the conveyor belt is lowered, waiting for the glass II to enter.

Glass II process:

When the output end of the cleaning machine has a glass through the signal, the input section drive wheel begins to act, glass II enters the input section, when the sensor detects that the glass end enters the input section and moves Δx, the drive wheel stops, the positioning block pops out, the drive wheel moves in reverse, the glass backsteps, and is positioned by the positioning block;

Aluminum frame positioning system action (side aluminum frame positioning mechanism, bottom aluminum frame positioning mechanism), the operator first check the cleanliness of the glass, after passing the manual installation of aluminum frame, positioning, after aluminum frame installation, step on the foot switch, drive wheel action, glass forward to the detection section, glass thickness, height, width and aluminum frame thickness detection;

When glass I enters the main plate press (main drive belt), the end passes a certain distance Δx from the sensor on the right side of the main machine, the conveyor belt stops, the small guide wheel retreats, the right baffle closes, the conveyor belt moves in reverse, and the glass is positioned by the baffle;

The front box is pressed to the station I (the main body drives the glass I and the aluminum frame to maintain a certain gap, about 6 mm), and the conveyor belt is lifted; According to the glass area automatically confirm the partition (A, B, C three zones), the Z-direction position of the conveyor belt is adjusted, so that the middle porosity of the conveyor belt is aligned with the gap between the glass I and the glass II aluminum frame, and the inert gas begins to fill;

After confirming the filling, press the front box to station II and keep the pressure;

The baffle opens and the small guide wheel resets.

Glass III process:

When the closing machine is in standby mode, the two pressing plates are open (default width), and there is no glass signal at the output end of the cleaning machine, the closing machine is always in standby state;

When there is glass at the output end of the cleaning machine, the transmission wheel of the input section of the closing machine begins to act, and the glass III enters the input section. When the sensor detects that the end of the glass enters the input section, the transmission wheel stops acting, the operator visually checks the cleanliness of the glass, and labels (labels can also be attached to the outside of the glass III), step on the foot switch, the transmission wheel action, and the glass continues to move forward;

When the sensor of the detection section detects that there is glass entering, the transmission wheel of the detection section stops, and at the same time, the glass thickness, height, width and aluminum frame thickness are automatically detected (the width detection is carried out during the glass entering process);

After the detection is completed, it is confirmed that when the host is idle, the glass II enters the plate press host (host drive belt), and when the end passes a certain distance Δx from the sensor on the right side of the host, the conveyor belt stops, the small guide wheel backs up, the right baffle closes, the conveyor belt moves in reverse, and the glass backs up by the baffle.

The baffle opens and the small guide wheel resets;

The front box of the main engine is closed (stroke I), the conveyor belt is lifted, and the front box is pressed (stroke II is in place); Baffle reset, guide wheel reset; Before the suction cup closes, the box opens and the conveyor belt drops.

5.6.3 *Anti-Interference Design of PLC System*

PLC has strong environmental adaptability and strong anti-interference ability, and has been widely used in industrial production. Even so, in order to improve the reliability and stability of the control, it is necessary to carry out anti-interference design during use, especially in the harsh and strong electromagnetic interference environment. The anti-interference design of PLC system needs to consider the following contents.

(1) Resistance to Power Interference

Power interference can cause PLC control system failure. The transmission grid covers a wide range, is subject to all spatial electromagnetic interference and generates induced voltage and current on the line. In addition, there are many factors affecting the power supply inside the workshop, such as switching operation surge, frequent start and stop of large power equipment, harmonics caused by AC and DC transmission devices, and short-circuit transient impact of the power grid, which are transmitted to the power supply through the transmission line. In order to reduce the impact of power supply fluctuations on PLC operation, the following measures can be taken:

① The use of good performance of the power supply. The power grid interference is mainly coupled into the PLC system through the power supply of the PLC system (such as CPU power supply, I/O power supply, etc.), the transmitter power supply and the instrument power supply that has a direct electrical connection with the PLC system. The power supply of the PLC system generally uses a power supply with good isolation performance, while the power supply of the transmitter power supply and the power supply of the instrument directly connected with the PLC system are often designed by the user themselves, and have not received enough attention. For the transmitter and the shared signal instrument power supply, the distributor with small distributed capacitance and large suppression band (such as using multiple isolation and shielding and leakage sensing technology) should be selected to reduce the interference of the PLC system.

② Use uninterruptible power supply. In order to ensure the uninterrupted power supply of the grid, modern control system needs to use the uninterruptible power supply (UPS) with strong interference isolation performance to improve the safety and reliability of power supply. UPS has strong anti-interference isolation performance.

③ Hardware filtering measures. In the case of strong interference or high reliability requirements, the isolation transformer with a shielding layer should be used to supply power to the PLC system.

④ Correctly select the ground point and improve the grounding system.

⑤ Cable laying should be reasonable and standardized. Strong and weak electricity strictly separate.

(2) Grounding Design of the Control System

Grounding refers to the neutral point of the power system and the electrical device, the exposed conductive part of the electrical equipment and the external conductive part of the device connected to the ground through the conductor. According to different purposes, it can be divided into working grounding, lightning protection grounding and protection grounding. Good grounding can avoid accidental voltage shock hazards, is an important condition to ensure the reliable operation of PLC. Perfect grounding system is one of the important measures for PLC control system to resist electromagnetic interference. The grounding method of the grounding system can be generally divided into three forms: series single point grounding, parallel single point grounding and multi-branch single point grounding.

The ground wire of PLC control system includes system ground, shielding ground, alternating ground and protection ground. The disturbance of the grounding system to the PLC system is mainly the uneven distribution of potential at each ground point, and the ground potential difference between different ground points, which causes the ground loop current and affects the normal operation of the system. The cable shield layer must be grounded at one point. If both ends of the cable shield layer are grounded, there is a ground potential difference and a current flows through the shield layer. When an abnormal state (such as lightning strike) occurs, the ground current will be larger. In addition, the shielding layer, the ground wire and the earth may form a closed loop, and under the action of changing magnetic fields, the induced current will appear in the shielding layer, which interferes with the signal loop through the coupling between the shielding layer and the core wire. If the system is confused with other grounding, the resulting ground circulation may produce unequal potential distribution on the ground wire, affecting the normal operation of the logic circuit and analog circuit in the PLC. The logic voltage interference tolerance of PLC is low, and the distribution interference of the logic ground potential can easily affect the logic operation and data storage of PLC. The distribution of simulated ground potential will lead to the decrease of measurement accuracy and the distortion of measurement and control signal.

(3) Anti-I/O Interference

Signal interference will cause I/O signal abnormal work and measurement accuracy decline, serious will cause damage to components. For the system with poor isolation performance, the signals will interfere with each other, causing the bus backflow of the common ground system, resulting in logical data changes, misactions or crashes. Anti-I/O interference can choose the anti-interference performance of the I/O module, in addition, you can also take the following measures.

① When wiring, the input and output of PLC are routed separately, and the switching quantity and analog quantity are also laid separately. The analog signal should use a shielded wire, the shielding layer should be grounded at one end, and the grounding resistance should be less than 1/10 of the shielding layer resistance. Power line, control line and PLC power line and I/O line should be distributed separately, the isolation transformer and PLC and I/O should be connected by twisted pair. Separate the PLC I/O line and the high-power line, if it must be in the same slot, you can add a partition, the best slot routing. Stay away from strong interference sources, such as welding machines, high-power silicon rectifier devices and large power equipment, avoid installing high-voltage appliances in the same switch cabinet, according to the installation instructions, the PLC in the cabinet should be away from the power line. Inductive loads close to the PLC, such as high-power relays, contactor coils, etc., should be connected to the RC circuit. The AC output line and the DC output line should not be collinear, away from the high voltage line and the power line, and avoid parallel.

② The I/O input cable should not be too long. If the distance is long, the cable needs to be shielded, and the input/output lines should be separated. The output terminal is divided into independent output and common output, and different types and voltage levels of output voltage can be used in different groups. Use inductive load should be reasonable choice, or add isolation relay.

③ Correct selection of the ground point, improve the grounding system and the suppression measures of the inverter interference.

5.6.4 *Debugging of PLC System*

On the basis of hardware and software design, PLC system debugging should be carried out. Electromechanical joint adjustment is the necessary step before the system is put into use. The PLC system needs to debug both the hardware and the software. The hardware debugging of PLC system is relatively simple, mainly the preparation and debugging of PLC program.

Compile the application program offline, check the hardware of the control system after the offline debugging is passed, and download the application program for online debugging when there is no problem, and then modify the program online until the logic is correct and stable, and finally summarize and organize the relevant information.

5.7 Electrical Control System Design

5.7.1 Overview

The electrical control system is composed of a number of electrical components, used to achieve the control of one or some objects (automatic control, protection, monitoring), so as to ensure the safe and reliable operation of the controlled equipment. The purpose of the process design of electrical control system is to meet the requirements of the manufacture and use of electrical control equipment. Whether it is PLC system, or other such as open CNC system, electrical control is essential, is an important part of intelligent manufacturing equipment control system, so it is necessary to carry out correct and reasonable design of electrical system.

General electrical control system to carry out three parts of the design.

① Input part design. This part mainly includes the design of sensors, switches, buttons and other hardware and wiring.
② Logical part design. This part mainly includes the design of relays, contacts, etc.
③ Execution part design. The main design content of this part includes electromagnetic coil, indicator light and other execution parts.

Draft electrical design task book; Determine the electric drive scheme and control mode; Select motor capacity, structure form; Design electrical control schematic, calculate the main technical parameters; Select electrical components and make a list of electrical components; Write the design calculation specification.

After the completion of the electrical schematic design and the selection of electrical components, you can carry out the overall configuration of the electrical control equipment, that is, the design of the general assembly drawing and the general wiring diagram, and then design the electrical assembly drawing and wiring diagram of each part, and list the component catalog of each part, the input and outlet line number and the main material list and other technical data, and finally write the instruction manual.

The design of the electrical control system should meet the requirements of the production machinery and the process of the electrical control line to the maximum extent, and strive to make the control line simple, economic, safe and reliable, convenient operation and maintenance under the premise of meeting the production requirements. Electrical schematic diagram is the theoretical basis for electrical line installation, debugging, use and maintenance, is the basis for process design and formulation of other technical data, and is the central link of the entire design. The electrical control system diagram mainly includes electrical schematic diagram, electrical installation wiring diagram and electrical component layout diagram.

The premise of electrical control system design is to be able to read diagrams. Read the machine first, then the electricity. First understand the basic structure, operation, process requirements and operation methods of mechanical and electrical equipment, and have an understanding of the mechanical structure, controlled quantity and operation of the equipment, so as to clarify the requirements for electrical control and prepare for the analysis of the circuit. In addition, read the Lord first, then read the auxiliary. Start from the main circuit to read the diagram, find out how many axes (motor) drive the electromechanical equipment and the function of each axis, combined with the processing technology and the main circuit, analyze the braking mode of the motor, and find out the electrical components of the electrical equipment.

The same is true when designing. After the first opportunity to re-electricity, the first master and then auxiliary, and finally the overall inspection, analysis of the working principle of each local circuit and the control relationship between each part, and then take into account the entire control line, from the overall point of view to further check and understand the connection between the control links.

In order to ensure the reliability and safety of primary equipment operation, it is necessary to have many auxiliary electrical equipment for service, that is, it is necessary to have a combination of several electrical components that can achieve a certain control function, called the control loop or secondary loop. These devices should have the following functions:

① Protection function. When the electrical control system fails (overcurrent or overload), it needs a set of fault detection circuit and automatic protection equipment that can disconnect or switch circuits for equipment and lines.
② Automatic control. In some high-voltage and high-current switchgear applications, people generally do not operate directly, and automatic control devices need to be designed. When the equipment is running, it automatically realizes the actions such as suction and closing; When there is a fault, the circuit is automatically cut off and the power supply equipment is automatically controlled.
③ Monitoring function. Electromechanical equipment needs to set up a variety of audio-visual signals, such as warning lights, buzzers, sound, monitors, etc., for electrical monitoring of a device.
④ Measurement function. During the operation of mechanical and electrical equipment, various instruments are required to measure various parameters of the line, such as voltage, current, frequency and power.

5.7.2 *Basic Circuit of Commonly Used Control Line*

① Power supply circuit. The power supply of the power supply circuit has a variety of AC 380 V, 220 V and DC 24 V.

② Protection circuit. The working power supply of the protection (auxiliary) circuit has a variety of single-phase 220 V (AC), 36 V (DC) or DC 220 V, 24 V, etc., for short circuit, overload and voltage loss protection of electrical equipment and lines, which is composed of fuses, thermal relays, voltage loss coils, rectifier components and voltage regulator components.
③ Signal loop. The circuit that can reflect or display the normal and abnormal working status information of the equipment and line in time, such as different color signal lights, different sound audio equipment, etc.
④ Automatic and manual circuits. In order to improve work efficiency, electrical equipment is generally equipped with automatic links, but in the installation, debugging and emergency handling, the control line also needs to set manual links for debugging. Automatic and manual conversion is realized through combination switch or transfer switch.
⑤ Brake stopping circuit. Cut off the power supply of the circuit, and take some braking measures to control the motor to stop quickly, such as energy consumption braking, power reverse braking, reverse pull braking and regenerative power generation braking.
⑥ Self-locking and blocking circuit. After the start button is released, the line remains energized, and the electrical link that the electrical equipment can continue to work is called the self-locking link, such as the dynamic contact of the contactor is connected in series in the coil circuit. Two or more electrical devices and components, in order to ensure the safety and reliability of the operation of the equipment, only one power start, the other can not power start the protection link, called the lock link. For example, the breaking contacts of two contactors are connected in series in each other's coil circuit.

5.7.3 Common Protection Links

The electrical control system must meet the requirements of the production process under the premise of safety and reliability. When the electrical control system is designed and operated, all kinds of faults and abnormal conditions that may occur in the system must be fully considered, and the corresponding protection devices should be set up. The protection link is an integral part of all electrical control systems. The common protection links of low-voltage motors are:

(1) Short Circuit Protection

A short circuit occurs when the insulation of the electrical appliance or line is damaged, the load is short circuited, the wiring is wrong, etc. The instantaneous fault current generated by short circuit can reach ten to dozens of times the rated current, so that electrical equipment or distribution lines are damaged due to overcurrent, and even cause fire due to arc. Short-circuit protection requires instantaneous characteristics, that is, it requires cutting off the power supply in a very short time. The common methods of short circuit protection include fuse protection and low voltage circuit breaker protection.

(2) Overcurrent Protection

Overcurrent protection is a kind of current protection which is different from short circuit protection. Overcurrent refers to the operating state of a motor or electrical component exceeding the rated current. Transient overcurrent does not cause immediate damage to the electrical component, as long as the current value returns to normal before the maximum allowable temperature rise is reached. But too large impact load will cause too large impact current and damage the motor. Excessive electromagnetic torque of the motor will also cause damage to the mechanical rotating parts, so it is necessary to cut off the power supply instantaneously. The possibility of generating overcurrent in the operation of the motor is greater than the occurrence of short circuit, especially in the condition of frequent starting and positive and negative rotation, and repeated short-time work.

Overcurrent protection is usually realized by over-current relay combined with contactor. The overcurrent relay coil is connected in series in the protected circuit, and the normally closed contact of the overcurrent relay is connected in series in the contactor coil circuit. When the circuit current reaches the limit value, the overcurrent relay operates, the normally closed contact is disconnected, the contactor coil is powered off and released, and the contactor main contact is disconnected to cut off the motor power supply. This overcurrent protection link is often used in the control circuit of DC motor and three-phase winding rotor induction motor.

(3) Overload Protection

Overload protection is a kind of overcurrent protection. Overload means that the operating current of the motor is greater than its rated current, but within 1.5 times the rated current. Overload can be caused by sudden increase of load, lack of phase operation or decrease of power supply voltage during operation. Long-term overload operation will damage electrical equipment. Thermal relays are usually used for overload protection. When there is more than 6 times the rated current overheat relay, it needs to be operated after 5 s, so that before the thermal relay does not operate, the heating element of the thermal relay may be burned out first, so when using the thermal relay for overload protection, it must also be equipped with a fuse or low-voltage circuit breaker short-circuit protection device. It is worth pointing out that overcurrent protection cannot be used for overload protection.

(4) Loss of Voltage Protection

The motor stops because the power supply voltage drops or disappears, and once the power supply voltage recovers, it may start by itself, which is easy to cause production safety accidents. The protection set to prevent the motor from starting on its own or the electrical components from putting themselves into work when the voltage is restored is called the loss of voltage protection. The starting and stopping device controlled by contactor and button has the function of voltage loss protection. This is because when the power supply voltage disappears, the contactor will automatically release and cut off the motor power supply; When the power supply voltage is restored, the contactor self-locking contact is disconnected and will not start by itself.

(5) Undervoltage Protection

When the power supply voltage is reduced to 60–80% of the rated voltage, the motor power supply needs to be cut off and stop working, this protection is called undervoltage protection. In addition to the use of contactor and button control, that is, the use of contactor itself undervoltage protection, undervoltage relay can also be used for protection. The voltage relay coil is connected to the power supply, and the normally open contact is connected to the contactor coil circuit in series. When the power supply voltage is lower than the release value, the voltage relay action releases the contactor coil, and the main contact is disconnected from the motor power supply to realize the undervoltage protection.

(6) Overvoltage Protection

Large inductive loads and DC electromagnetic mechanisms, DC relays, etc., will produce higher induced electromotive force when the current is broken, so that the electromagnetic coil insulation breakdown and damage, this situation requires the use of overvoltage protection measures. The method is to parallel a resistor at both ends of the coil, the resistor and the capacitor in series or the diode and the resistor in series to form a discharge loop to achieve overvoltage protection.

(7) Weak Magnetic Protection

The large decrease of the magnetic field of the DC motor will cause the motor to overspeed and require weak magnetic protection. The weak magnetic protection is realized by inserting the undercurrent relay in the circuit of the motor excitation coil. When the motor is running, if the excitation current is too small, the undercurrent relay is released, its contact disconnects the motor armature circuit contactor coil circuit, the contactor coil power off release, the contactor main contact disconnects the motor armature circuit, the motor disconnects the power supply, to achieve the purpose of protecting the motor.

(8) Other Protection

In addition to the above protection, there are overspeed protection, stroke protection, pressure protection, etc., which are connected to a normally open contact or normally closed contact controlled by these parameters in the control circuit to achieve the control of the control circuit. These devices are mechanical and electrical, such as centrifugal switches, tachometer generators, stroke switches, pressure relays, etc.

5.7.4 Fault Maintenance

Electrical control circuit failure, light so that electrical equipment can not work, affect production, etc., serious will cause personal injury accidents. Therefore, it is required that in the event of a failure, the cause must be identified in time and quickly eliminated. Troubleshooting can generally be performed in the following steps:

(1) Observe and Investigate the Fault

Electrical faults are diverse, the same fault may have different fault phenomena, different kinds of faults may also appear the same fault phenomenon. The uniformity and diversity of fault phenomena make it difficult to find faults. However, the fault phenomenon is the basic basis for finding electrical faults and the starting point for finding electrical faults, so it is necessary to carefully observe and analyze the fault phenomenon, find out the most important and most typical aspects of the fault phenomenon, and figure out the time, place, and environment of the fault.

(2) Analyze the Cause of the Failure

Analyzing the cause of the fault according to the fault phenomenon is the key to find the electrical fault. State analysis, graph analysis, unit analysis, loop analysis, inference analysis, simplified analysis, tree analysis, computer aided analysis and so on are commonly used in the analysis of electrical equipment faults.

(3) Determine the Fault Location

Determining the fault location is the ultimate purpose of finding the fault of electrical equipment. Determining the fault location can be understood as determining the fault point of the equipment, such as short circuit points, damaged components, etc., and can also be understood as determining the variation of some operating parameters, such as voltage fluctuations, three-phase unbalance. The fault location is determined on the basis of careful investigation and detailed analysis of the fault phenomenon. In this process, often to use a variety of means and methods, can be used investigation and research method, power test method, measurement method, analogy method and so on.

5.8 Summary of This Chapter

This chapter mainly introduces the intelligent manufacturing equipment control system. Firstly, the basic knowledge of intelligent manufacturing equipment control system is described, and then the hardware platform design and software design method of intelligent manufacturing equipment control system are introduced. Taking PLC as the main means of modern industrial automatic control, the design method is described. Combined with the development trend of fieldbus control, the related knowledge of fieldbus is supplemented. Finally, the design of the electrical control system is briefly introduced.

References

1. Yang Xianqi, Chang Zongyu. Fundamentals of Mechatronic Engineering Control[M]. Beijing: National Defense Industry Press, 2005.
2. Chu Yanfang, Zhang Ruihua. Overview of Fuzzy Control Theory[J]. Science & Technology Information, 2009, 000(020):161–162.

3. Zhang Yudong. Overview of Expert System Development[J]. Computer Engineering and Applications, 2010, 46(19):43–47.
4. Zhao Bo, Wang Hongyuan. Hydraulic and Pneumatic Technology[M]. Beijing: China Machine Press, 2017.
5. Zhang Guozhong. Intelligent Control Systems and Applications[M]. Beijing: China Electric Power Press, 2007.
6. Tian Maosheng, Tang Xiaoqi, Meng Guojun, et al. Implementation of Interactive Control in Open Control System of Industrial Robots Based on Embedded PC[J]. Journal of Computer Applications, 2010, 30(11):3087–3090.
7. Li Lan, Mei Lifeng. Electric Drive and Control[M]. Beijing: China Machine Press, 2016.
8. Gao Anbang, Shi Lei, Zhang Xiaohui. Self-Study Manual for Siemens S7-200/300/400 Series PLC[M]. Beijing: China Electric Power Press, 2012.
9. Ma Ding. Detailed Explanation of Common Modules of Siemens PLC and Design Examples of Industrial Systems[M]. Beijing: Electronic Industry Press, 2009.
10. Su Shaojing. Digital Testing Technology[M]. Beijing: National Defense Industry Press, 2015.
11. Feng Dongqin, Shi Yiming, Chu Jian. "Foundation Fieldbus (FF) Technology" Lecture - Lecture 1: Development and Characteristics of Foundation Fieldbus (FF)[J]. Process Automation Instrumentation, 2001(06):54–56.
12. Wang Junjie, Zhang Wei, Xie Chunyan. "LonWorks Technology and Its Application" Lecture - Lecture 1: Development of Fieldbus and LonWorks Technology[J]. Process Automation Instrumentation, 1999(07):3–5.
13. Tian Min, Gao Anbang. New Development of LonWorks Fieldbus Technology[J]. Journal of Harbin University of Science and Technology, 2010, 15(01):33–39.
14. Gao Jie, Gao Yan, Jiang Dengke, Shi Hongjun, et al. Current Situation and Development of Field bus[J]. Coal Mine Machinery, 2006(06):915–916.
15. Fan Kai. Development Trend of Fieldbus[J]. Process Automation Instrumentation, 2000(02):1–4.
16. Zhou Kai. PC Numerical Control: Principle, System and Application[M]. Beijing: China Machine Press, 2007.

Chapter 6
Design of Intelligent IoT Electromechanical Equipment System

The Internet of Things (IoT) is the Internet connecting all things. It is an extended and expanded network based on the Internet. Through information sensors, radio frequency identification technology, global positioning system, infrared sensors, laser scanners and other devices and technologies, it can collect any object or process that needs to be monitored, connected, and interacted in real time, and collect all kinds of information needed, such as sound, light, heat, electricity, mechanics, chemistry, biology, and location, Through various possible network connections, achieve ubiquitous connections between objects and people, and achieve intelligent perception, recognition, and management of objects and processes. The Internet of Things is an information carrier based on the Internet, traditional telecommunications networks, etc. It enables all ordinary physical objects that can be independently addressed to form an interconnected network. The ultimate goal of the development of the Internet of Things is to make extensive connections between things, people, and reality, in order to facilitate relevant information exchange, management, control, and identification. The application of the Internet of Things is shown in Fig. 6.1, and the IoT network based on intelligent sensors is shown in Fig. 6.2.

Future intelligent manufacturing equipment should enhance its core value and create value for users through connectivity and intelligent features. In fact, now intelligent manufacturing equipment like common items has entered our daily lives.

Not to mention the emergence of smart cars and autonomous vehicles, small products in daily life also reflect intelligence everywhere. For example, when the water in the smart kettle (tea making kettle) boils, it can automatically keep warm and can choose the "tea making" function according to needs; The household toaster can automatically mix, feed, and bake, and can also make appointments. It can also set various flavors and cooking conditions on its own; The intelligent washer can automatically contact for repair before it malfunctions, and can also coordinate the cleaning dosage and water temperature. Smart homes and appliances can rely on IoT technology to connect all household appliances in the home, allowing consumers to control and monitor their devices at anytime and place.

M. Xu et al., *Design and Case of Intelligent Electromechanical Equipment System*, Advanced and Intelligent Manufacturing in China,
https://doi.org/10.1007/978-981-95-3833-1_6

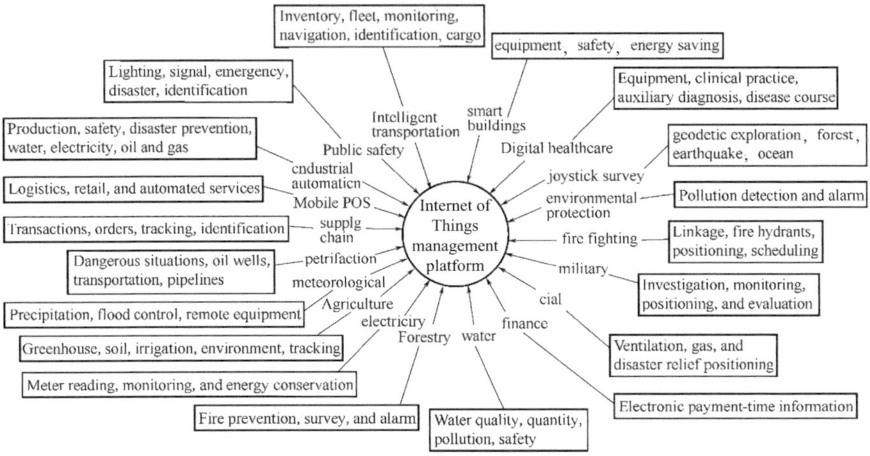

Fig. 6.1 Application of Internet of Things

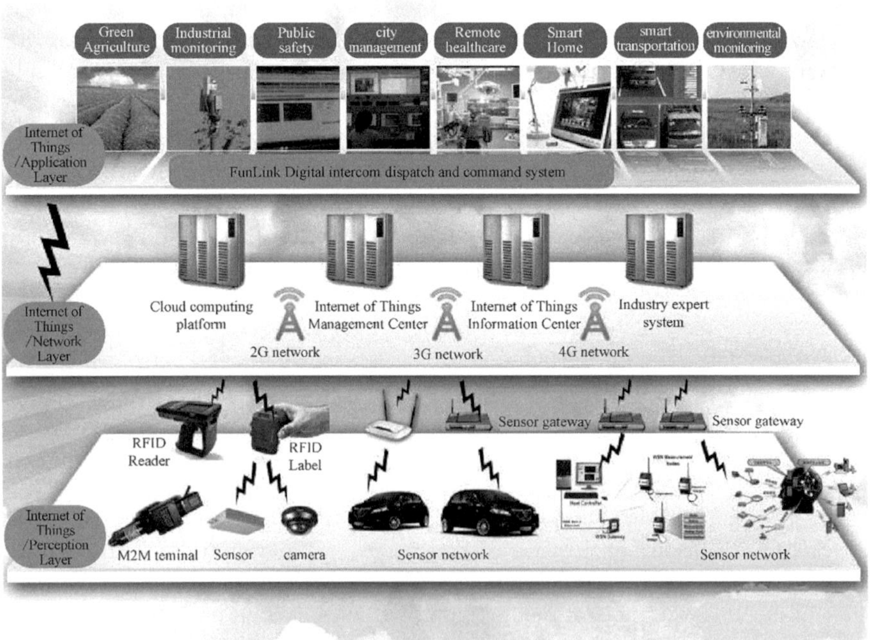

Fig. 6.2 IoT network based on intelligent sensing

6.1 Overview of the Internet of Things

In the 1990s, the concept of the Internet of Things emerged and aroused people's interest. The Internet of Things (IoT) is based on the computer Internet and uses radio frequency identification, wireless data communication, computer and other technologies to build a physical Internet covering everything in the world. The Internet of Things is the integrated part of future networks, which is a global dynamic network infrastructure based on standard and interoperable communication protocols and has self configuring capabilities. In this network, all physical and virtual items have specific encoding and physical characteristics, and are seamlessly linked through intelligent interfaces to achieve information sharing. Industrial Internet is an Internet that links goods, machines, computers and people. It is a key infrastructure that links the whole industrial system, the whole industrial chain and the whole value chain, and supports the development of industrial intelligence. Industrial Internet includes three elements: data collection (edge layer), industrial PaaS (platform layer) and industrial APP (application layer). The data acquisition layer is the foundation to build an accurate, real-time and efficient data acquisition system, which is responsible for collecting field data and performing protocol conversion and edge computing. Some of the collected data is processed on the edge side and returned directly to the machine equipment, and the other part is transferred to the cloud for comprehensive utilization analysis, further optimizing and forming decisions. The platform layer is the core, used to build a scalable operating system and provide a basic platform for the development of industrial APP applications. The application layer is crucial, forming application services that meet different industries and scenarios, and presenting them in the form of apps. Automation equipment is the source of industrial data generation and the foundation of industrial Internet. Robots are a typical representative of automation equipment, reflecting the level of automation development in a country. The Internet of Things can be seen as the integration of information space and physical space, digitizing and networking everything, achieving efficient information exchange between items, between items and people, and between people and the real environment, and integrating various information technologies into social behavior through new service models. This is a higher level of comprehensive application of information technology in human society. The application of the Internet of Things is shown in Fig. 6.3.

The innovation of the Internet of Things is reflected in at least the following four aspects [1].

Firstly, the Internet of Things proposed a panoramic framework for the socialization of information technology. The information industry is classified as the fourth industry after industry, agriculture, and service. The Internet of Things has stimulated the enthusiasm of various industries in society to apply information technology to change production and lifestyle. Liu Haitao of China proposed the "Perceived Society Theory", which suggests that the socialization of information technology will enable IoT technology to enter various levels of social production and life.

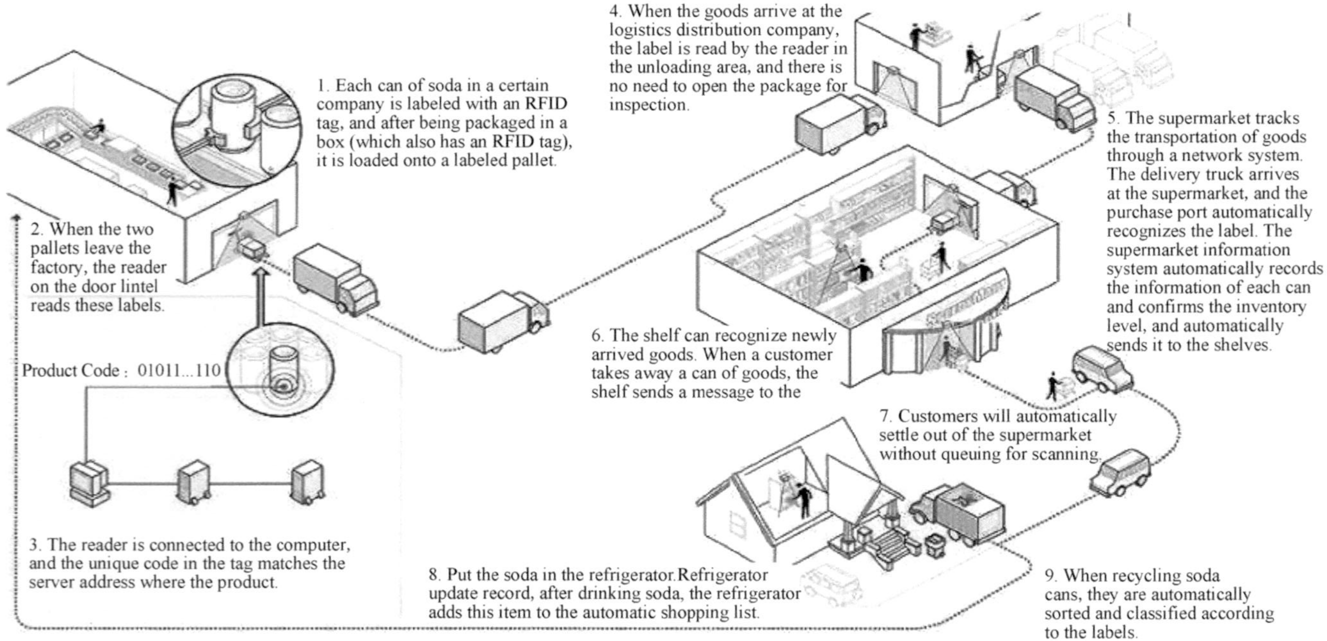

Fig. 6.3 IoT applications

Secondly, the Internet of Things is characterized by perception. The addition of sensors to the communication network allows the network's tentacles to extend longer towards objects. The biggest difference between the Internet of Things and the traditional Internet lies in perception. Perception is realized through large-scale and diverse sensors increasingly deployed in human living space, and data information of social individual behavior is obtained through perception. It can be foreseen that the scope of perception is global, and the perceived information will be seamlessly integrated on a global scale, forming an intelligent network.

Thirdly, the Internet of Things will generate massive amounts of data. The data generated by various sensors will form an ocean of data. The era of the Internet of Things is the era of big data. Facing massive amounts of data, how to store, transmit, and analyze data will be a new topic.

Fourthly, the Internet of Things is fundamentally based on intelligence. A microprocessor is a normalized intellectual core. It endows all nodes, terminals, and servers of the Internet of Things with infinite intelligent capabilities in the form of general-purpose computers and embedded systems. The infinite number and infinite intelligence of microprocessors highlight the intelligent characteristics of the Internet of Things.

The Internet of Things user end realizes the information exchange and communication between any things, which is an extension and expansion of the Internet foundation. Its core and foundation is still the Internet. The basic characteristics of the Internet of Things are comprehensive perception, reliable transmission, and intelligent processing.

① Comprehensive perception. By utilizing technologies such as Radio Frequency Identification (RFID), QR codes, and sensors, comprehensive real-time information collection and acquisition of the state of objects or people can be achieved through perception, capture, and measurement.
② Reliable transmission. Connect connected objects to information networks, rely on various communication networks, and enable reliable information exchange and sharing anytime, anywhere on a global scale.
③ Intelligent processing. Using computing and database technologies to analyze and process massive perceptual data and information, achieving intelligent decision-making and control [2].

In order to describe the key links of the Internet of Things more clearly, according to the perspective of information science, the information function model of the Internet of Things is abstracted around the process of information flow, as shown in Fig. 6.4.

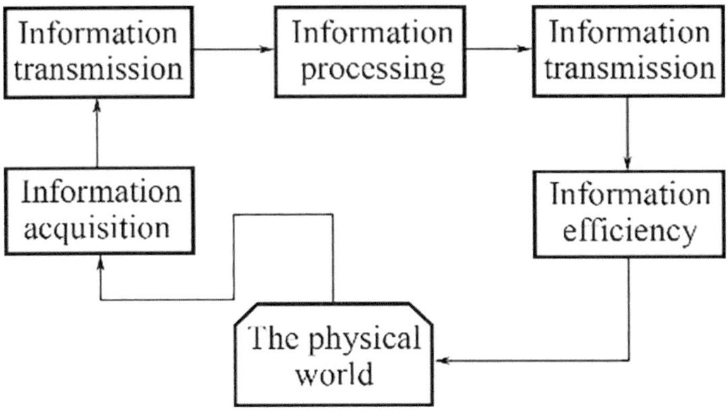

Fig. 6.4 IoT information function model

6.1.1 The Main Functions of the Internet of Things

The main functions of the Internet of Things include:

① Information acquisition. Including information perception and information recognition. Information perception refers to the perception of the state of things and their sensitivity and perception of change patterns; Information recognition refers to the representation of the motion states and changes of perceived objects in a table show.

② Online monitoring. Including information transmission, and reception, ultimately changing the state of things.

The localization method can also be achieved through GPS and Beidou, which is one of the most basic functions of the Internet of Things.

③ Information processing. This refers to the process of processing information, which involves acquiring knowledge, achieving cognition, and understanding how some past information is used for data mining and statistical analysis to generate new information, resulting in decision support and statistical reports.

④ Command and dispatch. The process by which information ultimately becomes effective, with many different forms of expression the most important aspect is to regulate the state of the object and its transformation method, based on pre-set rules or regulations dispose of events generated by objects to keep them in the expected state of motion.

⑤ Alarm linkage. Mainly providing event alarms and prompts, sometimes also based on workflow or rules the linkage function of the engine.

⑥ Remote maintenance. This is a service that IoT technology can provide or enhance, mainly applicable to enterprise production product after-sales networking services.

6.1.2 Key Technologies of the Internet of Things

As shown in Fig. 6.5, the key technologies of the Internet of Things include radio frequency identification, QR code sensing network, cloud computing cloud storage, etc.

(1) Radio Frequency Identification Technology

The basic principle of Radio Frequency Identification (RFID) technology is to utilize wireless radio frequency automatic recognition of recognized objects through spatial coupling of signals. RFID systems generally consist of RFID and readers, It is a highly concerned technology in the development of the Internet of Things. RFID is the beginning an automatic recognition technology developed in the 1990s, utilizing radio frequency signals achieve contactless information transmission through spatial coupling, and the technology of identifying through the information transmitted [3].

A complete RFID system consists of one reader and multiple applications answer device (or tag) and application software composition, reader and tag contactless data communication between signatures. Label is composed of components and chips,

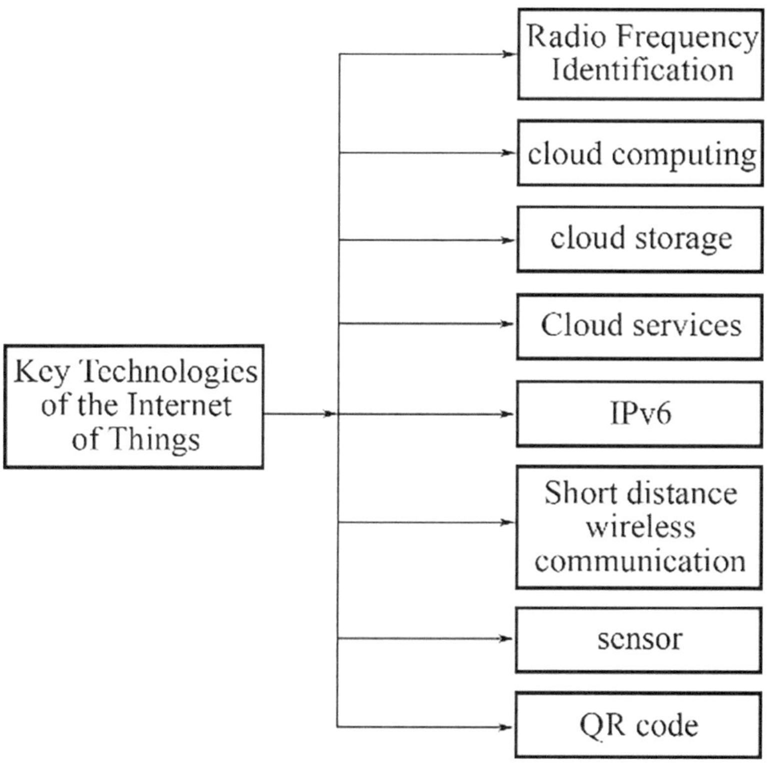

Fig. 6.5 Key technology of IoT

each label has a unique electronic code attached to an object to identify the target object. This code transmits radio frequency information to the reader through an antenna, the reader a device that reads information. Radio frequency technology in logistics, transportation, identity recognition, anti-counterfeiting, asset management, and food products, information statistics, data retrieval, and security control have all been widely applied. The characteristics of radio frequency identification technology are:

① Wide adaptability. RFID technology relies on electromagnetic waves to transmit information in a non-contact manner, unaffected by dust, fog, plastic, paper, wood, and various obstacles, and directly completes communication.
② High transmission efficiency. The read and write speed of RFID systems is extremely fast, usually less than 100 ms. RFID readers can even recognize and read the content of multiple tags simultaneously, greatly improving information transmission efficiency.
③ Uniqueness. Each RFID tag is unique, and is linked to the product one by one through RFID tags correspondence can clearly track the 'past and present' of each product.
④ Simple structure. RFID tags have a simple structure, high recognition rate, and require simple reading devices. Especially with the gradual popularization of NFC technology on smartphones, every user's phone will become the simplest RFID reader.

(2) QR Code

QR code, also known as two-dimensional barcode, is a popular encoding method on mobile devices in recent years. Compared to traditional barcodes, QR codes have a large information storage capacity and a wide coding range. They can encode digitized information such as images, sounds, text, signatures, fingerprints, etc. They are low-cost, durable, and have good security. The error rate is less than one in ten million, and they are highly reliable.

A QR code is a graphic that uses a specific geometric shape distributed in a plane dimension according to a certain pattern, with alternating depths, and is used to record symbol information of data. In coding, cleverly utilizing the concept of "0" and "1" bit streams that form the basis of computer internal logic, using several geometric shapes corresponding to binary to represent textual numerical information, automatically recognized by image input devices or photoelectric scanning devices to achieve automatic information processing. It has some commonalities of barcode technology: it has a specific character set; Each character occupies a certain width; Has certain verification functions, etc. At the same time, it also has an automatic recognition function for information from different peers and can handle the rotation and change points of graphics.

The popularization of intelligent terminals has promoted the application field of QR codes. QR code technology has penetrated into every aspect of life and is widely used in various fields such as mobile payment, product traceability, electronic ticketing, anti-counterfeiting traceability, healthy travel, information acquisition, website redirection, advertising push, etc. It will have a broader development in the future.

(3) Sensor Network

A sensor network is a randomly distributed wireless network that integrates sensors, data processing units, and communication units with small nodes, formed through self-organization. Wireless Sensor Network (WSN) is a multi hop self-organizing network system formed by a large number of sensor nodes through wireless communication. Its function is to collaboratively perceive, collect, and process information of perception objects in the network coverage area. It can achieve data collection, quantization, processing, fusion, and transmission applications.

Sensor Network (WSN) is an emerging technology that integrates multiple fields of technology such as computer, communication, networking, intelligent computing, sensors, embedded systems, microelectronics, etc. It forms an autonomous network with a large number of types of sensor nodes to achieve dynamic intelligent collaborative perception of the physical world. Wireless sensor networks originated from military applications such as battlefield monitoring, and today they are used in many civilian fields such as environmental monitoring, agricultural monitoring, health monitoring, medical care, and traffic control.

(4) Cloud Storage, Cloud Computing

"Cloud" is a popular concept in recent years. Cloud computing aims to integrate multiple relatively low-cost computing entities into a perfect system with powerful computing power through the network. A core concept of cloud computing is to continuously improve the processing capabilities of the cloud, reduce the processing burden on user terminals, and ultimately simplify it into a simple input/output device, allowing users to enjoy the powerful computing processing capabilities of the cloud on demand. The perception layer of the Internet of Things obtains a large amount of data information, which is transmitted through the network layer and then stored on a standard platform, by reusing high-performance cloud computing to process this data, we can endow it with intelligence and ultimately transform it into useful information for end users.

Cloud storage (Fig. 6.6) is an extension and derivative development of the concept of cloud computing. Through cluster applications, grid technology, or distributed file systems, cloud storage brings together a large number of different types of storage devices accross the network through application software to work together and provide data storage and business access functions to the outside world, ensuring data security and saving storage space. Simply put, cloud storage is an emerging technology that places storage resources on the cloud for people to access. Users can easily access data on the cloud at any time and anywhere through any networked device.

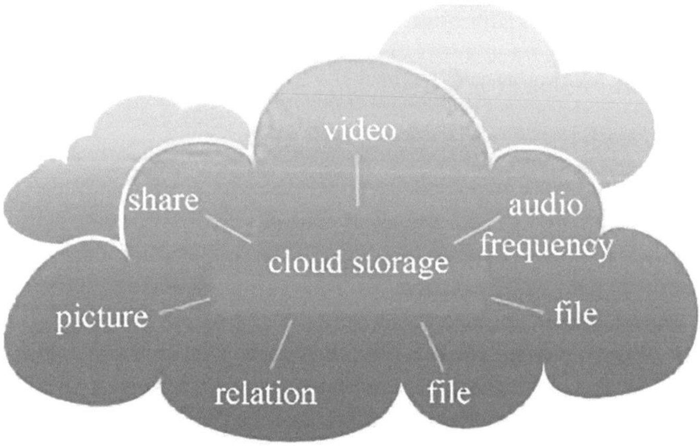

Fig. 6.6 Cloud storage schematic diagram

(5) IPv6

IPv6 is the abbreviation of Internet Protocol version 6. It is an IP protocol designed by the Internet Engineering Task Force (IETF) to replace IPv4. The length of the IP address in IPv6 is 128 bits, that is, the total number of possible addresses is up to 2^128. It can effectively solve the problem of insufficient IPv4 network address resources, removes the obstacles for access devices to connect to the Internet, and promote the application and development of the Internet of Things.

6.1.3 Application of the Internet of Things

The current world is the world of the Internet of Things. The Internet of Things has a wide range of applications, covering various fields such as intelligent transportation, environmental protection, government work, public safety, safe home, intelligent fire protection, smart healthcare, intelligent education, industrial monitoring, environmental monitoring, lighting control, elderly care, personal health, flower cultivation, water system monitoring, food traceability, enemy investigation, and intelligence collection. It effectively promotes the intelligent development of these fields, makes resource allocation more reasonable, and thus improves industry efficiency. In the industrial field, the application of the Internet of Things has greatly improved the networking rate of equipment, and has performed excellently in equipment status monitoring, fault diagnosis, maintenance, and other aspects. In the field of life services, the Internet of Things has greatly improved the scope, methods, and quality of services, enhancing people's quality of life. In the field of national defense and military affairs, from equipment systems such as satellites, missiles, airplanes, submarines, to individual combat equipment, the embedding of IoT technology has effectively improved military intelligence, informatization, and precision, greatly enhancing military combat effectiveness, and is the key to future military transformation. The application fields of the Internet of Things are shown in Fig. 6.7.

(1) Smart Water

Water management (Fig. 6.8) refers to the full application of emerging information technology in urban water management, embedding and equipping sensors into natural water and social water circulation systems, and real-time perception of the status of urban water supply and drainage systems, surface water, and water sources through online monitoring equipment such as data acquisition instruments, wireless networks, water quality, water level, and water pressure. It also integrates and shares information related to water related fields such as meteorology, hydrology, water environment, urban landscaping, and construction transportation, Forming a "perceptual

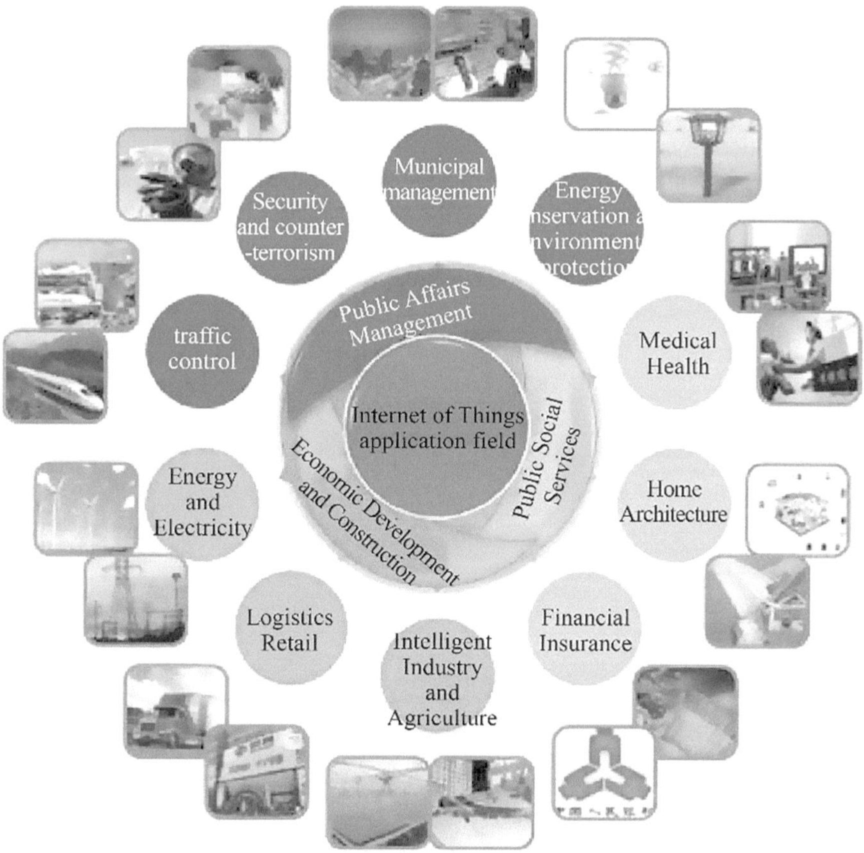

Fig. 6.7 Application field of IoT

Internet of Things" through universal connectivity; Then, through supercomputers and cloud computing, the "Water Internet of Things" is integrated to form the "Urban Water Internet of Things". With the support of multi-source coupled binary water cycle simulation, water resource regulation, water virtual reality platform, etc., the seamless integration of digital city water facilities and physical city water facilities is completed, and corresponding processing results are made and decision suggestions are proposed to achieve the entire production of the water system The precise management of management and service processes enables the planning, design, and management of urban water affairs in a more refined, dynamic, flexible, and efficient manner, achieving the state of smart water affairs, providing intelligent support for various fields such as e-government, water business management, cross industry coordination and management of water related affairs, and social public services.

The construction goal of smart water management is to build an intelligent water management system that integrates high-tech applications through years of implementation, basically achieving information digitization, control automation, and decision-making intelligence, so that the perception content is fully covered, the collection of information is fully mastered, the transmission time is all-weather, and the application runs through the entire process.

(2) Smart Transportation

The application of Internet of Things technology in road traffic is relatively mature. With the increasing popularity of vehicles in society, traffic congestion and even paralysis in big cities have become a major problem. People spend more and more

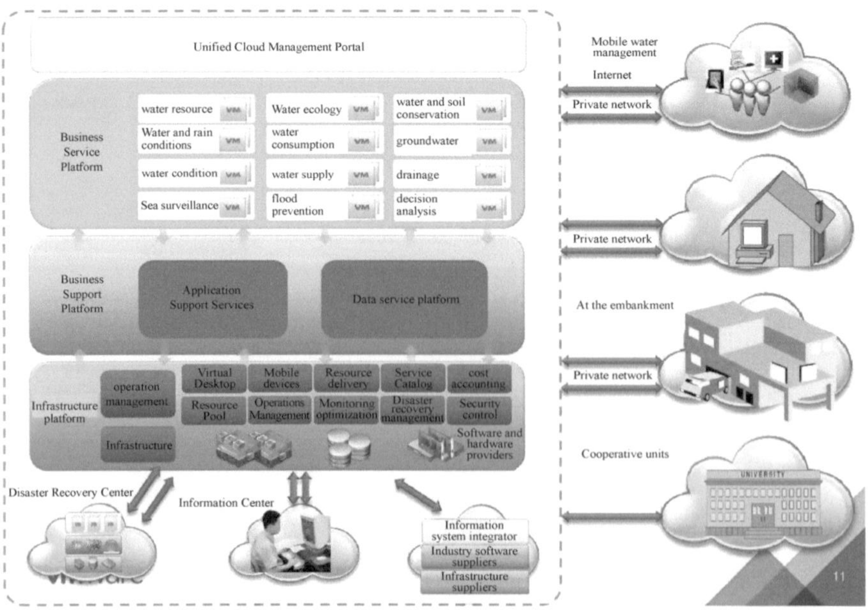

Fig. 6.8 Smart water platform

time driving and commuting, seriously affecting their quality of life. At the same time, it causes great energy waste, air pollution, noise pollution, and exacerbates the urban heat island effect. Smart transportation system refers to a system that utilizes cameras and various wireless sensing technologies to collect, organize, and analyze traffic flow data, enabling the road system to adjust traffic volume to the optimal state based on its own intelligence, ensuring traffic safety, energy conservation, and efficiency. Real time monitoring of road traffic conditions and timely transmission of information to drivers, enabling them to make timely travel adjustments and effectively alleviate traffic pressure; The establishment of an automatic toll collection system (ETC) at highway intersections eliminates the time required for card collection and return at entrances and exits, improving the efficiency of vehicle traffic. Installing a positioning system on buses can provide timely information on the bus route and arrival time. Passengers can determine their travel routes based on the route they take, avoiding unnecessary time waste. The increasing number of vehicles in society not only brings traffic pressure, but also makes parking difficult increasingly prominent as a problem. Many cities have launched smart roadside parking management systems, which are based on cloud computing platforms, combined with Internet of Things technology and mobile payment technology, to share parking space resources, improve parking space utilization and user convenience. This system can be compatible with both mobile phone mode and radio frequency identification mode. Through the mobile app software, it can timely understand parking space information and location, make reservations in advance, and achieve payment and other operations, greatly solving the problem of "difficult parking" [4]. An example of smart transportation is shown in Fig. 6.9, and urban intelligent supervision is shown in Fig. 6.10.

The last mile problem is also a concern for citizens. With the development of Internet of Things technology, multiple "shared bicycle" resources have been launched in recent years. Shared travel based on Internet of Things technology has brought great convenience to people's lives, and shared travel has become the trend. Remote unlocking, mobile payment, positioning and navigation based on mobile

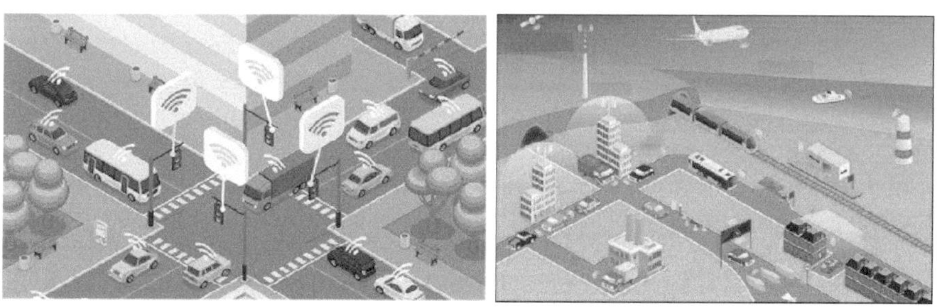

Fig. 6.9 Example of smart transportation

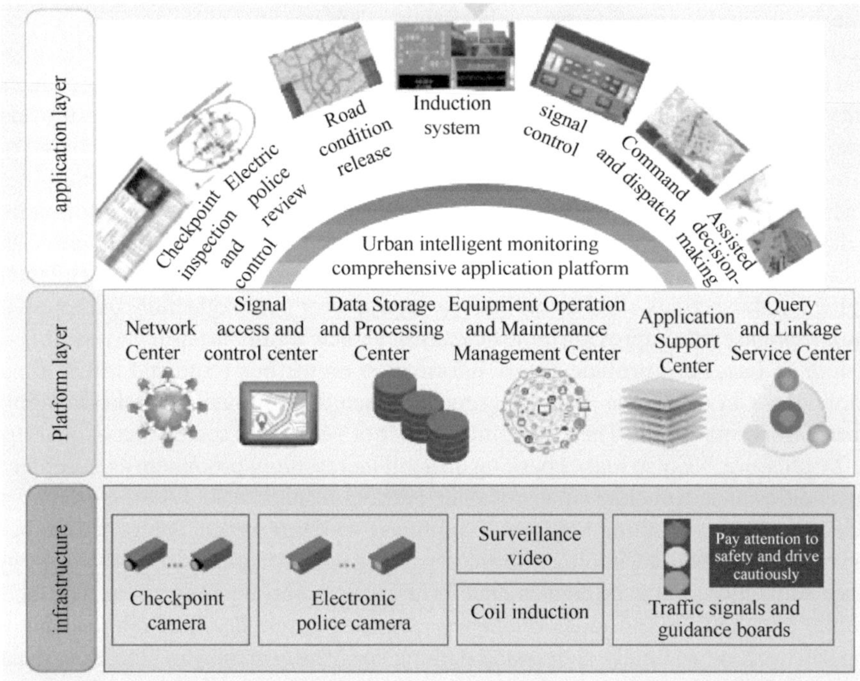

Fig. 6.10 Urban intelligent supervision

Internet and Internet of Things have brought great convenience to people's lives. Figure 6.11 shows the principle of shared bicycles.

(3) Smart Home

Smart home is the fundamental application of the Internet of Things in households. With the popularization of broadband services, smart home products involve various aspects. When there is no one at home, you can use product clients such as mobile phones to remotely operate the smart air conditioning, adjust the room temperature, and even learn the user's usage habits, thereby achieving fully automatic temperature control operation; Realize the switch, brightness, and color adjustment of smart lights through the client; The smart socket is equipped with built-in WiFi, which can achieve timed on/off of remote control sockets, monitor device power consumption, generate electricity consumption meters, etc.; The intelligent weight scale is equipped with sensors that can monitor blood pressure and fat, and has data analysis functions, which can provide health suggestions based on the body's condition; The intelligent toothbrush is connected to the client, providing reminders on brushing time and position. It can also produce charts based on brushing data and monitor the oral health status in real-time. Smart cameras, window sensors, smart doorbells, smoke detectors, smart alarms, etc. are essential safety monitoring devices for homes. Users can view the real-time situation of their homes at any time and place, eliminating

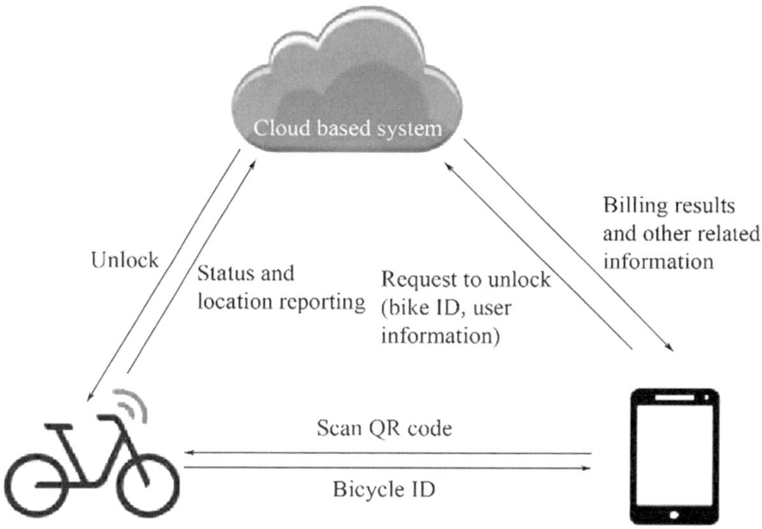

Fig. 6.11 Principles of shared bicycles

safety hazards. With the promotion of 5G technology, smart homes will usher in a broader market. Figure 6.12 show home automation solutions, and Fig. 6.13 show examples of smart homes.

(4) Public Safety

In recent years, global climate anomalies have occurred frequently, and the suddenness and harmfulness of disasters have increased further. The Internet can monitor the potential safety hazards in the environment in real time, so as to prevent in advance, give early warning in real time, and take timely response measures to reduce the threat of disasters to human life and property. In 2013, the University at Buffalo in

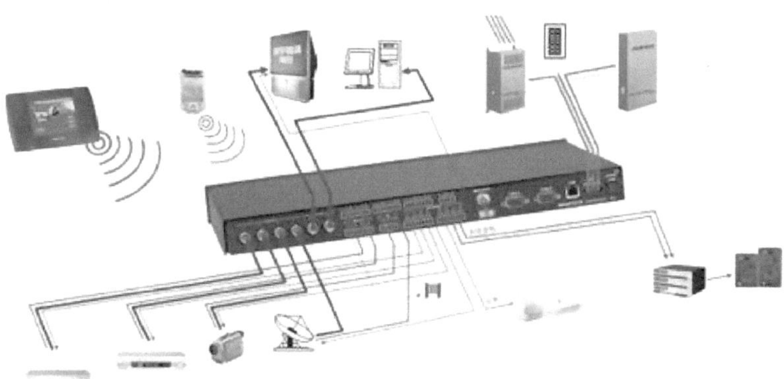

Fig. 6.12 Home automation solutions

Fig. 6.13 Example of smart home

the United States proposed the Deep Sea Internet project, which involves placing sensors in the deep seabed to analyze underwater conditions, so as to prevent and control marine pollution, detect seabed resources, and even warn of tsunamis. The use of IoT technology can intelligently perceive indicator data of the atmosphere, soil, forests, water resources, etc., playing a significant role in improving the human living environment. By utilizing remote sensing and IoT technology, natural disasters such as floods, forest fires, landslides, and mudslides can be predicted, benefiting humanity.

The sudden outbreak of novel coronavirus pneumonia at the end of 2019 has posed a serious threat to the health of people around the world, and the "health code" technology has played an important role in the prevention and control of the epidemic in China. Health codes are a means of prevention and control using the Internet of Things and big data technology. According to the principle of voluntary processing, citizens fill in information online, and 303 machines automatically compare and review the data. Subsequently, QR codes of varying colors are generated based on these review results. During the usage phase, the principle of differentiated management is adopted to scientifically and safely guide people to resume work, production, and daily travel based on different colored health codes. The health code has the characteristics of accuracy, scientificity, and dynamism. "One person, one code" monitors the health status of each person, and the backend data generates results based on objective evaluation standards and indices. The color of the health code will dynamically change according to one's movement trajectory and body temperature. Real time monitoring is used to minimize the risk of epidemic spread. Based on big data technology and Internet of Things technology, when a sudden cluster infection occurs in a city, the source can be identified in the first time, high-risk populations can be targeted, precise prevention and control can be carried out, and the spread of the virus can be contained in a short period of time, greatly reducing social risks.

(5) Smart Agriculture

In traditional agriculture, people obtain mainly agricultural information through manual measurement, which often consumes a lot of manpower. The Internet of Things can effectively reduce labor consumption and impact on the agricultural environment by utilizing wireless sensor networks to obtain accurate crop environment and information. The scenario of facing the yellow soil and facing the sky will no longer exist.

With the application and promotion of Internet of Things technology in the agricultural field, our agricultural production has become more intelligent and automated, and smart agriculture (as illustrated in Fig. 6.14) will also be widely adopted. By leveraging the information provided by advanced technology, farmers can actively choose an intelligent system that suits their agricultural production needs, thereby increasing agricultural product yields and revenues.

The agricultural expert intelligent system is a system that takes the development and utilization of intelligent expert systems as the guide, analyzes and judges environmental data such as climate, soil, and water quality, systematically plans the distribution of parks, reasonably selects and matches agricultural product varieties, and scientifically guides ecological rotation practices. The agricultural production Internet of Things (IoT) control system is based on agricultural Internet of Things technology, which collects real-time parameters such as light, temperature, humidity, and the growth status of agricultural products at the agricultural production site through various wireless sensors. It is a system that remotely monitors the production environment [5–7]. The organic agricultural product safety traceability system adds

Fig. 6.14 Smart agriculture

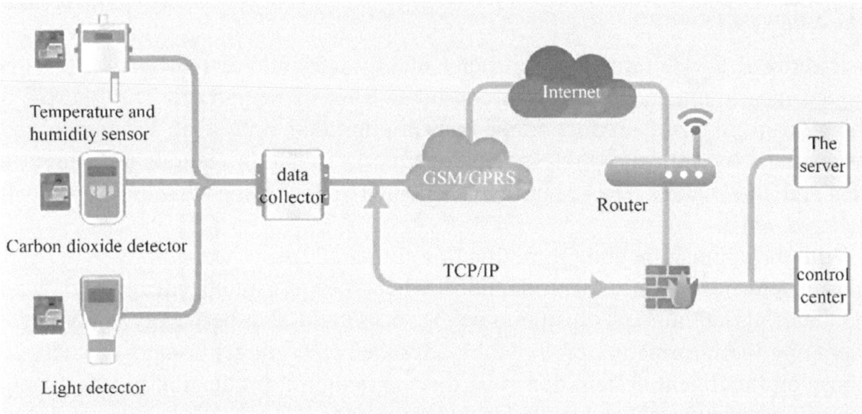

Fig. 6.15 Example of agricultural IoT

RFID electronic labels to the organic agricultural products themselves or freight packaging during the production process, and continuously adds them in transportation, warehousing, sales, and other links. Thus establishing a safety traceability system for organic agricultural products [8, 9]. An example of an agricultural IoT application is shown in Fig. 6.15.

The application of IoT technology in agriculture is mainly manifested in:

① Real-time monitoring. Real-time collection of air temperature, air humidity, and oxygen in the greenhouse through sensing devices that also gather data on carbon sequestration, lighting, soil moisture, soil temperature, outdoor temperature, and wind speed; This data is then transferred via mobile communication networks. The communication network transmits the data to the service management platform for analysis and processing.

② Remote control. Greenhouses with better conditions are equipped with electric rolling shutters, exhaust fans, electric irrigation systems and other mechanical and electrical equipment, all of which enable remote control functions. Farmers can log in to the agricultural Internet of Things (IoT) system platform via their mobile phones or computers to control the opening and closing of water valves, exhaust fans, and rolling shutters inside the greenhouse. The control logic can also be set, and the system can automatically turn on or off the roller shutter, water valve, fan, and other devices based on internal and external conditions.

③ Cloud based queries. After logging into the system using their mobile phones or computers, farmers can access real-time information on various environmental parameters, historical temperature and humidity curves, historical operation records of mechanical and electrical equipment, and historical photos inside the greenhouse. After logging into the system, you can also query local agricultural policies, market conditions, supply and demand information, expert advice, and other notices, thereby achieving targeted and comprehensive information services.

④ Remote warning. The warning function requires pre-set upper and lower limits that are suitable for the conditions, and the set values can be modified according to crop types, growth cycles, and seasonal variations. When a certain data exceeds the limit, the system promptly sends a warning message to the corresponding farmers, reminding them to take timely measures [10].

(6) Smart Healthcare

The Medical Information System (Fig. 6.16) is an efficient information support system, standardized information standard system, normalized information security system, scientific government management system, professional business application system, convenient medical service system, and a humanized health management system. It is built on the foundation of medical big data, electronic medical records, resident health records, as well as advanced information technology, automation, and intelligence. It integrates technologies such as the Internet of Things, radio frequency identification, wireless sensors, and cloud computing. The hospital information integration platform is illustrated in Fig. 6.17.

The health and hygiene industry is an important development direction for IoT applications. Smart wearable devices such as smart bracelets are connected to applications that share data with other applications, establishing a complete product and service ecosystem in the field of personal health, covering everything from exercise to nutrition. Calories and nutritional information do not need to be manually recorded in tables. These devices measure activity levels through accelerometers and use barcode scanners to comprehensively monitor calorie, nutritional, and exercise levels. These data are analyzed in the cloud and transmitted to individuals through terminals such as web pages or smartphones in the form of charts, illustrations, and images.

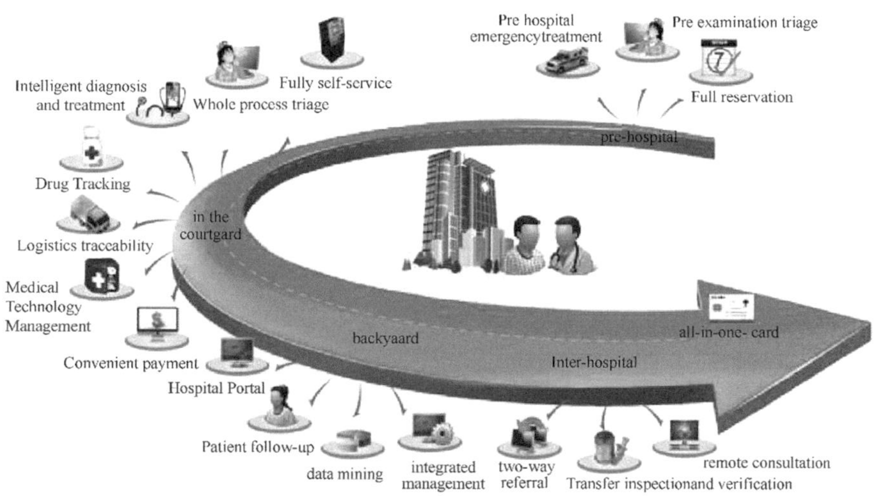

Fig. 6.16 Smart healthcare

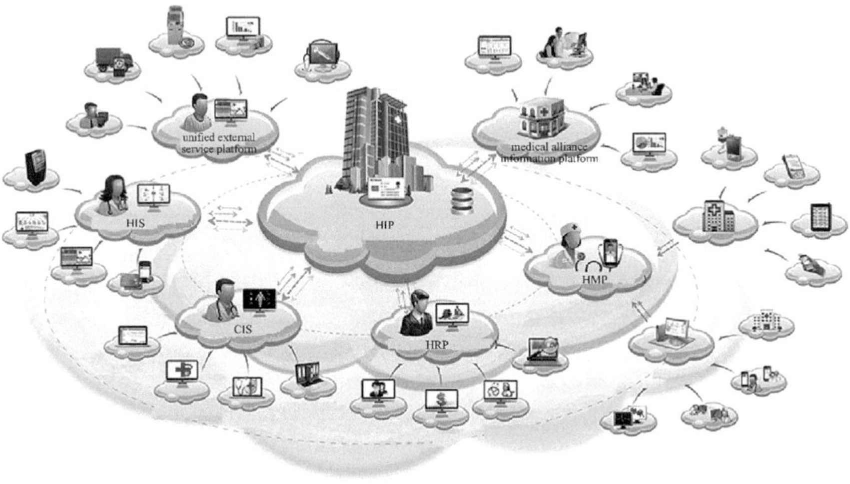

Fig. 6.17 Hospital information integration platform

In addition to wearable devices, daily household activities also collect various information about people. The networked weighing system transmits data to cloud servers, which then relay the data to the personal information dashboard on web pages or smartphone applications. The sleep tracker system records environmental data such as noise levels, indoor temperature, and brightness, and works in conjunction with sensors placed under mattresses to provide detailed information about nighttime sleep patterns and cycles. Through big data analysis, it provides guidance and recommendations for human health. These systems integrated with smartphone applications have produced an increasing number of personalized programs for assisting with sleep and early waking. In addition, there are now systems for measuring and correcting posture, devices for measuring the intensity of exercise and oxygen consumption during exercise, and trainers who provide timely feedback via smartphone applications (Fig. 6.18).

Fig. 6.18 Smart wearable devices

Fig. 6.19 Magnetic controlled micro capsule gastroscopy robot

More and more medical devices have emerged in the internet field, including sphygmomanometers, blood glucose meters, and home dispensing systems that provide reminders, configure appropriate dosages, and alert nursing staff and medical personnel in case of abnormalities. In the near future, doctors may embed miniature sensors and nanorobots in our bodies to monitor our organs and tissues, determine when medication is needed, and administer it at the optimal dosage. The magnetically controlled microcapsule gastroscopy robot has entered our lives. It can eliminate the discomfort of gastroscopy examinations by simply 'swallowing' a robot (Fig. 6.19), which completes an examination of the entire digestive tract without blind spots, providing detailed information that is transmitted to clinical doctors for auxiliary diagnosis.

The internet is likely to spark a revolution in the pharmaceutical industry. People will no longer need to go to the hospital for brief annual check-ups, and nurses will not need to constantly monitor high-risk patients. Sensors will provide monitoring data around the clock, every day of the year. With the use of next-generation software and sophisticated algorithms, smart medical instruments can analyze detailed data streams, identify potential problems and trigger points in the early stages, and enable doctors and other practitioners to develop more proactive and comprehensive treatment plans for illnesses. The future smart ward management system is shown in Fig. 6.20.

(7) Smart Education

With the rapid development of education informatization, especially during the COVID-19 pandemic in 2020, colleges and universities have adopted the online teaching mode, and education informatization has ushered in greater development space.

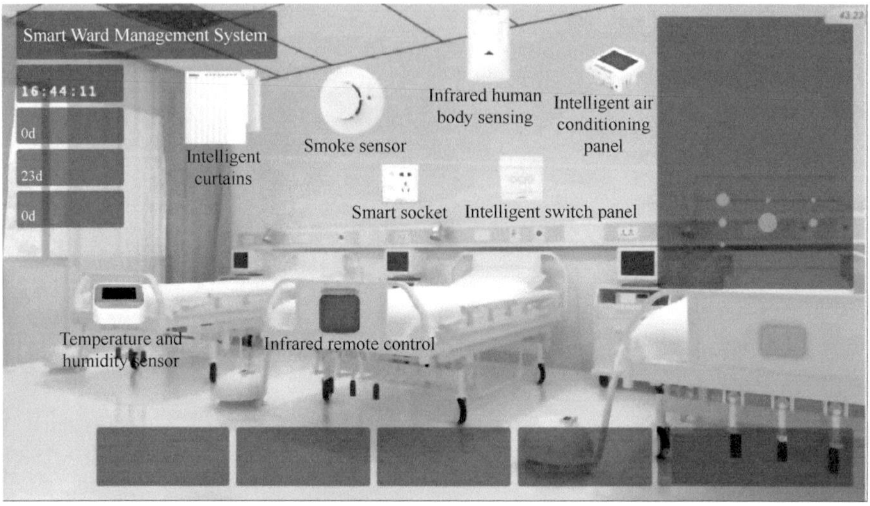

Fig. 6.20 Smart ward management system

Smart education (Fig. 6.21) is a virtual education environment based on digital information and networks, established on computer and network technology. It collects, processes, integrates, stores, transmits, and applies campus information such as teaching, scientific research, management, technical services, and life services, optimizing the utilization of digital resources, expanding the time and space dimensions of real education, and achieving comprehensive informatization of the education process.

(8) Smart Community

In the context of the acceleration of a new round of information technology reform, such as the Internet of Things, the next generation Internet and cloud computing, countries and government organizations around the world have proposed plans to change the future development blueprint of cities with information technology, that is, to build smart cities. Smart city is a new concept that has emerged as an evolution of digital cities. It is a deep integration of industrialization, informatization, and urbanization construction, and an advanced form of urban informatization. Smart community is an organic component of smart city, which is an extension of virtual government, public services, and security monitoring systems involved in smart city. Its basic components include sensor layer, public data dedicated network, application system, comprehensive application interface, and database.

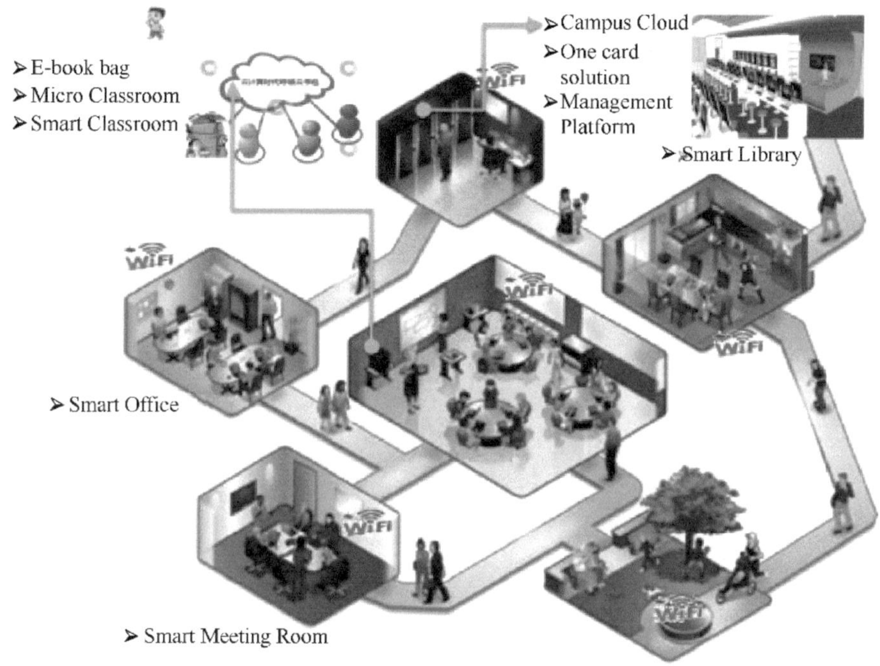

Fig. 6.21 Smart education

6.2 Internet of Things Architecture

The application of the Internet of Things is extremely extensive. With the continuous evolution of application requirements, various new technologies will gradually be incorporated into the Internet of Things system. When planning and designing the system, it is necessary to establish a framework supporting system architecture, which determines the technical details, application modes, and development trends of the Internet of Things. The perception link of the Internet of Things has strong heterogeneity. In order to achieve interconnection, interoperability, and interoperability between heterogeneous information, the future Internet of Things requires an open, layered, and scalable network architecture as the framework.

Domestic researchers mostly use the USN high-level architecture as the foundation, which is hierarchically divided into five levels from bottom to top: underlying sensor networks, ubiquitous sensor access networks, ubiquitous sensor network basic backbone networks, ubiquitous sensor network middleware, and ubiquitous sensor network application platforms.

The technical framework of the Internet of Things (Fig. 6.22) includes perception layer technology, network layer technology, application layer technology, and public technology.

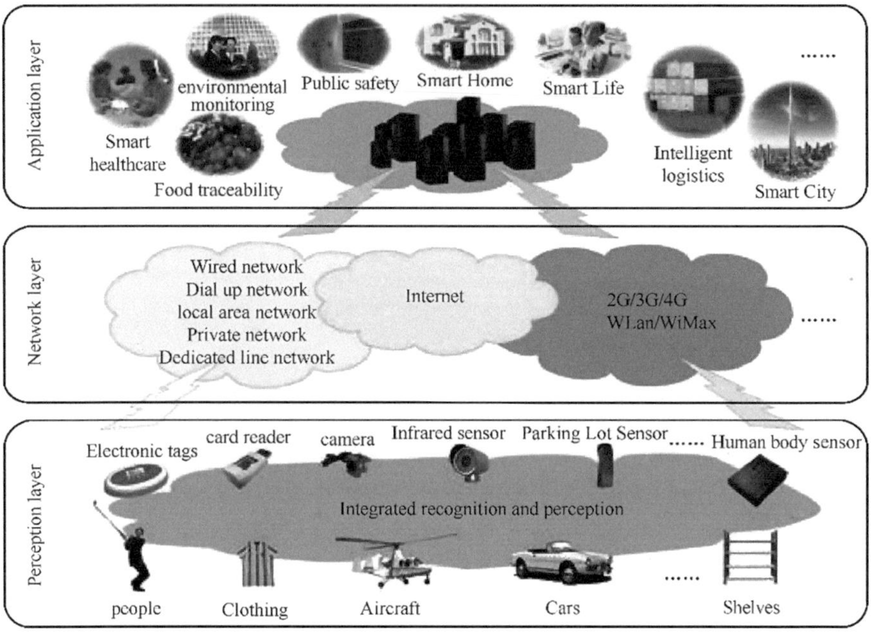

Fig. 6.22 IoT architecture

(1) Perceptual Layer

The perception layer is responsible for data collection and perception, mainly utilizing technologies such as sensors, RFID, multimedia information collection, QR codes, and real-time positioning to collect physical events and data that occur in the physical world. This includes various physical quantities, labels, audio and video data. The networking of sensor networks and collaborative information processing technology for data collection in the Internet of Things enables short distance transmission of data obtained by sensors, RFID, etc., autonomous networking, and collaborative information processing among multiple sensors.

(2) Network Layer

To achieve a wider range of interconnection functions, the perceived information can be transmitted without obstacles, with high reliability and high security, but it requires the integration of sensor networks, mobile communication technology and Internet technology. The maturity of mobile communication, Internet and other technologies, especially the development of 5G communication technology, can meet the needs of data transmission for the Internet of Things.

(3) Application Layer

The application layer mainly includes the application support platform sublayer and the application service sublayer. The application support platform sublayer is used to support information collaboration, sharing, and interoperability across industries, applications, and systems. The application service sublayer includes industry applications such as intelligent transportation, smart healthcare, smart home, smart logistics, and smart electricity.

(4) Public Technology

Public technology does not belong to a specific level of IoT technology, but is related to all three layers of IoT technology architecture, including identification and parsing, security technology, network management, and quality of service (QoS) management [11].

6.3 Basic Principles and Functions of IoT Terminals

6.3.1 Concept of IoT Terminals

IoT terminals are devices that connect the sensor network layer and the transmission network layer in the Internet of Things, enabling data collection and transmission. They possess various functions such as data collection, preliminary processing, encryption, and transmission. The overall terminal equipment can be categorized into situational awareness layer, network access layer, network control layer, and application layer, with each layer corresponding to the control equipment on the network side. The terminal should have the ability to perceive changes in the scene and select the best service channel for users.

6.3.2 Basic Principles and Functions of IoT Terminals

The IoT terminal serves as the intermediary equipment between the sensor network layer and the transmission network layer, and is also crucial equipment within the IoT. Through its conversion and collection capabilities, various external perception data can be gathered, processed, and transmitted to the Internet via diverse network interfaces. Without the IoT terminal, sensor data would be unable to reach its designated destination, effectively rendering the Internet of Things non-existent.

The IoT terminal consists of three parts: peripheral sensor interface, central processing module, and external communication interface. It is connected to sensing devices (such as RFID card readers, infrared sensors, environmental sensors, etc.) through peripheral sensing interface. The data from these sensing devices is read and processed by the central processing module. According to the network protocol, it is connected to external communication interfaces (such as GPRS module, Ethernet interface, WiFi, etc.) and sends to the designated central processing platform of Ethernet.

6.3.3 Classification of IoT Terminals

(1) By Industry Application

According to industry application scenarios, IoT terminals can be divided into industrial equipment detection terminals, agricultural facilities and parameter detection terminals, logistics RFID recognition terminals, power system detection terminals, security video monitoring terminals, etc.

① Industrial equipment testing terminal. Industrial equipment detection terminals are mainly used to collect data from displacement sensors, position sensors, vibration sensors, liquid level sensors, pressure sensors, temperature sensors, etc. on large equipment in workshops or large machinery in industrial and mining enterprises, and send them to the upper computer for data processing through the wired or wireless network interface of the terminal, achieving timely tracking and confirmation of the operating status of large or important equipment in the factory workshop, this achieves the goal of ensuring safety in production.

② Agricultural facilities and parameter detection terminals. Agricultural facilities and parameter detection terminals are used to collect data from air temperature and humidity sensors, soil temperature sensors, soil moisture sensors, light sensors, gas content sensors, etc. The data is packaged, compressed, encrypted, and sent to the central processing platform for data aggregation and processing through the terminal's wired or wireless network interface. These types of terminals are generally installed in greenhouses, which can promptly detect abnormal environmental factors in agricultural production and ensure the safety of agricultural production.

③ Logistics RFID identification terminal. Logistics RFID identification terminal equipment is categorized into fixed, vehicle-mounted, and handheld types. Fixed terminal devices are generally installed at the entrance of warehouses or other cargo channels to track the entry and exit of goods; Vehicle mounted terminals are installed on transport vehicles and handheld terminals are used for handheld use. These two types of terminals have GPS positioning function and basic RFID tag scanning function, which are used to identify the status, position, performance and other parameters of goods. Transfer location information and basic cargo information to the central processing platform through wired or wireless networks to improve logistics efficiency. For example, when we purchase goods online, we can keep track of the logistics information status on the platform at any time, greatly improving the shopping experience.

(2) By Usage

According to the usage situation, IoT terminals mainly include fixed terminals, mobile terminals, and handheld terminals.

① Fixed terminal. Fixed terminals are used in stationary situations, fixed throughout the year, with a reliable external power supply and wired data links, to detect information about various fixed equipment, instruments, or environments. Examples include terminals used in logistics warehousing, facility agriculture, and industrial equipment, as mentioned earlier.

② Mobile terminals. Mobile terminals are used in situations where the terminal and the device being tested are moving simultaneously. Data transmission is generally carried out through wireless data links, mainly detecting images, positions, and device statuses, and they need to have good seismic and electromagnetic interference resistance capabilities. They are found in various applications, such as in-car instruments, in-car video surveillance, and GPS positioning for trucks and buses. Nowadays, buses and trucks generally come equipped with positioning systems and driving recorders, which utilize car power and are equipped with large capacity memory cards to effectively record driving images. In the event of an accident, it can be traced back to the source.

③ Handheld terminal. Handheld terminals are compact, lightweight, portable, and easy to operate. They are the transformation and upgrade of mobile terminals, with interfaces that can connect to external sensing devices. The collected data can generally be transmitted in a timely manner through wireless, or connected to wired transmission after accumulating a certain degree. Most of these types of terminals are used in fields such as logistics radio frequency identification, factory parameter table inspection, and crop pest and disease surveys.

(3) By Transmission Method

According to transmission methods, IoT terminals can be divided into Ethernet terminals, WiFi terminals, 3G terminals, 4G terminals, 5G terminals, and so on. Some intelligent terminals have multiple interfaces as mentioned above.

① Ethernet terminal. This type of terminal is typically used in situations where there is a need for large data transmission and good Ethernet conditions. It is easy to wire on site and has the capability to connect to the Internet. It is generally used in fixed equipment testing in factories, smart buildings, smart homes, and other similar environments.

② WiFi terminal. This type of terminal is generally used in situations where data transmission volumes are large and Ethernet conditions are good, but the wiring of the terminal is not easy or impossible. Networking is achieved by setting up WiFi routers or WiFi gateways around the terminal. It is generally used in wireless cities, intelligent transportation systems, and other scenarios that require wireless transmission of large data volumes, or in situations where high data transmission rates are necessary but wiring around other terminals is not practical.

③ 3G terminal. This type of terminal is applied in small data mobile transmission scenarios or field work scenarios with small data transmission, such as vehicle GPS positioning, logistics RFID handheld terminals, reservoir water quality monitoring. This type of terminal possesses the networking function in mobile or outdoor conditions, providing a broader market for the deep application of the Internet of Things.

④ 4G and 5G terminals. This type of terminal is an upgrade of 3G terminals, which improves the communication speed of uplink and downlink to meet the needs of mobile image monitoring, video distribution and other applications, such as the transmission of police car patrol images, dynamic real-time traffic information monitoring, etc. In some large data sensing applications, such as vibration acquisition or power signal monitoring, this type of terminal can also be utilized.

With the development of mobile Internet, more and more devices are connected to mobile networks. It is necessary to solve the problems of efficient management of various networks, ensuring interoperability, and enhancing user experience. 5G was born to address the aforementioned challenges and meet the growing demand for mobile traffic. The 5G era has arrived. Its advantages are that the data transmission rate is far higher than that of the previous cellular network, up to 10Gbps, faster than the current wired Internet.

(4) From Using Scalability

From the perspective of scalability, IoT terminals are mainly divided into two types: single function terminals and universal intelligent terminals.

① Single function terminal. The single function terminal has simple functions and fewer external interfaces, and is only suitable for specific occasions, meeting the needs of a single application and its partial expansions. At present, there are many such terminals in the market, such as image transmission service terminals for automotive monitoring, terminals for power monitoring, and RFID terminals for logistics, which have lower costs and are easy to standardize.

② Universal intelligent terminal. The universal intelligent terminal comprehensively considers the universality of industry applications, with multiple external interfaces that can meet two or more applications. It can meet different application needs through internal software settings, parameter modifications, or combinations of hardware modules. It has multiple wired and wireless interface modes for network connection, as well as Bluetooth, WiFi, ZigBee and other interfaces. In addition, a specific output interface is also reserved.

(5) From the Transmission Path

① Data transmission terminal. The data transmission terminal establishes a data transmission path between the input port and the application software. It allows data to be input through the module's input port and then output directly through the software, functioning as a "transparent" channel. This type of terminal, known as the data transmission terminal, has been widely used in IoT integration projects. Its advantage lies in the ease of constructing an IoT system that meets application requirements. However, its disadvantage is its limited functionality. Currently, the majority of general-purpose terminals available on the market belong to this category.

② According to transparent transmission terminals. This type of terminal generally combines data collected from multiple external interfaces and transmits them through the processor inside the terminal, thus having the advantage of multiple simultaneous transmissions and reducing the number of terminals. The disadvantage is that applications can only be selected based on the peripheral interfaces of the terminal. If all applications are met, the terminal requires many types of peripheral interfaces, which can cause waste of interface resources in the application. Therefore, the pluggable design of interfaces is a common feature of such terminals.

6.3.4 *Standardization of IoT Terminals*

The biggest obstacle to terminal promotion is standardization of terminals. The booming development of Internet of Things technology in China is estimated to be a trillion dollar market in the future. The main reason that currently hinders the large-scale promotion of IoT technology is the issue of terminal incompatibility. Devices and software from different manufacturers cannot be used on the same platform, and there is no unified standard for protocols between devices. Therefore, before the popularization of the Internet of Things and the large-scale promotion of terminals, standardization issues must be addressed, which are manifested in the following aspects:

① Standardization of hardware interfaces. The sensing devices of the Internet of Things are provided by different manufacturers. If each company has different interface or communication rules, it will lead to different terminal interface designs, and terminals cannot reserve interfaces for each manufacturer. Therefore, it is necessary for sensing device manufacturers and terminal manufacturers to jointly develop standard interface and communication specifications between IoT sensors and terminals to meet the hardware interoperability and interconnection requirements between devices from different manufacturers.

② Standardization of data protocols. Data protocol refers to the data flow interaction protocol between the terminal and the platform layer, which can be divided into business data flow and management data flow. The WMMP protocol jointly developed by China Mobile and Ericsson is a good management protocol, and its promotion and popularization will undoubtedly promote the standardization process of data protocols, facilitating the network access and management of newly developed terminals. The development of the Internet of Things requires the leadership of relevant national departments and the joint formulation of more comprehensive general protocols similar to WMMP by relevant industries to meet the interconnection issues of various applications and terminals from different manufacturers, and to expand the promotion of the Internet of Things in the future.

At present, the scale promotion of IoT terminals is mainly limited to the security, logistics, environmental monitoring and other fields of national key projects, and has not been widely used in other fields. The main reason is that the concept of IoT and its benefits are not fully known to people; In some industries, it is still difficult for promoters and users to find their own profit points and profit models, and the high cost and operating expenses of the system make customer enthusiasm low. Analyzing industry applications and reducing system costs (especially operating costs) is a necessary path for the large-scale promotion of the Internet of Things. Reduce terminal costs, sensor costs, and deployment costs. With the maturity of various technologies in the Internet of Things and the standardization of terminals, the costs of each link in the Internet of Things will be greatly reduced.

6.4 Application of Internet of Things Technology in Equipment

The early Internet realized the connection between people and static information. Now the Internet has evolved to connect people. The Internet of Things realizes the connection between people and things, as well as between things and things, involving all aspects of production and life. Through the interconnection of chips and sensors, it can perceive, capture, communicate and actively respond to the information of billions of intelligent devices in real time, truly realizing the "global village". The application of the Internet of Things is shown in Fig. 6.23.

Multiple market research reports show that every household in the United States has approximately 7 connected devices. Meanwhile, the Organization for Economic Cooperation and Development predicts that the number of connected devices per household in the United States will reach 20 by 2020. The survey data shows that 88% of mobile device users already know about home automation systems. More and more smartphone, e-book, Blu ray player and other device users say that Internet connectivity and the functionality of viewing content are the first considerations when purchasing products.

In the era of the Internet of Things, users are the controllers of devices. Devices in the network should be designed based on user experience and user-defined values, which is crucial for the widespread application of the Internet of Things. As more and more devices become intelligent and connected together, many existing physical products will transform into digital experiences. Many human–computer interactions will be replaced by interactions between machines, and new modes of interaction between machines and humans will emerge. Most communication between machines will become intangible, and human–machine communication will become more

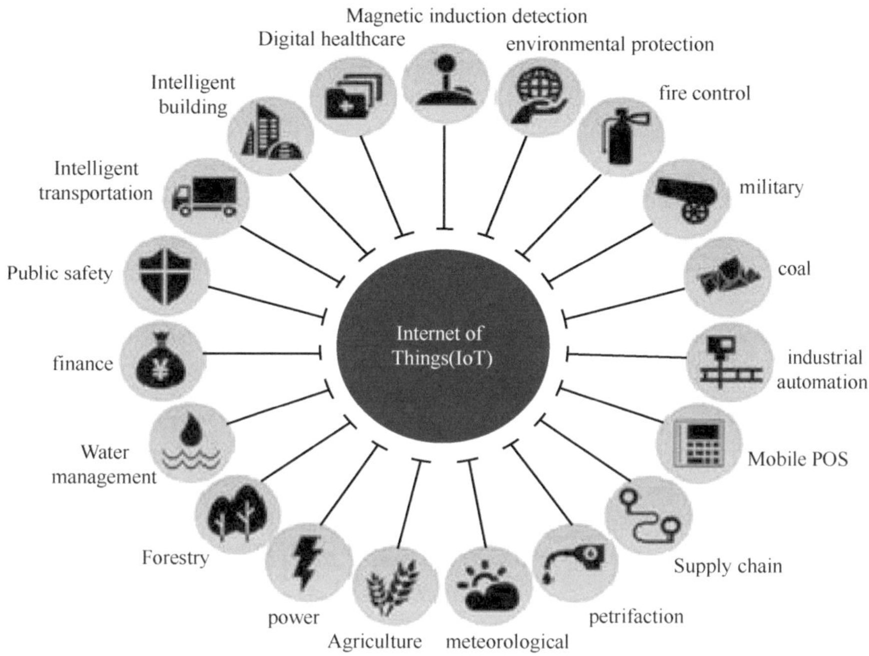

Fig. 6.23 Application of IoT

interactive. A successful IoT solution must be simple and reliable. When designing smart devices for the Internet of Things, the following factors need to be considered.

(1) Simplify Getting Started

The first step in introducing users into a new system is the most difficult when designing a new intelligent product. In the case of multi device interaction, it usually means duplicate authentication, such as gateway processes between different devices and switching to other services. Simplifying input, such as using code, gestures, facial expressions, or voice, instead of using the number 315 as a password for security verification, is a good measure. Considering that using IoT systems typically means switching devices between mobile terminal devices or embedded software, simple, secure, and intelligent verification is particularly important.

(2) Cross Platform Design and Interaction

The cloud is the key to achieving a consistent user experience across multiple IoT products. Cloud based applications and network devices provide users with seamless transformation between system elements in the simplest and most efficient way. Meanwhile, the responsive design of the Internet of Things encompasses all devices, platforms, and software, surpassing the consistency of experience on the web or mobile end.

(3) Personalization and Background

Freedom and individuality are the pursuits of modern people, and the same applies to products. Crowdsourcing and personalization are inevitable requirements for modern intelligent products. More and more digital tools can learn and recognize patterns from user behavior, providing a more refined experience. The user experience of IoT systems should be personalized, just like the personalized performance of network devices. With the development of the Internet and the Internet of Things technology, people can customize their own products anytime, anywhere, to achieve a true sense of "private customization".

(4) Integrated Experience

One of the most challenging tasks in IoT user experience design is to minimize the gap between network devices and the real world, and establish a smooth experience between all system elements. Designers and engineers of network products should make every effort to create a unified environment for IoT systems. According to the permission level, end users should be able to access data from devices, sensors, and integrated platforms, and perform high-quality data visualization and analysis on different terminal platforms.

(5) Innovative Experience Methods

Designers of IoT products targeting consumers have begun to pay attention to voice products, and more and more smart assistants are appearing at home. Smart home systems also possess smart assistant skills, but sound is not the only way of communication. A smooth user experience will become more contextualized and naturalized in the future. The co pilot of a smart car uses gestures to ensure safe driving. The activation function of biometrics can achieve faster and safer authentication. This is particularly important for medical IoT systems, industrial IoT, and other areas with limited access.

Taking smart wearable products as an example, smart wearable equipment is a key entry point in the era of the Internet of Things and a key link between people and things in the Internet of Things. Intelligent wearable devices not only refer to wearable devices for the human body, but also pertain to the intelligent future of various industries. With the driving force of medical demand and the driving force of medical intelligent wearables, experts predict that the intelligent wearable market will bring about a new wave of enthusiasm. The design requirements for smart wearables consider the comfort and position of wearing. From the perspective of human–computer interaction, pay attention to the factors between "human", "machine", and "environment", and adopt different experience design strategies at different stages of the product. The interaction of intelligent wearables has liberated hands, with voice interaction as the main approach and adopting a multi device and multimodal interaction mode.

6.5 Design of IoT for Mechanical and Electrical Equipment

6.5.1 Overview

Mechatronics integration involves multiple technologies such as machinery, electronics, control, and networking, and the role of the Internet of Things is becoming increasingly significant. Today's mechatronics products are not just lifeless mechatronics products, but also products with "intelligence".

The Internet of Things technology enables the interconnection of people, machine equipment, and system software. There is much mechanical and electrical equipment in the production workshop, and it is necessary to connect all the machinery and fixtures uniformly for network management, achieving the "Internet of Things" between people and mechanical and electrical equipment. The NC programming staff writes the program on the PC and sends it to the DNC network server. The actual operator of the machine equipment can download the required program for processing and production. After the task is completed, the NC machine tool program flow will be sent back to the network server through the DNC Internet, which will be sorted out by the program process management personnel or processing technicians. All production processes will remain digital and traceable management methods. Industrial production process control systems, the Internet of Things, ERP, CAD V CAM VCAE VCAI and other technologies can leave manufacturing companies for market applications. The production line of a manufacturing company runs at high speed, and the amount of information generated, collected, and solved by the production line equipment is much higher than the data collected by computers and human services in the company. The practicality of statistical data is also higher.

The workshop and workshop equipment are connected to different devices through barcodes, QR codes, RFID, industrial sensors, etc. Statistical data is collected every few seconds. Through these statistical data, we can analyze the operating rate of spindle bearings, the spindle bearing load rate, the operating rate, the repair rate, the output rate, the comprehensive utilization rate of machinery and equipment (OEE), the component compliance rate, and the quality percentage. At the same time, it can also make the production and manufacturing of text documents paperless, maintaining high efficiency and intelligent manufacturing.

6.5.2 IoT Platform Architecture Design Process

(1) Scheme Design Stage

The overall scheme design of the Internet of Things engineering includes the overall network scheme design of the engineering project and the overall system function scheme design, among which network design is divided into logical network design and physical network design.

The overall scheme design usually consists of the system design stage and the structural design stage, and the system design stage determines the overall architecture and logical network selection of the system; The specific implementation plan will be determined during the structural design phase, including the selection of physical networks and equipment, the selection and determination of data centers, implementation strategies for security, and detailed design of software module structures.

Design work should be carried out from top to bottom. Firstly, design the overall structure, framework and overview of the system, and provide detailed reports to user units and departments. Then, based on this, we delve deeper layer by layer until we proceed with the detailed design of each module.

In summary, the overall scheme design of the Internet of Things system is also based on the preliminary system analysis, including the division of the entire system, the configuration of machinery and equipment (including software and hardware devices), the storage scale of data, and the overall framework structure design of the entire system implementation plan.

(2) Core architecture Design and Network Communication Selection Stage

A complete IoT platform must have four core modules: device management, user management, data transmission management, and data management, and all other functional modules can be considered as extensions of these four modules. Figure 6.24 shows the IoT architecture of electromechanical equipment.

① Equipment management. Device management includes two parts: device type management and device information management. Device type management refers to defining the type of device, and its functions are generally defined by the manufacturer of the device. Device information management refers to

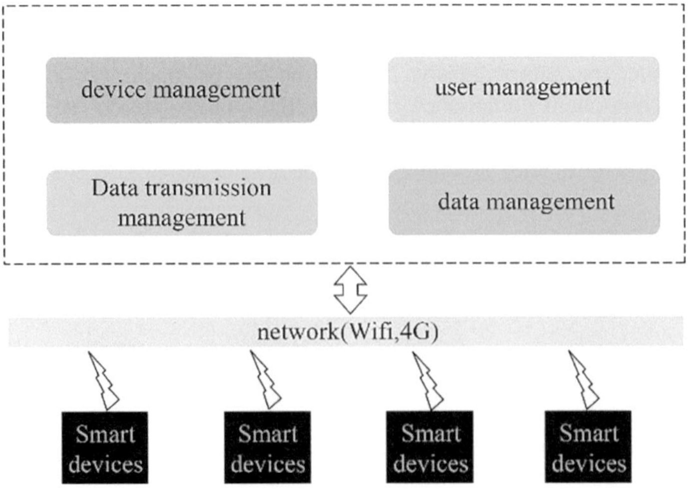

Fig. 6.24 Architecture of the Internet of Things for mechanical and electrical equipment

defining device related information, selecting device types based on defined user attributes, and allowing users to have complete control over the device upon activation.

② User management. User management includes four parts: organization management, user management, user group management, and permission management. Organization management pertains to the oversight of all devices, users, and data within the context of an organization. User management focuses on the composition of personnel within an organization. Administrators of each organization can add different individual users to their respective organizations and assign distinct permissions to each user. Additionally, a single user can belong to multiple organizations and serve in various administrative roles across these organizations.

③ Data transmission management. Data transmission management includes three parts: basic format, data parsing definition, and data storage. The basic format establishes a data transmission protocol for a type of device, which includes three components: device serial number, command code, and data. The definition of data parsing is that organizational administrators can define multiple commands using different parsing methods for the same device type according to their needs. Data storage should support a distributed architecture, where different storage locations can be defined for each device, and each data has a defined lifecycle. After the lifecycle ends, the system will automatically delete it.

④ Data management. Data management includes three parts: permission management, big data, and data export. Permission management refers to the ownership of data, and the concept of who data belongs to is very important. Only the owner of the device can define who data can be shown to. The permission of data is crucial in the Internet of Things platform. The role of big data is to rely on big data platforms to achieve visual analysis and processing of massive data in the Internet of Things, and to obtain valuable information. Data export refers to the ability of users to export partial IoT data for local analysis.

The selection of network communication is also a key aspect in the design of the Internet of Things. At present, the communication between cloud based IoT platforms and devices is essentially built on the TCP/IP protocol, with the only difference being the re-encapsulation of data packets. Based on this, WiFi and 4G are widely used to achieve communication between devices and cloud platforms. With the promotion of 5G, there is a broader field for communication selection. Communication between devices can be achieved through various methods such as WiFi, Bluetooth, ZigBee, and so forth. Common communication architectures are selected as shown in Table 6.1.

(3) Architecture Design Phase of Industrial IoT Projects

A typical IoT project consists of at least three parts: device side, cloud side, and monitoring side.

① Device side architecture design. The device end is mainly responsible for data collection, process logic execution, and control. From a functional perspective,

Table 6.1 IoT communication architecture

Communication architecture	Main features	Problems
Based on mobile network 3/4G communication	The simplest architecture is based on mobile communication to access the internet	Each device requires a SIM card
Based on WiFi LAN	Suitable for all IoT devices to operate in a local environment	Data traffic issues; Communication quality issues
Based on Bluetooth communication	A typical point-to-point communication method	Intelligent devices within a local area network do not have independent IP addresses for the public network, only one IP address within the local area network; Power consumption issue; Interference issue
Based on Zigbee	Zigbee is designed for networking between sensors and has strong low-power capabilities	Capacity issue of Bluetooth gateway; Bluetooth pairing issue

the device architecture can generally be divided into three layers: the first is the data acquisition and control output layer, the second layer handles the execution of process flows, and the third layer is the communication layer, responsible for data upload and command reception.

② Architecture design. The cloud generally includes three parts: web front-end, web back-end, and middleware. In an industrial-level IoT project, the web front-end usually displays four parts of content: process screens, various data reports, operation logs, and system diagnostic information. The web backend is relatively complex and typically involves handling GET and PUT requests, transmitting real-time data to the front-end interface, and establishing correspondences between device data and various reports, curves, and logs, all to facilitate the application's deployment across as many sites as possible. The main function of middleware is to communicate with on-site devices, obtain data, or send relevant control instructions. A middleware program is generally a service program or ordinary application program of a system, with a long lifecycle and can run continuously for a long time. Therefore, it can handle some relatively complex business logic, data conversion, and data dumping tasks.

③ Architecture design for monitoring end. The monitoring end generally includes PC, mobile or tablet monitoring. From a functional perspective, the architecture can be relatively simply divided into two layers: the UI interface display and operation layer, and the data communication layer to achieve information exchange with the server.

(4) Post Maintenance Phase

After the development of the Internet of Things platform, it is not a mature application platform, and maintenance work will take up a lot of time. Maintenance work is not only the maintenance of the development team, but more importantly, on-site maintenance, troubleshooting, timely positioning, and timely resolution. In response to the above challenges, it is necessary to consider a unified and configured architecture design at the beginning of the design.

6.6 Summary of This Chapter

This chapter introduces the concept, main functions, and key technologies of the Internet of Things, and elaborates on the IoT terminals and architecture. On this basis, the application of IoT technology in equipment and the design methods of IoT for electromechanical equipment were introduced. The rapid development of Internet of Things technology requires timely attention to databases and media to obtain the latest information on the latest developments and applications of the Internet of Things.

References

1. Yu Jianfeng 2013 Yu Jianfeng. Development and Practice of Internet of Things Engineering[M]. Beijing: Posts & Telecom Press, 2013.
2. Yun Liyu. Discussion on Internet of Things and Its Applications[J]. China Computer & Communication, 2017(3):3.
3. Wang Aiying. Smart Card Technology: IC Cards, RFID Tags and Internet of Things[M]. Beijing: Tsinghua University Press, 2015.
4. Li Yuze 2020 Li Yuze. Research on Development Status and Applications of Internet of Things[J]. Digital Communication World, 2020(03):234–23.
5. Li Daoliang Yang Hao 2018 Li Daoliang, Yang Hao. Research Progress and Development Trend Analysis of Agricultural Internet of Things Technology[J]. Transactions of the Chinese Society for Agricultural Machinery, 2018, 49(01):1–20.
6. Tian Hongwu Zheng Wengang Li Han 2016 Tian Hongwu, Zheng Wengang, Li Han. Application Status and Development Trend of Internet of Things Technology for Water-Saving in Field Agriculture[J]. Transactions of the Chinese Society of Agricultural Engineering, 2016, 32(21):1–12.
7. Li Jin Guo Meirong Gao Liangliang 2015 Li Jin, Guo Meirong, Gao Liangliang. Application of Agricultural Internet of Things Technology and Innovative Development Strategies[J]. Transactions of the Chinese Society of Agricultural Engineering, 2015, 31(S2):200–209.
8. Ge Wenjie, Zhao Chunjiang. Research and Application Status and Development Countermeasures of Agricultural Internet of Things[J]. Transactions of the Chinese Society for Agricultural Machinery, 2014, 45(07):222–230, 277.
9. Xu Gang Chen Liping Zhang Ruirui et al 2010 Xu Gang, Chen Liping, Zhang Ruirui, et al. Research on Application of Agricultural Internet of Things Based on Precision Irrigation[J]. Journal of Computer Research and Development, 2010, 47(S2):333–337

10. Application of Internet of Things Technology in China's Agricultural Production[OL]. http://www.qianjia.com/zhike/html/2020-09/15_28713.html.
11. Meng Guangjun. Overview of Basic Architecture of Internet of Things Technology[J]. Teaching of Forestry Region, 2013(3):64–65.

Chapter 7
Industrial Equipment Systems with Complex Processes and High Performance Sports Requirements

7.1 Case 1 Development of Fully Automatic Glazing Machine for Insulated Glass Units

Glass is widely used in industries such as construction, transportation, shipbuilding, aviation, and refrigeration. It is not only a good transparent material, but also a material with good thermal conductivity. Regardless of the field in which glass is applied, heat conduction through glass can lead to significant energy loss. With the development of the glass processing industry, more and more people have realized that insulating glass has significant energy-saving effects. The application of insulating glass in China is also increasing, especially with the increasing use of curtain wall structures in buildings, which further promotes the application of insulating glass. In addition, domestic and foreign practices have proven that improving the insulation performance of building envelope structures, especially enhancing the insulation performance of windows, is the most economical and effective method to prevent heat loss in buildings. Hollow glass plays a crucial role in related architectural applications.

Hollow glass is a product in which two (or more) pieces of glass are evenly separated by effective supports (usually hollow aluminum spacers or solid thermoplastic spacers filled with desiccants), and the edges are bonded and sealed to form a dry gas chamber between the glass layers, as shown in Fig. 7.1 [1].

The sealing process of glass involves two operational processes, namely the first sealing and the second sealing. The first sealing process mainly uses hot melt butyl adhesive to bond the bent aluminum partition frame and two or more pieces of glass into one. The second sealing process is to mix polysulfide adhesive and silicone adhesive and evenly apply them to the grooves formed by the glass and aluminum partition frame, ensuring the structural bonding between the glass and aluminum partition frame, as shown in Fig. 7.2. The insulated glass production line is a specialized equipment used to produce insulated glass. As a complete insulated glass production line, it should include the main processing equipment (such as glass slicing machine,

M. Xu et al., *Design and Case of Intelligent Electromechanical Equipment System*, Advanced and Intelligent Manufacturing in China,
https://doi.org/10.1007/978-981-95-3833-1_7

Fig. 7.1 Hollow glass structure

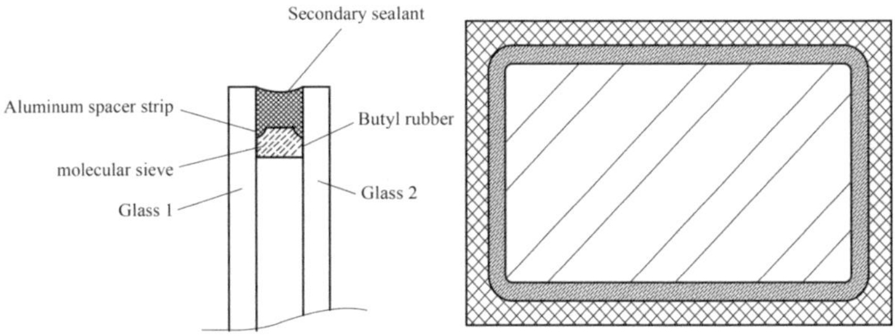

Fig. 7.2 Second sealing of insulated glass

edge grinding machine, glass cleaning machine, laminating pressing machine, gluing machine, etc.) and necessary auxiliary equipment (such as aluminum partition storage machine, glass loader, unloading machine, raw material conveyor, finished product conveyor, etc.).

The fully automatic gluing machine is a very important component of the fully automatic insulated glass production line, and it is also the core equipment of the insulated glass production line. It is responsible for gluing and sealing the glass periphery (i.e., secondary sealing), as shown in Fig. 7.2. The speed of gluing by the gluing machine directly affects the production speed of the entire production line, and the quality of gluing directly affects the quality of insulated glass. Through thorough market research and understanding of glass deep processing equipment both domestically and internationally, the author's team has developed a comprehensive set of equipment for insulated glass production lines, utilizing industrial computer and motion control card bus technology. The fully automatic glue coating robot is a representative large-scale intelligent manufacturing equipment that realizes the control of various movements (12 axes) from glue dispensing, and achieves the intelligence of the equipment at a relatively low cost, replacing imported equipment.

7.1.1 Design of the Mechanical Body of a Fully Automatic Adhesive Coating Machine for Insulating Glass

The overall mechanical simplified structure of the glue coating machine is shown in Fig. 7.3, and the actual body diagram is shown in Fig. 7.4. The equipment is mainly divided into four parts based on mechanical functions: the glue supply part, the mixing part, the glue application part, and the glass conveying part.

① Glue supply section. Responsible for providing colloids from the rubber barrel to the plunger pumps of component A and component B. Each rubber barrel has an air pump to squeeze the colloids inside the barrel to assist in glue supply. A.

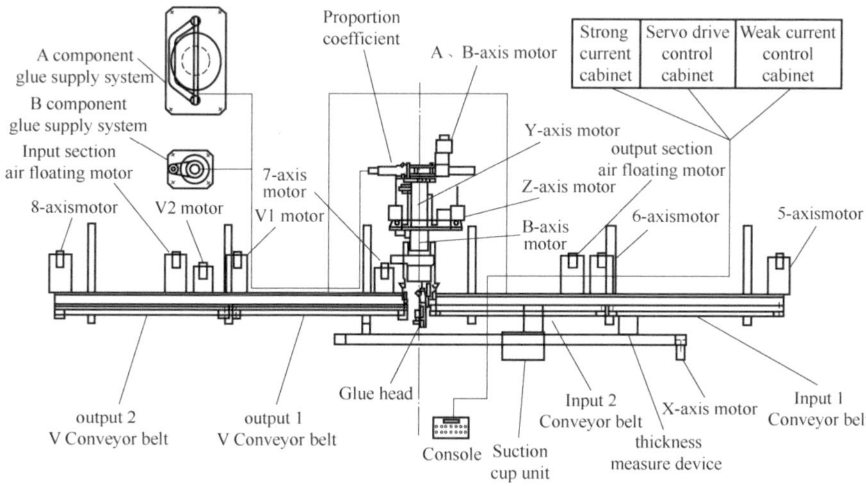

Fig. 7.3 Simplified overall mechanical structure of glue coating machine

Fig. 7.4 Solid diagram of fully automatic glue coating machine for insulated glass

Group is driven by motors A and B to operate the plunger pump. At the same time, in order to overcome the influence of colloid viscosity to the maximum extent, an auxiliary glue supply cylinder is installed at the front end of the plunger pump, which can suck the colloid into the storage cylinder at that cylinder in advance during glue application, for the next colloid suction use.

② Mixing adhesive part. A. The B plunger pump is driven by a motor to squeeze the colloids into the mixer of the mixing section in corresponding proportions (different colloids have different proportions, usually 10:1). After the two colloids are fully mixed in the mixer, they can be sent to the dispensing head of the dispensing section for dispensing.

The gluing part, which consists of a gluing head, a trowel, a crescent plate, etc., completes the relevant actions of the gluing process.

The conveying part consists of two sets of input synchronous belts, two sets of output synchronous belts, an X-axis, and thickness measurement units. The input segment 1 consists of a reducer, coupling, and synchronous belt structure, which is used to transport glass to the gluing area. At the end of this segment, there is a glass thickness detection device that detects the thickness of the glass when it is transported to that location. The structure of input segment 2 is the same as input segment 1, but there is no thickness measuring device and sensor function at its end. Glass length measurement is mainly completed in this segment. Input segment 2 and output segment 1 are the execution components for gluing in the X direction during the gluing process. The suction cup component is composed of suction cups, synchronous belt structures, guide rails, etc. It is used to suction and fix the glass during the gluing process, maintain the stability of the glass, and is also an execution component for gluing in the X direction. It always maintains synchronization with the 6 and 7 axes during its operation. Output segment 1 consists of two synchronous belts distributed in a V-shape. Besides being the execution component for gluing in the X direction, this segment also completes the glass's retraction action. The structure of output segment 2 is the same as that of output segment 1, and it is used to output the glued glass to the end of this segment and wait for unloading.

The gluing head mainly includes the gluing head, plastering board, gluing depth detection device (crescent plate), etc. The gluing head is one of the core components of the gluing machine, which applies glue to the periphery of the glass.

7.1.2 Design of Drive System for Fully Automatic Gluing Machine for Hollow Glass

The driving system of the fully automatic adhesive coating machine for insulated glass adopts a combination of motor drive and pneumatic drive. The system uses 12 motion control motors, 2 air floating motors, and 1 single end encoder.

The 9 servo motors are respectively A component, B component, input section 1, input section 2, output section 1, output section 2, glue head lifting shaft, suction cup unit (hereinafter referred to as X-axis), and motors on the glue head rotation function section. Three stepper motors are used: one for the output section 1 V with a V1 adjustment shaft, one for the output section 1 V with a V2 adjustment shaft, and one for the Y-axis adhesive head extension.

The incremental encoder, which occupies a shaft card channel but does not have a practical motor. It is connected to the rotation axis of the lunar plate, and the system measures the rotation of the lunar plate by reading the pulse number of the encoder. After calculation, the depth of the lunar plate measurement is obtained. The measurement value of the crescent plate is an important factor affecting the quality of glue application. Compared with the potentiometers used in the previous generation of equipment with low measurement accuracy and susceptibility to interference, this encoder greatly improves the measurement accuracy (theoretical accuracy of 0.1°/pulse). At the same time, due to the transmission of digital signals by the encoder, the anti-interference ability of the analog signals of the potentiometers is greatly enhanced, and the quality of glue application is significantly improved.

Two three-phase asynchronous motors are used to drive the air floating fan. The air floating structure is a newly adopted glass support technology for the mechanical part. The previous generation of glue machine was equipped with several rows of rollers on the backing plate, which supported the glass to always remain on a specific plane during transportation. However, due to the large number of required rollers, it is quite cumbersome and difficult to debug all rollers on the same plane, and there is a certain frictional resistance between the rollers and the glass during the glass conveying process. Therefore, the newly designed air flotation structure can form a uniform "air cushion" between the glass and the backing plate, with even gaps in the middle and much less resistance than the roller.

The partial action of the glue gun, the action of the glue supply part, and other auxiliary action driving mechanisms are achieved by a high-performance pneumatic system.

The incremental encoder is used to measure the glue depth of the crescent plate, occupying the 13th axis card channel. The 9 motor axes are the Yaskawa servo driver and Yaskawa servo motor, and the other 3 axes use the Start stepper motor. The servo motor adopts a pulse plus direction control method. On the Yaskawa servo drive, the parameter that needs to be set is the first bit of Pn000, which is 1, that is, the "position control (pulse column command)" mode is selected. In addition, according to the installation characteristics of the mechanical equipment, the parameter can be set to the 0th bit of Pn000 to select. "Rotation direction selection", some parameter settings can be found in Table 7.1

Table. 7.1 Part of the control parameters

Setting of motor steps per revolution (fine fraction setting)	Each model of driver has 16 selectable step counts (sub scores), which are set by the 1st, 2nd, 3rd, and 4th positions of the toggle switch on the driver. These 16 step counts basically cover the user's requirements for motor step pitch, and the step count setting must be effective when the driver is not powered or powered but the motor is not running
Driver output current setting	Each model of driver has 16 selectable output currents, which are set by the 7th, 8th, 9th, and 10th positions of the toggle switch on the driver. The driver outputs a sinusoidal current to the motor, with the current magnitude declared as the effective value
Control signal mode setting	Each model of driver has 2 control signal modes to choose from, set by the 5th bit of the toggle switch on the driver: Cpd mode: The rotation direction of the motor is controlled by the DIR commutation level, while the step signal depends on CP. For example, when DIR is high, the motor rotates clockwise, while when DIR is low, the motor rotates counterclockwise in the opposite direction. This commutation method is called the single pulse method. The 5th position of the toggle switch is set to the "0" position; CW/CCW mode: The driver receives two pulse signals (usually labeled as CW and CCW). When one of them (such as CW) has a pulse signal, the motor runs forward, and when the other (such as CW) has a pulse signal, the motor runs in reverse. We call this the dual pulse mode. The 5th position of the toggle switch is set to the "1" position
Automatic half current	It means that the driver will automatically enter the half-current state about 1 s after the pulse signal stops being applied. At this time, the motor phase current is half of that during operation to reduce power consumption and protect the motor. This function is set by the 6th position of the dial switch on the driver: 0—no function, 1—yes
Phase memory function (no time limit)	When the pulse signal stops being applied for about 1 s, the driver will automatically enter a half current state. At this time, the motor phase current is half of the running state to reduce power consumption and protect the motor. This function is set by the 6th position of the toggle switch on the driver: 0- no this function, 1- this function is available
Protection function	Over temperature protection, over-current protection, undervoltage protection, and protection signal output

7.1.3 Control System of Fully Automatic Gluing Machine for Hollow Glass

(1) The Main Process and Process of Gluing

A reasonable process flow is the key to completing the entire gluing action and is a prerequisite for efficient production of gluing machine equipment. Mastering the reasonable process flow and logical timing relationship of the glue coating machine can lay a solid theoretical foundation for developing the glue coating machine control program. The main process of CNC gluing machine is to transfer the insulating glass from the input section to the gluing area, apply glue on all sides of the glass, and finally output the finished product from the output section. The entire workflow of the glue coating machine is shown in Fig. 7.5.

Power on return to zero: Automatically return to zero for each axis in sequence, ensuring that there is no motion interference between the axes, and this operation is only performed once after power on.

Parameter initialization: Before processing the same batch of products, input parameters such as glass thickness, colloid ratio value, gluing speed, and glue gun flipping judgment value.

Glass input: When the glue source in the glue bucket is sufficient, the glue pressure is normal, there is no glass in each section of the equipment, the glue gun mixer is

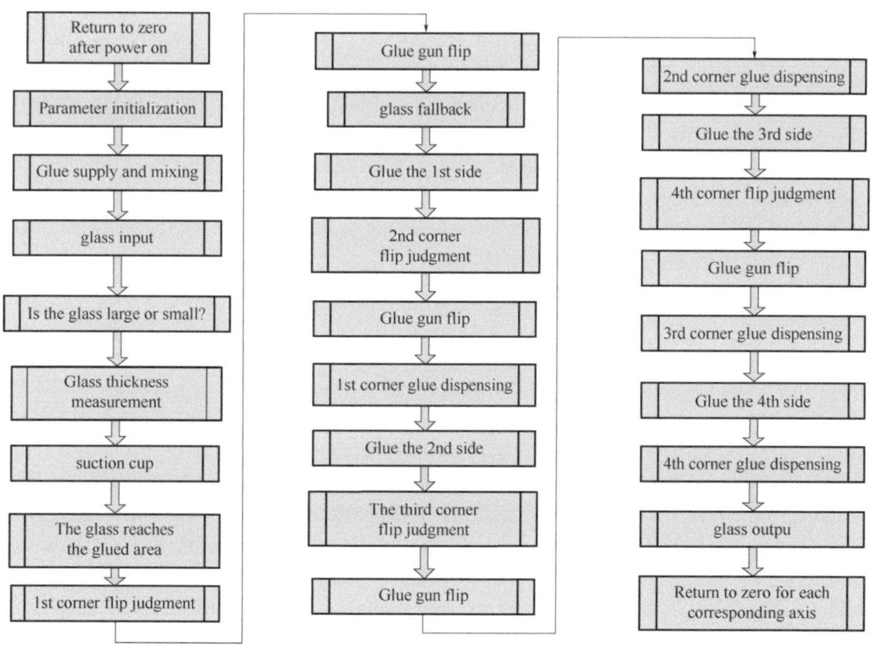

Fig. 7.5 Glue Application Flowchart

in position, and the foot switch is pressed, the glass input section motor starts to transport the glass. When the front edge of the glass passes through the photoelectric switch, it slows down the transmission. When the glass front passes through the next photoelectric switch, the servo motor brakes and measures the thickness of the glass.

Glass thickness measurement: The pneumatic solenoid valve is opened under the action of the photoelectric switch signal, driving the thickness measuring mechanism to press the glass tightly. The potentiometer reading changes, and the upper computer obtains accurate glass thickness values through analog-to-digital conversion and necessary filtering programs. The glass thickness value is the driving data obtained by adjusting the neutral belt with the glue gun.

Sucker suction: When the glass is transported to the glue coating area, the transmission stops. The supporting cylinder behind the suction cup assembly acts first, and the suction cup cylinder acts again. The vacuum generator starts and tightens the glass. When the vacuum degree of the vacuum sensor set inside the shaft reaches a certain value, a signal output is generated. At this time, it can be determined that the suction cup is tightly engaged, and then the cylinder lock is locked. The supporting cylinder is then retracted, and the suction cup moves synchronously with the input and output conveyor belt to transport the glass to the adhesive area [2, 3]

The gluing process mainly includes gluing on four sides, flipping the glue gun, and gluing on three corners. In the adhesive coating system, a plane that forms a 90° angle with the vertical plane is defined as the plane, and the axis parallel to that plane is set as the Z-axis. As shown in Fig. 7.6, the adhesive system moves along the axis in a positive direction to apply adhesive to the first side of the insulating glass. At the end of the first side of glue application, the mixing section stops supplying glue and instead uses a glue bucket to supplement the glue to the mixer. The glue gun stops applying glue, and the glue gun component rotates counter clockwise 90° relative to its rotation center to align the glue gun mouth with the second side of the glass. At this point, the plastering board is pressed against the first side, and the glue gun is applied to the corners of the first side. After the plastering board leaves the glass, apply glue to the second side. Use the same operating process to sequentially coat the other edges of the glass with adhesive.

(2) The Control System of the Glue Coating Machine

The intelligent manufacturing equipment fully automatic gluing machine uses industrial control computer and Delta motion control card as hardware platform, Windows operating system as software operation platform, and realizes real-time high-speed communication between digital controller and driver, I/O module, etc. through Delta DMC-NET bus technology. An open CNC system for fully automatic gluing machine is developed. The hardware architecture of the gluing machine control system is shown in Fig. 7.7, and some control parameters are shown in Table 7.1. The overall framework of the control system is shown in Figs. 7.8.

The upper level software used in this case is Visual Studio2005 based on the Windows operating system, and the lower level debugging software includes Pewin32Pro, PmacPlotPro, and PmacTurningPro. For the convenience of users, a dynamic link library Pcomm32.DLL for Windows is also provided.

Fig. 7.6 Apply glue on the first side of the glue machine

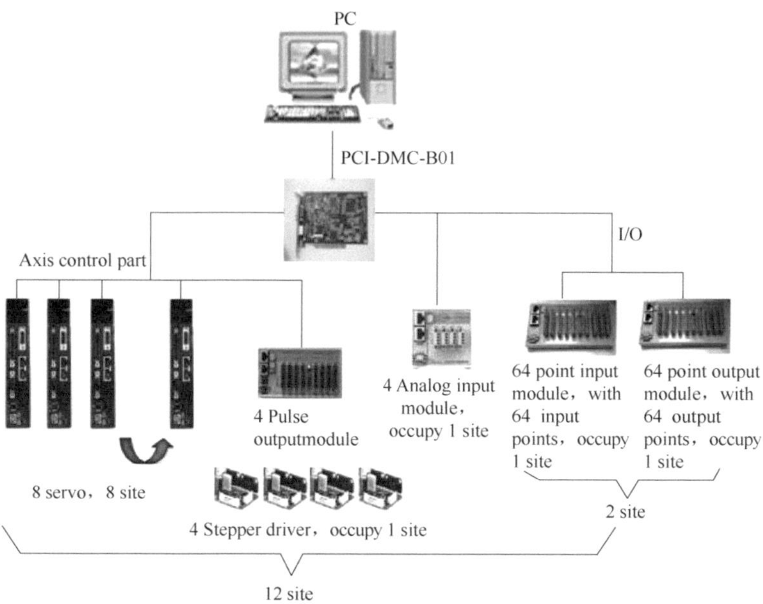

Fig. 7.7 Hardware composition of control system

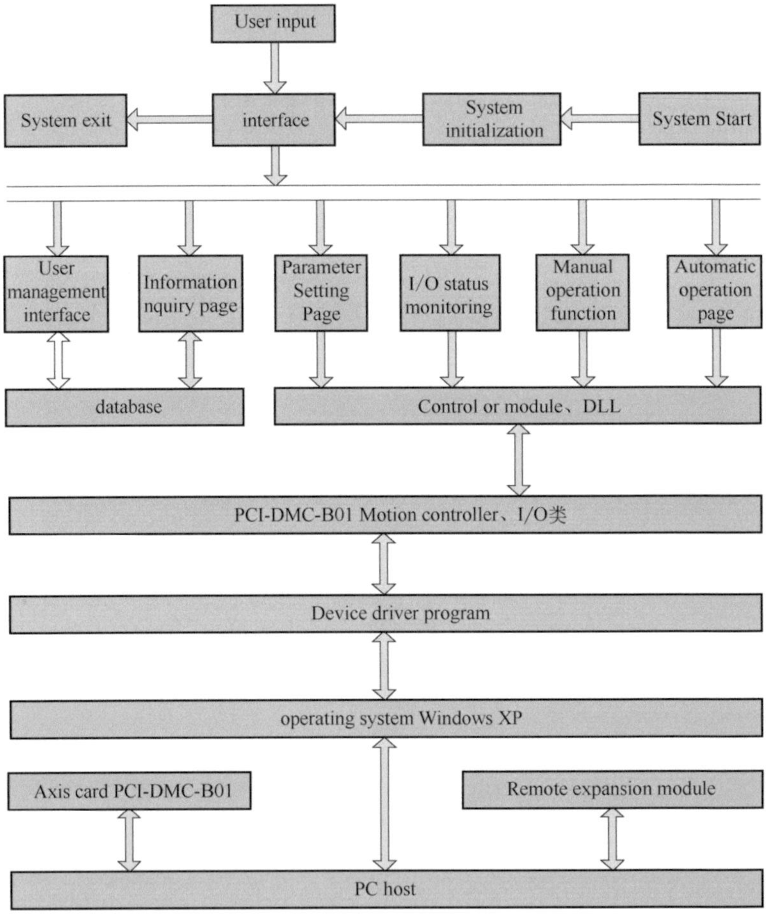

Fig. 7.8 Overall framework of control system

Pewin32Pro is a Windows system executable program for the DeltaTauUMAC controller, which is a development and debugging tool for lower level programs on the upper computer. According to functional requirements and drawing on the experience of previous generation development, this control program has designed a set of fully functional and easy to operate human–machine interaction interfaces on the upper level (Fig. 7.9). Through this interface, combined with the buttons on the control panel, almost all the work of gluing can be completed. This interface is divided into four main parts in terms of content: production management, glue setting, manual operation, and user assistance.

Production management includes production queries and operator management. Through production queries, detailed records of the equipment's operation can be found. Through the operator management item, accounts and login passwords can

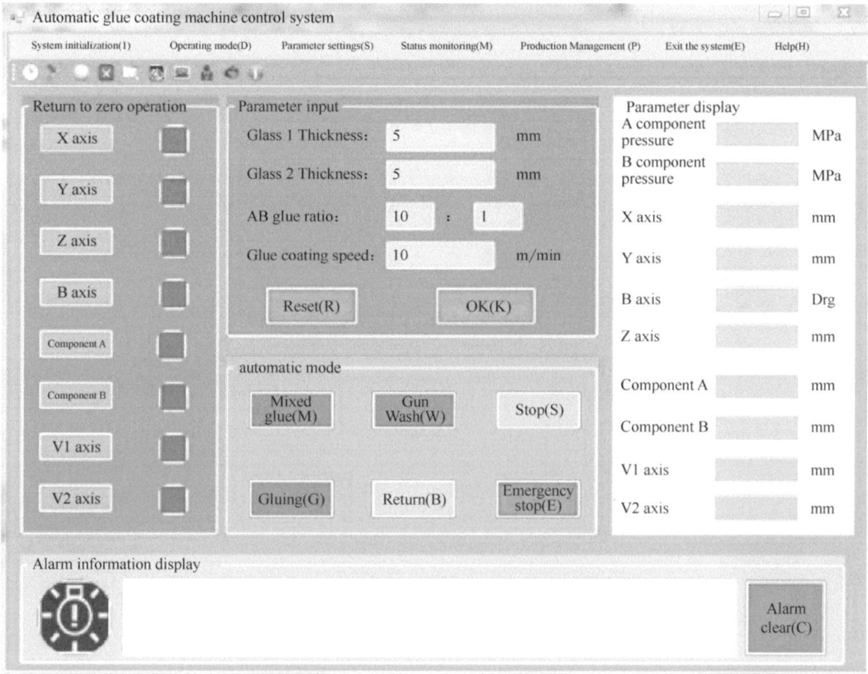

Fig. 7.9 Main interface of the control system

be set for operators to avoid non operators from operating the device. The production management interface is shown in Fig. 7.10.

The glue setting interface allows users to modify the amount of glue applied at each edge and corner, as shown in Fig. 7.11.

The basic settings correspond to the initialization of glass parameters mentioned earlier. The initial glass thickness, colloid ratio, and aluminum spacer fillet area are set through the corresponding options on the interface. The basic settings interface is shown in Fig. 7.12.

The starting position of gluing on each side can be adjusted through the upper computer. The starting setting is used to adjust the proportion of starting glue output, which can be adjusted at any time based on the current glue application effect. The opening position of the glue gun is also directly related to the quality of glue application. Based on the theoretical recommended values and combined with actual situations, users are allowed to make adjustments. The starting setting interface is shown in Fig. 7.13

The on/off timing of the glue gun also has a significant impact on the starting section and corner dispensing of glue, especially affecting the quality of dispensing. When the gluing environment (such as workshop temperature), colloid or its ratio changes, in addition to adjusting the relevant parameters in the above interface, the

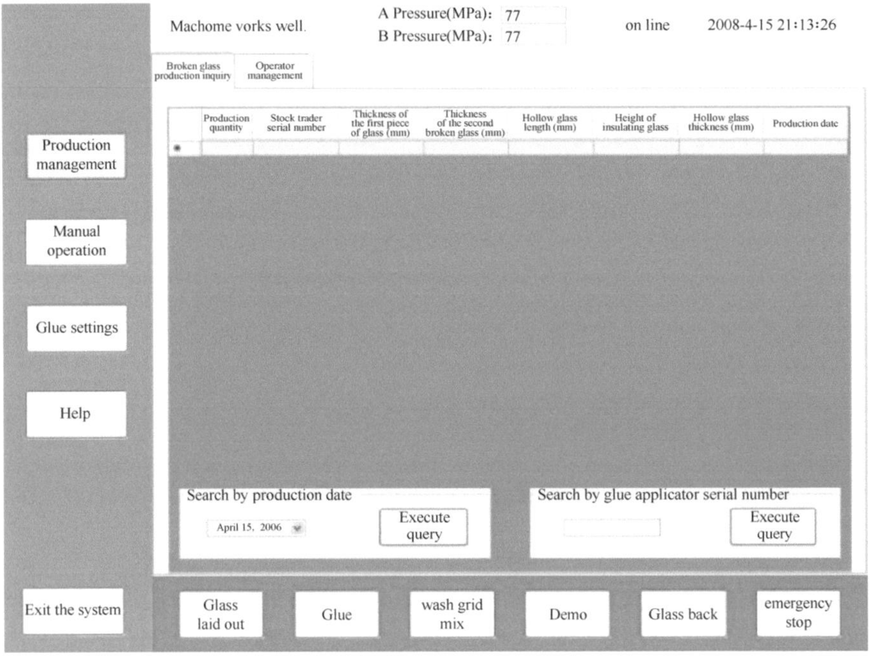

Fig. 7.10 Production management interface

closing delay of the glue gun at the end of each side and the firing delay at the beginning of each side can also be adjusted, as shown in Fig. 7.14.

The glue setting also includes control and adjustment of relevant parameters such as position, speed, and acceleration during the glue application process. The position of the switch nozzle, glass transportation position, glue coating speed, transportation speed, and acceleration have all been parameterized, and can be easily adjusted on the upper interface, as shown in Figs. 7.15, 7.16, 7.17, 7.18 and 7.19.

Pewin32Pro provides comprehensive and powerful features for user development, with the most representative being the manual operation interface. Therefore, for systems controlled by UMAC controllers, the manual operation interface is particularly important and indispensable for users in terms of functionality.

Although Pewin32Pro is very powerful, its English interface and operation, which targets only variables, inevitably cause inconvenience for users. Furthermore, the interface is overly professional and may not be suitable for industrial production environments. Therefore, our team has developed a concise yet fully functional interface specifically for use in industrial production.

This interface includes a lot of manual setting content, almost involving all aspects that need to be operated, including setting manual functions for motors that require manual functions during the operation process. Through the selection function of the interface, it can be operated in conjunction with the “forward” and “reverse” buttons

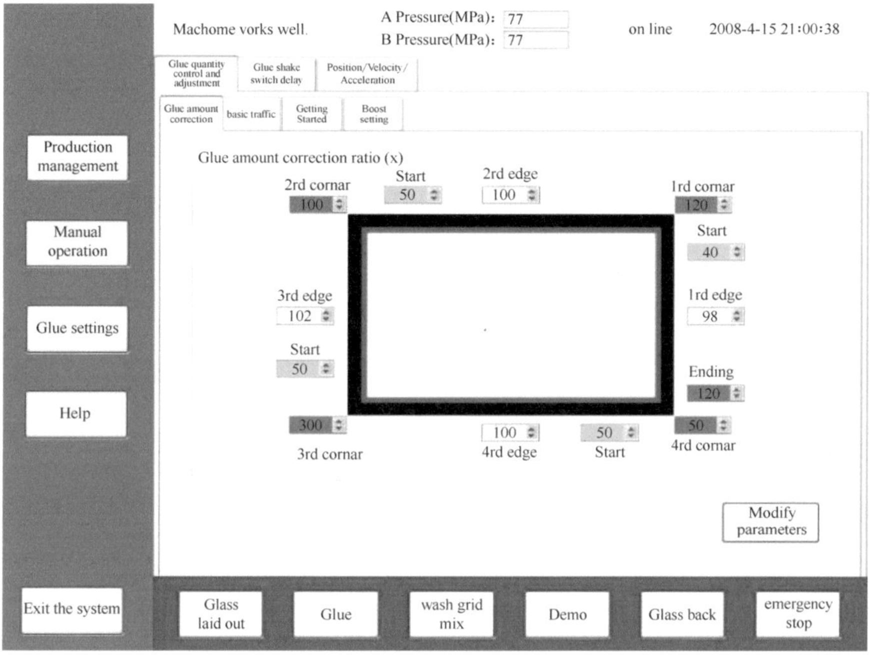

Fig. 7.11 Glue quantity correction interface

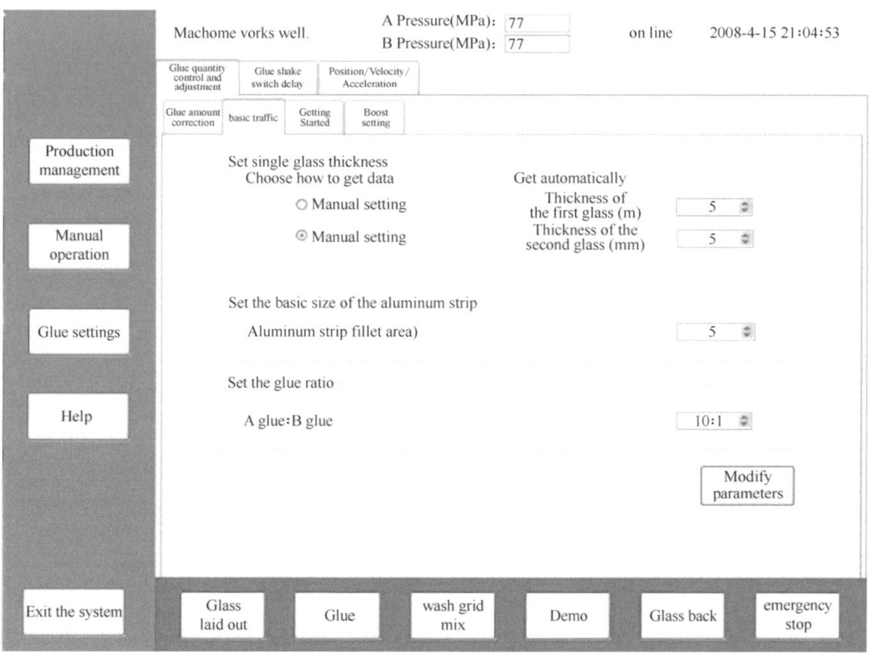

Fig. 7.12 Basic settings interface

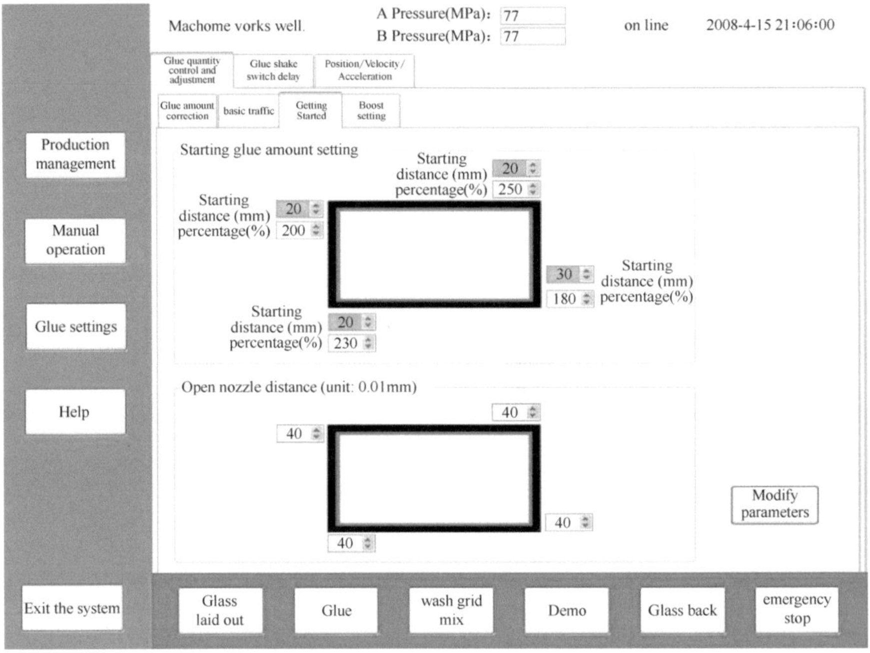

Fig. 7.13 Start up settings interface

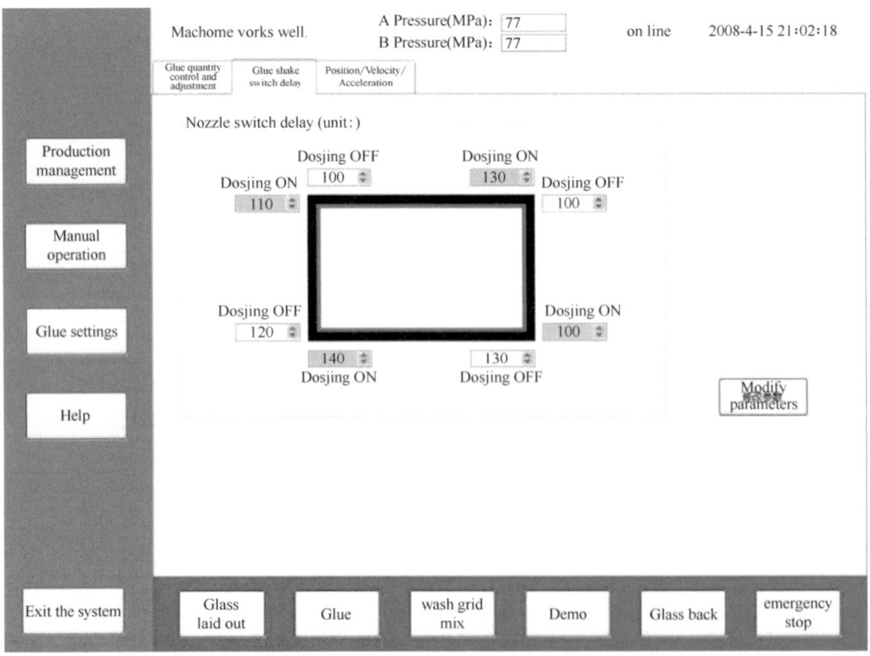

Fig. 7.14 Adhesive gun switch delay

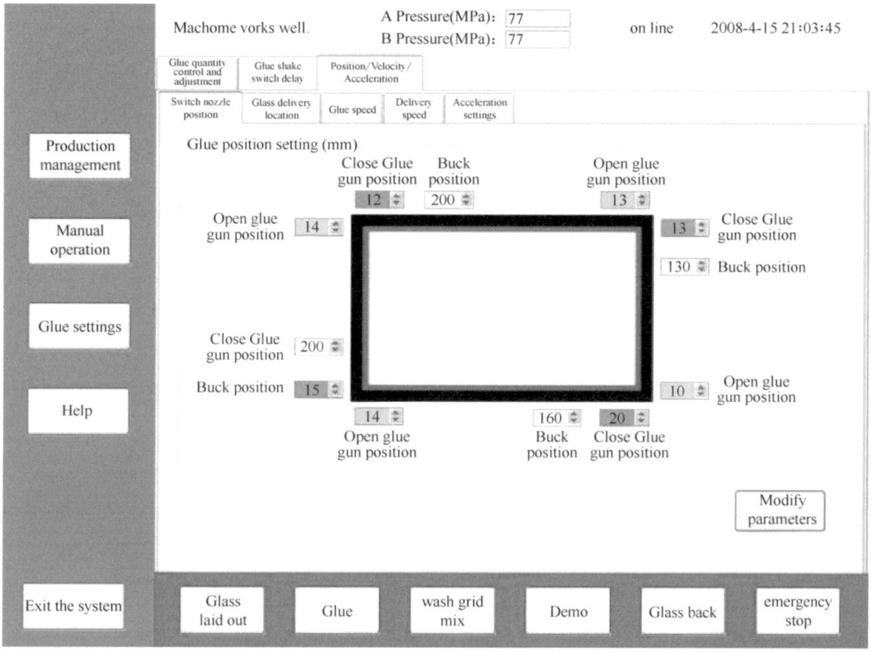

Fig. 7.15 Position of the switch nozzle

on the control panel. The interface also has an adjustment bar for motor speed, as shown in Fig. 7.20.

The zeroing process is the foundation for the gluing program, and the corresponding function buttons for each axis zeroing operation are set in the gluing initialization interface. At the same time, the initial pressure value for gluing is also set on this interface, as shown in Fig. 7.21.

In terms of manual functions, status control buttons for the output ports of each I/O port are also set up, making it very convenient to operate during equipment debugging and maintenance. The digital input interface can display the status of each input port in real time, and users can monitor the status of the concerned state variables in real time, as shown in Figs. 7.22 and 7.23.

(3) Using Help

① Shortcut button. At the bottom of any interface, there are some of the most commonly used buttons, which can quickly enter the corresponding functional interface for operation without the need for multi-level menus, improving the convenience of operation and work efficiency.

② Status display bar. The most important or commonly used information during the gluing process needs to be displayed at all times. The upper part of the program is set to display the connection status between the upper and lower positions, the pressure of A and B colloids, the current time, and

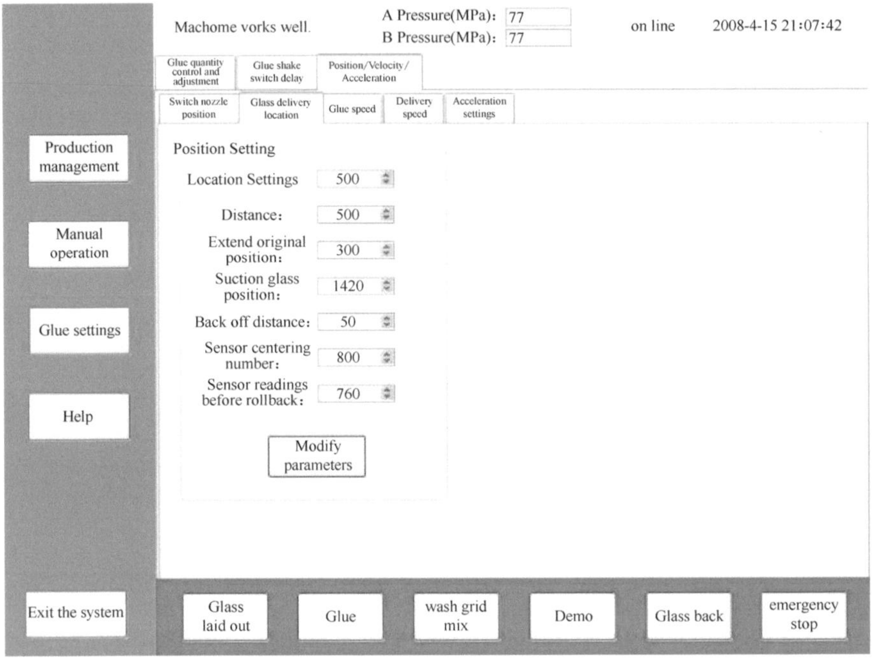

Fig. 7.16 Glass transportation location

other information. When the system encounters an abnormality or prompts the user to perform a certain operation, relevant prompt or alarm information will also be displayed on the status bar.

③ Console panel. The operation panel provides users with convenient and fast operations, and a reasonable functional layout will make the operation of workers simple, providing efficient and reliable safety protection.

The operation panel of the glue coating machine adopts the arrangement shown in Fig. 7.24. In addition to implementing corresponding functions through the lower program, each button can also achieve more flexible functions through the combination of states between the buttons.

Power off/on: No indicator light rotary switch is used to control the power supply of the computer and UMAC controller. Frequent operation is generally not allowed.

Forward: A non self-locking button without an indicator light, which is effective when held down when switched from ‘manual/automatic’ to ‘manual’ by rotating it, and works in the forward direction with the selected motor on the program interface.

Reverse: Non self-locking button without indicator light, effective when pressed and held when switched from ‘manual/automatic’ to ‘manual’ by rotating it, coordinated with the selected motor on the program interface for negative direction action.

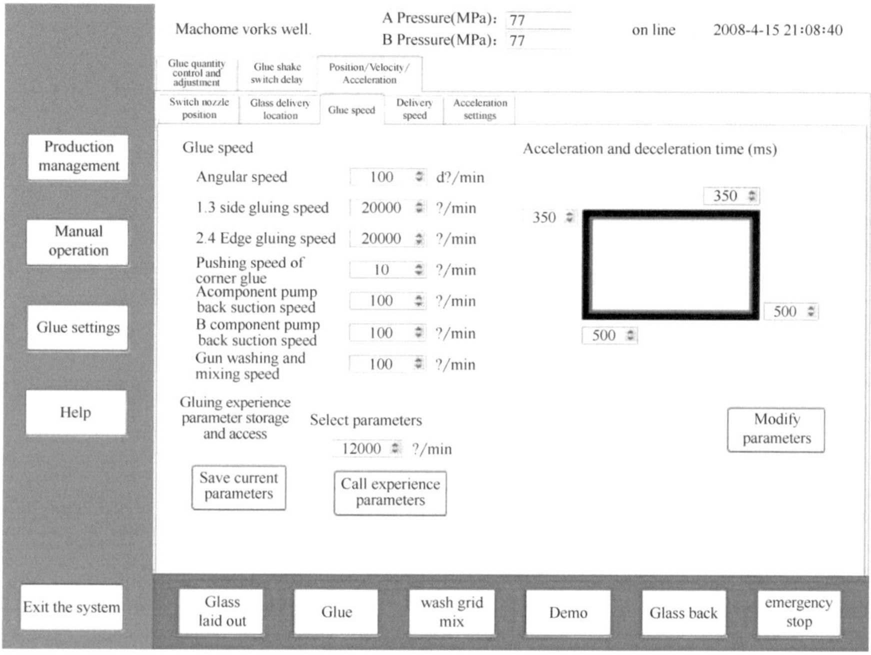

Fig. 7.17 Glue speed

Automatic/Manual: Select switch with indicator light to select whether the control mode of the glue coating machine is automatic or manual. The light will light up when in the manual position. If in manual mode, the upper program will have interference collision prevention judgment.

Joining waiting: A dual state non self-locking button with an indicator light. After pressing it (the light is on), the glass will stop at the position where it has been completed. Wait until it is pressed again (the light is off) before continuing to run the program.

Thickness measurement waiting: A dual state non self-locking button with an indicator light. After pressing it (the light is on), the glass stops at its current position after measuring the thickness. Wait until it is pressed again (the light is off) before continuing to run the program.

Mixing glue/white glue: a rotary switch with an indicator light, the light goes out when the glue is mixed. When the white glue is in the position, the light will be on and only the white glue will be released. The glue release button can be used to manually remove the glue or wash the gun.

Normal/Forced: A rotary switch with a key is usually in a normal state during normal operation. Forceful manual operation is only allowed for professional technicians, and the function of forced operation is used when the machine does not respond to

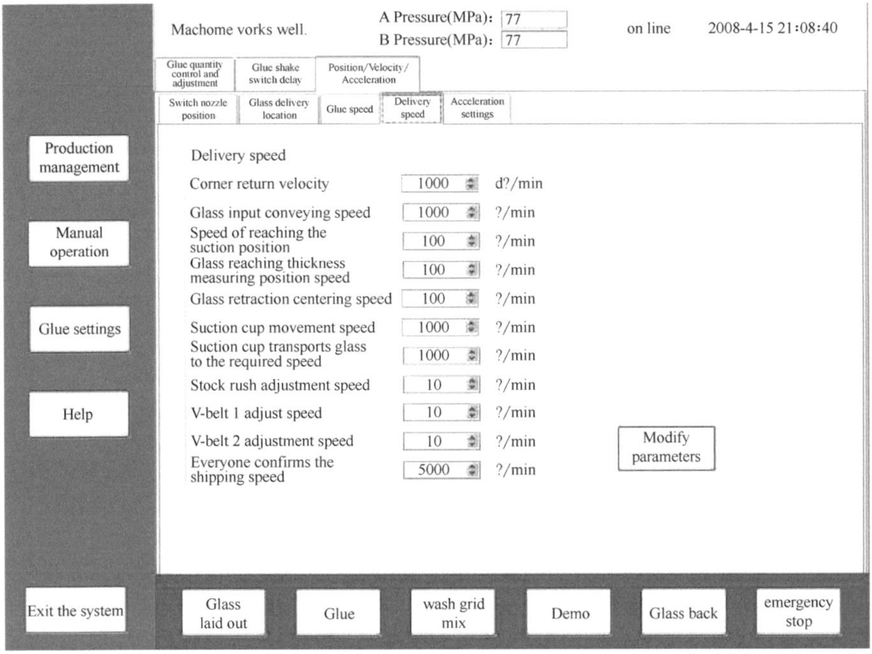

Fig. 7.18 Transport speed

the program or manual operation fails. To ensure safety, the machine speed is very slow at this time. The mandatory function is rarely used because the interference situation is not judged at this time. Therefore, the operator needs to understand the possible consequences of this operation and ensure absolute safety before using it.

Alarm: The indicator light flashes when there is an error or prompt message in the program or action, and the buzzer in the operation panel cooperates with intermittent sound reminders to give an alarm.

Start/Reset: Non self-locking button without indicator light, a. Automatic gluing program start button, automatic/manual knob rotating to automatic mode is effective; b. Resume button after pause; c. Clear the alarm indicator light alarm or reminder message.

Pause: A dual state non self-locking button with indicator lights, a stop or pause operation button. When in an effective state, the light is on, and the stop or pause state is restored through the "Start/Reset" button.

Glue dispensing: Non self-locking button with indicator light, the indicator light shows the on/off status of the glue gun nozzle during automatic gluing, the light is on when the glue gun nozzle is open, and the light is off when the glue gun nozzle is closed. Press and hold (light on) in manual mode, the glue gun nozzle will dispense

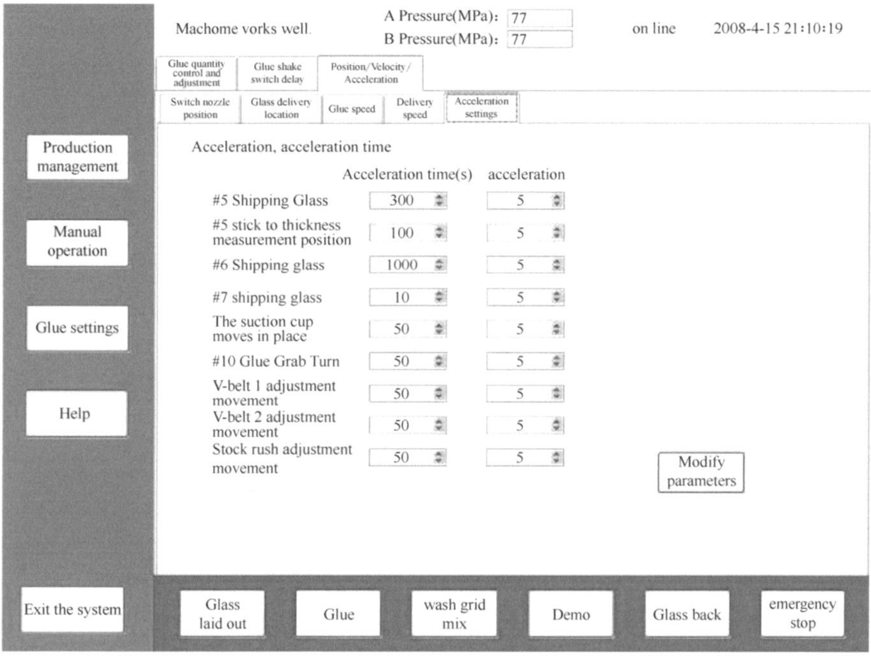

Fig. 7.19 Acceleration

glue, and use the "Mix Glue/White Glue" button to manually remove the mixed glue or wash the gun.

Setting 1: A dual state non self-locking button with indicator lights, combined with the relevant settings in Custom Option 1 on the upper interface of the program to perform gluing, and the light will be on when it is effective.

Setting 2: A dual state non self-locking button with indicator lights, combined with the relevant settings in the custom options on the upper interface of the program to perform gluing. When effective, the light will be on.

7.1.4 *Lower Control of Fully Automatic Gluing Machine for Hollow Glass*

To ensure reliable operation and good response speed of the glue machine, UMAC provides users with PLC and motion control programming languages similar to BASIC language. UMAC allows up to 256 motion programs (PROG) and 32 PLC programs to run simultaneously. The program written by the user can be debugged, monitored for status, and provided with download function on the upper computer through the Pewin32Pro software package equipped with UMAC under Windows,

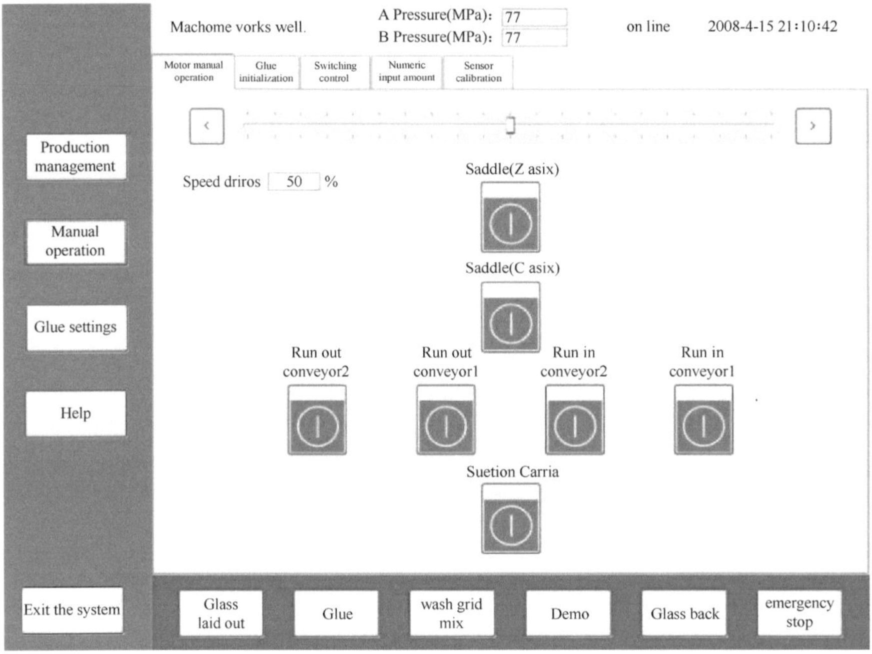

Fig. 7.20 Motor manual operation

allowing users to download (write) the written lower program into the EEPROM of UMAC and complete the writing of the lower program. In addition, as long as users write the corresponding lower computer program and download it to UMAC, UMAC can achieve independent operation.

UMAC internal variables (I, M, P, Q variables) are four times more convenient and flexible to use than PMAC internal variables.

The glue coating machine requires a lot of logical judgment during its operation, so the lower program is written using PLC program. In the process of programming the lower computer, the corresponding subroutine modules are mainly written based on the requirements of the gluing process and the functions of the hardware.

DeltaTau company provides dynamic link libraries based on Windows and other operating systems for user convenience. The dynamic link library under Windows is Pcomm32.DLL, which supports the development of upper level programs in languages such as VB, VC, Dephi, and C++ Builder. This dynamic linking library contains 499 library functions, which users can easily call according to their needs, such as:

OpenPmacDevice()—Open the UMAC data exchange channel.
ClosePmacDevice()—Close the UMAC data exchange channel and release system resources.

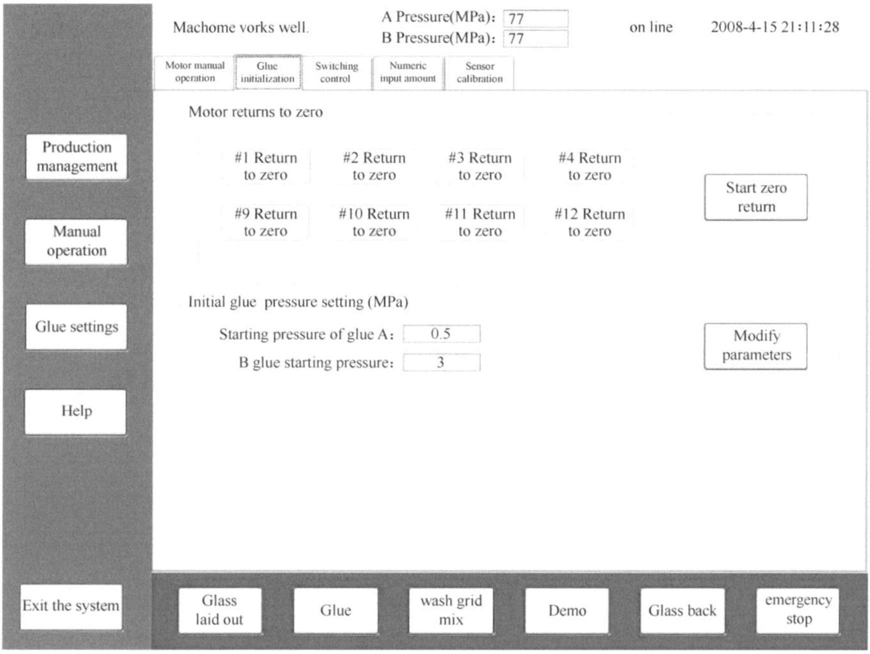

Fig. 7.21 Glue initialization

PmacGetResponse ()—Used to send an online instruction to PMAC, this is the most widely used function, which can be various instructions recognized by UMAC and can also obtain a response from UMAC.

Through various functions, users can access almost all the memory and register address spaces on the card, achieve communication between the upper computer and the lower computer, write the required configuration data to the lower computer through the upper interface, and display various lower information on the upper computer, such as the status of each control axis, the switch status of each I/O port, the collected values of A/D sensors, and various alarm information of the lower computer. This dynamic link library makes it very convenient for users to develop upper programs. The communication principle between the upper and lower positions of the glue coating machine is shown in Fig. 7.25.

① Lower PLC program. According to the glue coating machine process, the lower program is written using a modular structure. Write sub PLC program blocks based on different functions, call the corresponding PLC according to the glue process, and close the PLC program for the already used functions to save system resources to the maximum extent.

The PLC program corresponding to the lower program of this glue machine is shown in Table 7.2.

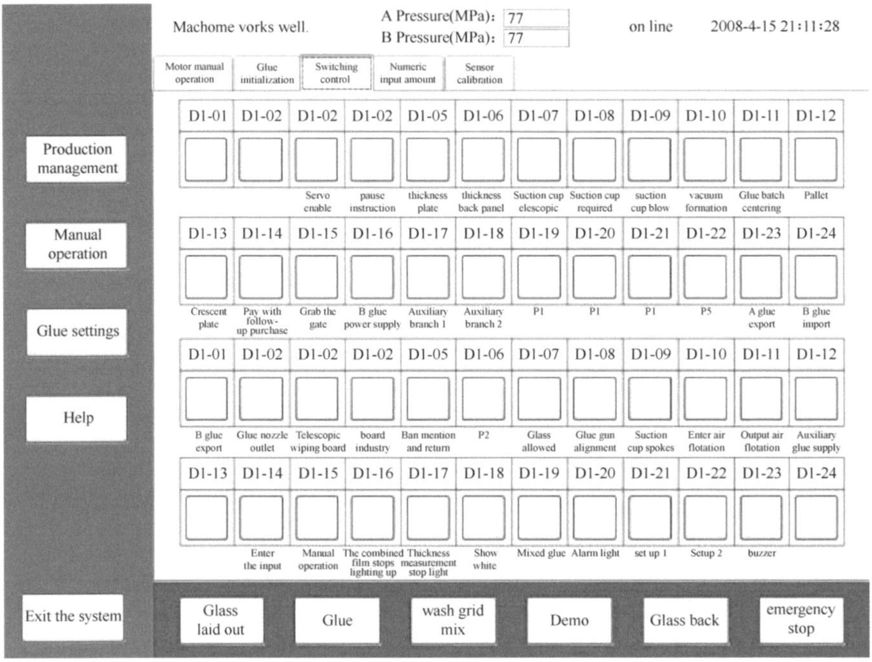

Fig. 7.22 Switch control

② Follow the implementation. According to the following process of gluing, this equipment has effectively controlled the amount of glue produced. During the gluing process, relevant data was collected (using PmacTurningPro software tool) to verify the implementation of this follow-up process based on the relevant data.

During the glue application process on the second side, the glue application speed, the following speed of motor A, the following speed of motor B, and the glue application depth were collected.

Figure 7.26 shows the sampling curve of glue application speed. Curve 1 represents the following speed of motor A, curve 2 in the figure represents the following speed of motor B, and curve 3 represents the movement speed of axis 6, which is the user set adhesive speed.

Figure 7.26 shows the glue following speed. It can be seen from the figure that the following speed ratio of component A to component B is about 2.5:1. The actual speed ratio of components A and B is calculated to be:

$$n_{\mathrm{b}}\xi_{\mathrm{b}} = \frac{5.01}{N} n_{\mathrm{a}}\xi_{\mathrm{a}} \tag{7.1}$$

Among them, $\varepsilon_b = 39.322$, $\varepsilon_B = 31.667$, $N = 10$, substitute into formula (7.1) to get:

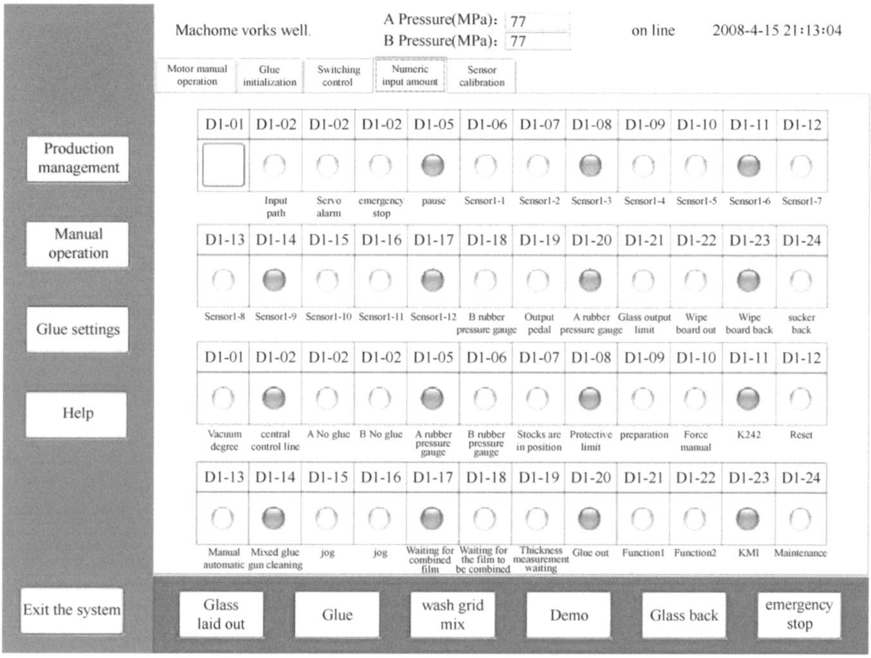

Fig. 7.23 Digital input

$$\frac{n_a}{n_b} = \frac{10 \times 39.322}{5.01 \times 31.667} = 2.48 \tag{7.2}$$

It can be seen from formula (7.1) that the actual data collected by the A and B motors are consistent with the theoretical values calculated. Curve 3 shows that the gluing speed is a constant value during the gluing process, that is, the gluing speed set by the user.

Because during the gluing process, the gluing speed U remains constant, the gluing width b remains constant, and the gluing ratio N remains constant, the rotational speed na of component A motor is linearly proportional to the gluing depth h. According to Eq. (7.2), it can be seen that there is a good linearity between the motor speeds of components A and B, and the speed corresponds to the glue application speed, as shown in Fig. 7.27.

③ Monitoring of alarm information. A complete upper level interface not only needs to make the operation of operators simple, but also needs to provide users with complete information, among which alarm information is an important item. The device should achieve real-time monitoring of various states of the lower computer. Based on the characteristics of the device and the severity of various information, situations that require user intervention should be set as alarm messages. Table 7.3 shows the alarm information statistics used in this process, which can fully meet the needs of users.

Fig. 7.24 Console panel

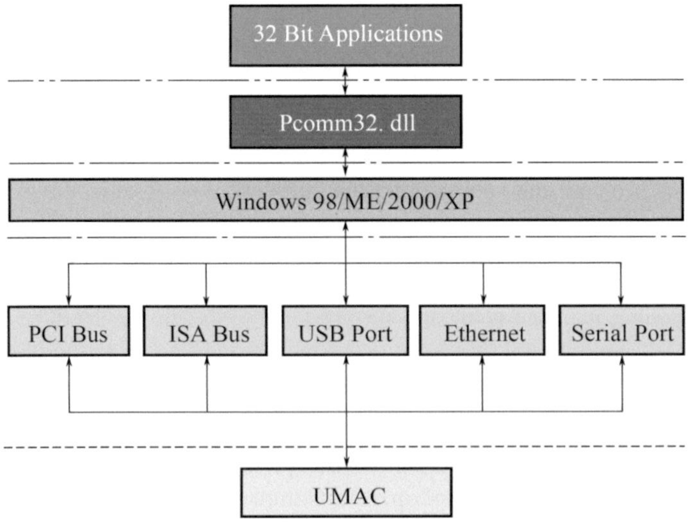

Fig. 7.25 Functional Block Diagram of Host-Lower Computer Communication

Table. 7.2 PLC program summary table

S. No.	Program corresponding functions
PLC0	Pause program
PLC1	Glass glue depth measurement, filtering program
PLC2	Manual zero return procedure for each axis
PLC3	Lower I/O port detection scanning program
PLC4	Forward and reverse operation in the console operation panel
PLC5	Parameter initializer
PLC6	IP1 Glass Shipping Procedure
PLC7	IP2 Glass Shipping Procedure
PLC8	Glue gun Y-axis and V1 adjustment procedure
PLC9	V2 Adjustment Program
PLC10	First side gluing, height measurement, second corner turning and back suction
PLC11	Glue dispensing, second side glue application, third corner and back suction
PLC12	Glue dispensing, third side glue application, fourth corner and back suction
PLC13	Glue dispensing, glue dispensing on the fourth side, glue dispensing on the fourth corner, and zero return of the X, B, Z, A, B axes
PLC14	Dispensing procedure
PLC15	A, B motor suction program
PLC16	After gluing, the Z-axis and B-axis return to zero
PLC17	OP1 Glass Delivery Procedure
PLC18	OP2 Glass Delivery Procedure
PLC19	Glass thickness measurement, filtering procedures
PLC21	Operation panel button operation scanning program
PLC22	Alarm information program
PLC29	A, B glue pressure collection, filtering program
PLC30	Fallback sensor position acquisition, filtering program Program corresponding functions

The alarm information should be set to the highest level and should be issued to the user in the shortest possible time. Therefore, in order to improve the real-time performance of system alarms in this process, the detection of alarm information is placed in an independent PLC22 program, and its scanning cycle is ensured to ensure the priority of system alarms. The relevant alarm information is shown in Table 7.3.

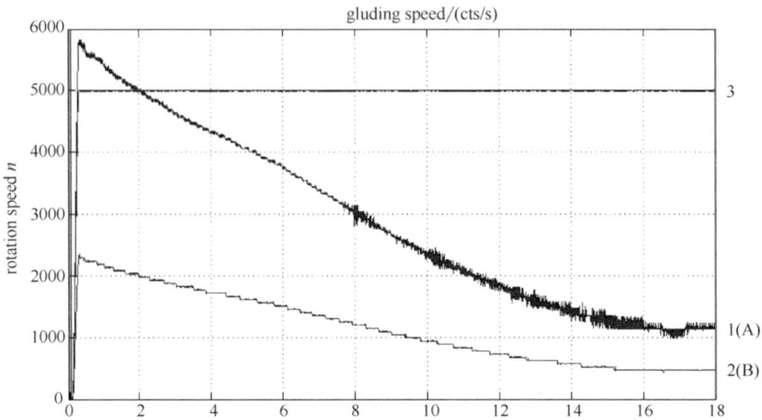

Fig. 7.26 Glue following speed

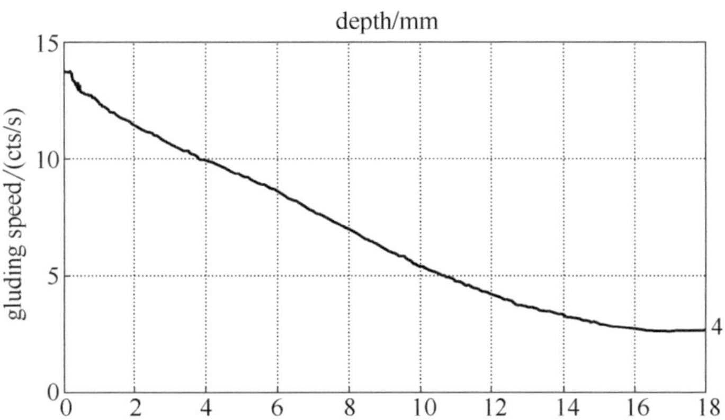

Fig. 7.27 Depth curre

7.2 Case 2 Development of Fully Automatic Vertical Glass Grinding Machine

The key equipment in energy-saving glass deep processing equipment—the large vertical glass edge grinding machine, has gained the trust and recognition of various glass processing enterprises due to its unique structural characteristics.

Compared with the horizontal glass edge grinding machine, the fully automatic vertical glass edge grinding machine (Fig. 7.28) has three advantages: firstly, the vertical edge grinding machine equipment is developing towards space, reducing the footprint [4–6];Secondly, based on the theory of material mechanics, when thin and brittle objects are erected, their strength is high, resulting in a low rate of glass

Table. 7.3 Alarm information table

S. No.	Alarm content	Treatment measures
1.	Communication failure between upper and lower computers	Stop all programs and require the user to re-establish the connection
2.	Emergency stop or not enabled	Check the emergency stop status and whether it is enabled
3.	The servo drive has an alarm	Stop all programs and check the servo drive
4.	Open loop alarm for motors 1–12	Stop all programs, troubleshoot, and close the open-loop motor loop
5.	1–12 positive limit alarm	Stop all programs, check motor position, and remove limit
6.	Negative limit alarm No. 1–12	Stop all programs, check motor position, and remove limit
7.	Abnormal thickness	Stop all programs, check the glass, and check the measurement environment
8.	Abnormal length	Stop all programs, check glass, measure environment, check sensors
9.	Depth anomaly	Stop all programs, check the glass depth, check the thickness encoder
10.	Component A is missing glue	Stop all programs and replace the glue of component A
11.	Component B is missing glue	Stop all programs and replace the glue of component B
12.	Abnormal pressure of component A	Stop all programs, check the pressure of A glue pipeline, and check the pressure sensor
13.	Abnormal pressure of component B	Stop all programs, check the pressure of B glue pipeline, check the pressure sensor
14.	KM1-2 not engaged	Check KM1-2 contactor
15.	Input section maintenance 1 is on	Stop all programs and confirm the status of the first stage of maintenance
16.	Input section maintenance 2 is on	Stop all programs and check the status of stage 2
17.	Abnormal return position	Stop all programs, monitor the retraction position, sensor status
18.	Trigger glass output limit switch	Stop all programs and check the glass output position
19.	The wiper plate is not extended in place	The program is waiting in the current state, checking the wiper, checking the rubber head extension sensor, and checking the corresponding cylinder solenoid valve
20.	The wiper board has not returned to its proper position	The program is waiting in the current state, checking the wiper, checking the rubber head extension sensor, and checking the corresponding cylinder solenoid valve

(continued)

Table. 7.3 (continued)

S. No.	Alarm content	Treatment measures
21.	The suction cup has not retracted into place	The program waits in the current state, detects the suction cup, checks the suction cup extension sensor, and checks the corresponding cylinder solenoid valve
22.	The suction cup does not reach vacuum	The program is waiting in the current state, checking the suction cup, checking the vacuum generator, and checking the vacuum sensor
23.	The glue gun is not in place	The program is waiting in the current state, checking the glue gun position
24.	Z axis servo not ready	Check the Z-axis motor status, enable and brake status
25.	Force manual operation	High risk, general users are not allowed to operate
26.	Y-axis adjustment abnormality	Stop all programs and check the Y-axis adjustment position
27.	V1 adjustment abnormality	Stop all programs and check the V1 axis adjustment position
28.	V2 adjustment abnormality	Stop all programs and check the V2 axis adjustment position
29.	Corner 1 does not meet the reversal condition	Stop all programs and check the angle reversal conditions and status
30.	Corner 2 does not meet the reversal condition	Stop all programs and check the angle reversal conditions and status
31.	Corner 3 does not meet the reversal condition	Stop all programs and check the angle reversal conditions and status
32.	Corner 4 does not meet the reversal condition	Stop all programs and check the angle reversal conditions and status
33.	The glue mixing state exceeds 15 min without glue application	Issues alarm information and buzzer to remind users to perform glue spraying or gun cleaning

damage during processing operations; The third is to fully ensure the convenience of loading and unloading for batch processing.

7.2.1 *Mechanical Body Design of Fully Automatic Vertical Glass Edge Grinding Machine*

The large vertical automatic grinding wheel glass edging machine is primarily used for grinding the four sides of glass, therefore, its main structural components can be categorized into a conveying section and a grinding section. The conveying section also includes a roller conveying part and a clamping mechanism wrapped in polyester ethylene material. The roller conveying part is mainly used to feed the glass to be ground and the ground glass into and out of the grinding section. The clamping

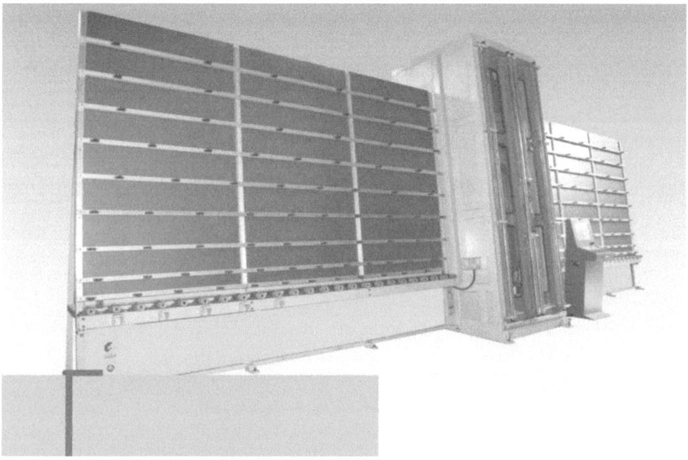

Fig. 7.28 Fully automatic vertical glass edging machine

mechanism is primarily used for clamping and conveying glass during the grinding process. The main components of the grinding section include the frame, grinding head mechanism (including upper and lower grinding heads), and support frame. In order to ensure the reliability of glass delivery and grinding, the frame is fixed with anchor bolts and inclined at a 7° angle to the vertical direction. By using a ball screw pair to drive the upper and lower grinding head mechanism to move up and down and rotate its inner rotating cylinder, the four sides of the glass are ground. The support frame is used for accurate positioning of the clamping mechanism and the upper and lower grinding heads. The specific structure is shown in Figs. 7.29 [7].

The frame of the vertical glass edging machine is a frame structure welded from standard hollow square tube profiles, with a length of 1780 mm, a width of 1360 mm,

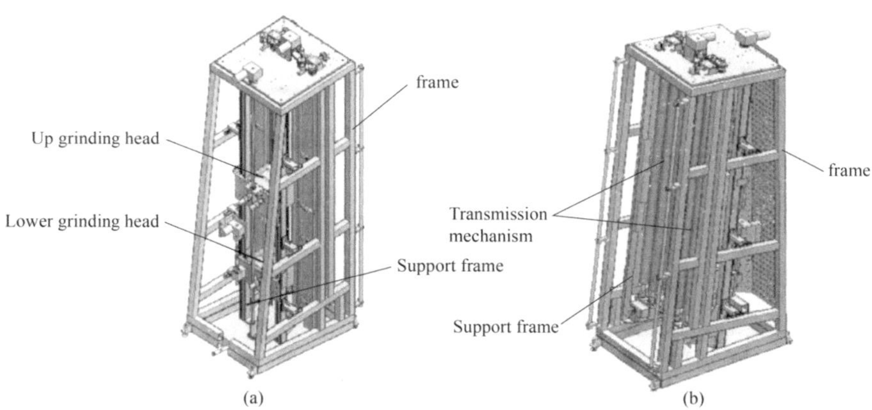

Fig. 7.29 Structure of large vertical automatic grinding wheel glass edging machine

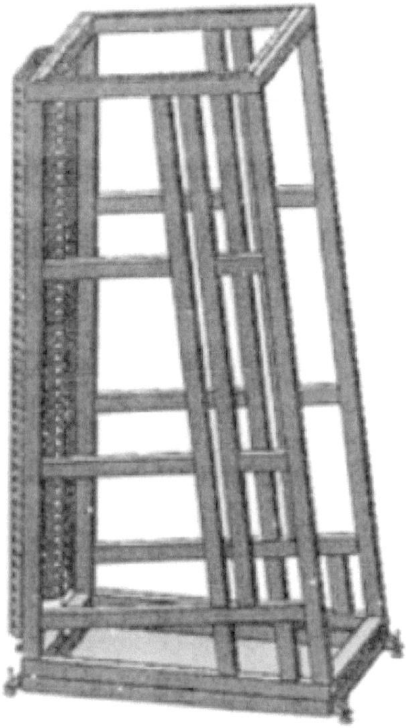

Fig. 7.30 Rack structure diagram

and a height of 3800 mm. The side length of the hollow square tube is 100 mm, and the wall thickness is 5 mm. The rack is mainly composed of a frame body welded with multiple square tubes, two rear doors, and four support legs. The specific structure is shown in Figs. 7.30.

The frame is the external frame structure of a vertical glass edging machine, mainly used to fix the support frame, lower grinding head, and other auxiliary mechanisms. The support frame is fixed to the frame through six support plates on both sides and four bolts below, while the screw connected to the lower grinding head is fixed to the top plate of the frame. Therefore, the rack mainly bears the pressure of the support frame and the gravity of the lower grinding head.

The material used for the rack is Q235 carbon steel, with an elastic modulus of 2.06×1011 Pa and a Poisson's ratio of 0.3.

The support frame of the vertical glass edging machine is a structure welded from standard hollow square tube profiles and hollow rectangular tube profiles, with a total length of 900 mm, width of 800 mm, and height of 3700 mm. The size of the hollow square tube is 100 mm $\times$ 100 mm $\times$ 5 mm, and the size of the hollow rectangular tube is 100 mm $\times$ 50 mm $\times$ 5 mm. The structure of the vertical glass edging machine support frame mainly consists of two front pillars, two rear pillars, two crossbeams, and four reinforcing ribs. The top and bottom panels of the support frame are connected to the frame through bolts, and the specific structural form is shown in Fig. 7.31.

Fig. 7.31 Support frame structure

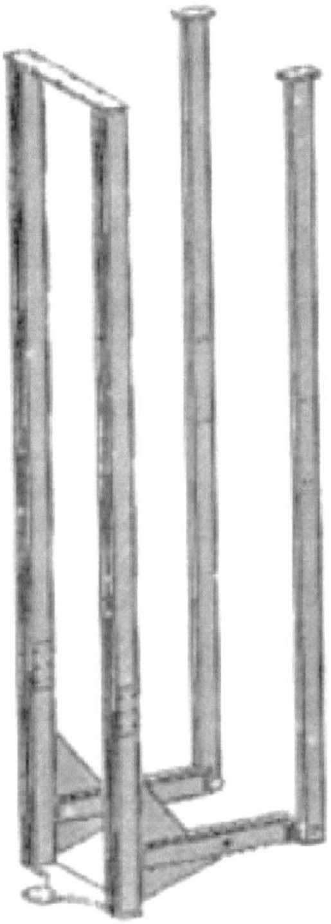

The frame is the core support component of the glass edging machine, used to support the clamping roller and upper grinding head mechanism. At the same time, the upper and lower grinding head mechanisms are fixed on the support frame through guide rails. The deformation of the support frame will directly affect the movement accuracy of the upper and lower grinding heads and the deformation of the clamping roller.

The material used for the support frame is Q235 carbon steel, with an elastic modulus of 2.06×1011 Pa and a Poisson's ratio of 0.3.

The pinch roller of the vertical glass edging machine is a hollow steel pipe with an outer layer of polyurethane rubber attached, with an outer diameter of 120 mm and a length of 2870 mm. The thickness of the polyurethane rubber layer is 9 mm, and the wall thickness of the steel pipe is 15 mm. The pinch roller is a mechanism used to clamp and transport glass during the glass grinding process. The conveying action of glass is performed by two identical clamping roller components. A pinch roller assembly contains four pinch rollers, which are divided into two active rollers and two driven rollers according to their functions. The position of the driven rollers is adjustable for clamping, while the active and driven rollers rotate in opposite directions to transport glass. These two pinch roller components are installed on the two crossbeams of the support frame, and the specific structural form of the pinch roller components is shown in Fig. 7.32.

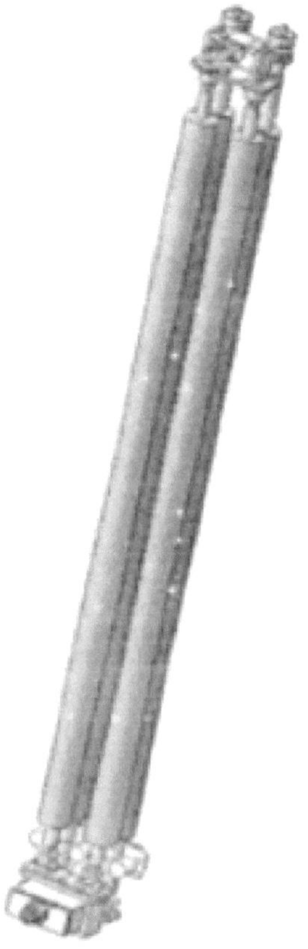

Fig. 7.32 Structure of the pinch roller component

The steel pipe material for the clamping rod is 45 high-quality carbon structural steel, with an elastic modulus of 2.1 × 1011 Pa and a Poisson's ratio of 0.3. The outer rubber of the pinch roller is polyurethane rubber, with an elastic modulus of 8.0 × 107 Pa and a Poisson's ratio of 0.47.

The transmission system of the lower grinding head and the upper grinding head of the vertical glass edging machine cooperate with each other, completing high-precision grinding of the four edges of the glass in three steps. The lower grinding head transmission system is responsible for grinding the bottom edge of the glass, involving main movements including up and down motion and rotational motion. The working principle is shown in Fig. 7.33.

Before the lower grinding head starts grinding the glass, there is a distance between the lower grinding head and the bottom edge of the glass. In order to make the lower grinding head move faster to the bottom edge of the glass within a shorter distance, it is required that the lower grinding head transmission system has good starting response characteristics. The rotational inertia of the entire transmission system has an important impact on the starting response characteristics. Therefore, The structural design aims to minimize the moment of inertia of the entire transmission system as

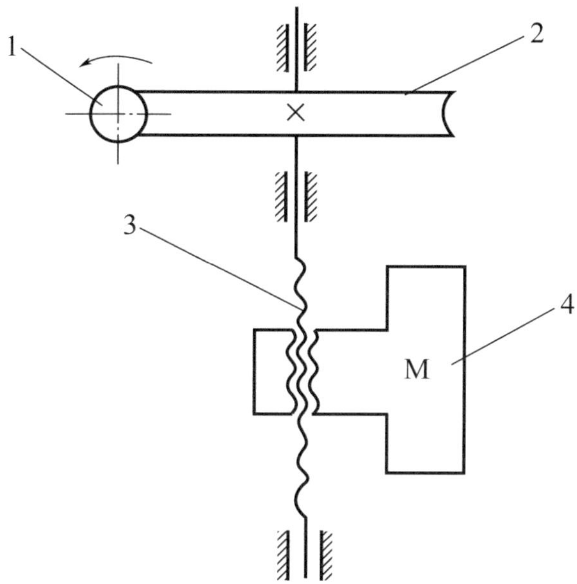

Fig. 7.33 Schematic diagram of the lower grinding head transmission system. 1 Worm gear. 2 Worm gear. 3 Lead screws. 4 Grinding head mechanism

the optimization objective to improve the starting response characteristics of the lower grinding head transmission system.

The glass conveying system of the vertical glass edging machine includes 8 clamping rollers -4 active rollers and 4 driven rollers. The position of the driven rollers is adjustable for clamping, and the active and driven rollers rotate in opposite directions to transport glass. The working principle of the glass conveying system is shown in Fig. 7.34.

As shown in Figs. 7.34, the driving roller rotates automatically under the drive of the motor and its position remains fixed. The position of the driven roller can be changed under the action of the screw, and it can clamp the glass under the action of the cylinder connected to the synchronous belt. During the glass grinding process, the active and driven rollers are responsible for clamping the glass or driving the glass forward and backward through opposite rotation. The synchronous belt connected to the cylinder can maintain consistent motion between the driving roller and the driven roller [8].

According to engineering practice, it is known that the starting response characteristics of the glass conveying system of a vertical glass edging machine have a significant impact on the grinding efficiency of glass. The better the startup response characteristics of the system, the higher the glass grinding efficiency and accuracy of the vertical glass grinding machine. The pinch roller is the core component of the conveying system, and its rotational inertia has a significant impact on the starting response characteristics of the conveying system, with its degree of influence far

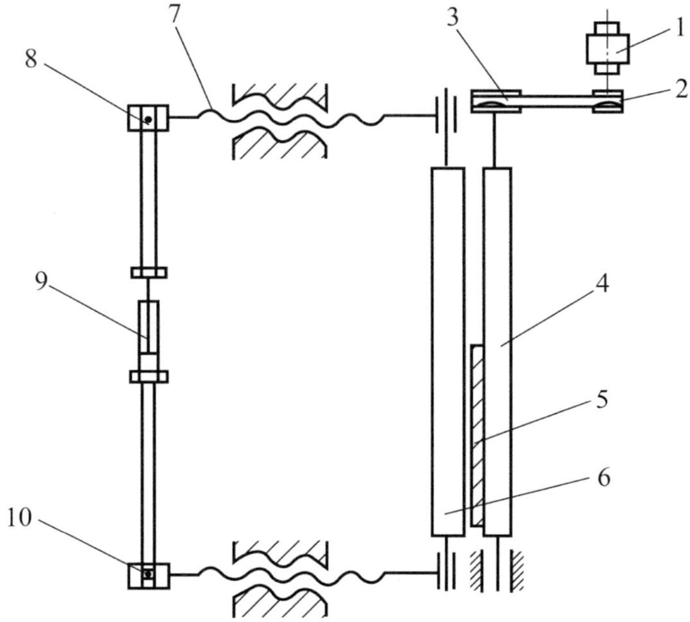

Fig. 7.34 Schematic diagram of glass conveying system. 1. Motor. 2 Pulleys 1. 3 With wheels 2. 4 Active rollers. 5 Glass. 6 Driven roller. 7 Lead screw. 8 Synchronous belt pulley 1. 9 Cylinder. 10 Synchronous pulleys 2

greater than that of other mechanisms. Therefore, during the design process, the starting response characteristics of the glass conveying system are mainly improved by optimizing the rotational inertia of the pinch roller. The structural form and size of the pinch roller are shown in Fig. 7.35.

As illustrated in Fig. 7.35, the pinch roller's structure consists of a hollow square tube coated with a layer of polyurethane rubber. To facilitate optimization, the stepped shaft at the end of the pinch roller will be excluded from the design considerations. The simplified structure of the pinch roller is a hollow steel pipe, 3000 mm in length,

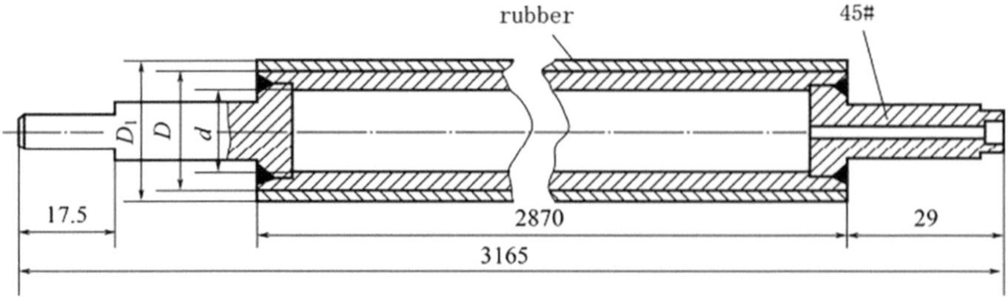

Fig. 7.35 Mechanism form and size of pinch roller

Fig. 7.36 Key components of grinding head and clamping roller

coated with a layer of polyurethane rubber. The key components of the grinding head and pinch roller are shown in Fig. 7.36.

7.3 Design of Control System for 2 Fully Automatic Vertical Glass Grinding Machine

The development process of the control system for a fully automatic vertical glass edging machine is similar to that of a fully automatic gluing machine, which is no longer detailed here.

The control system software interface is shown in Fig. 7.37.

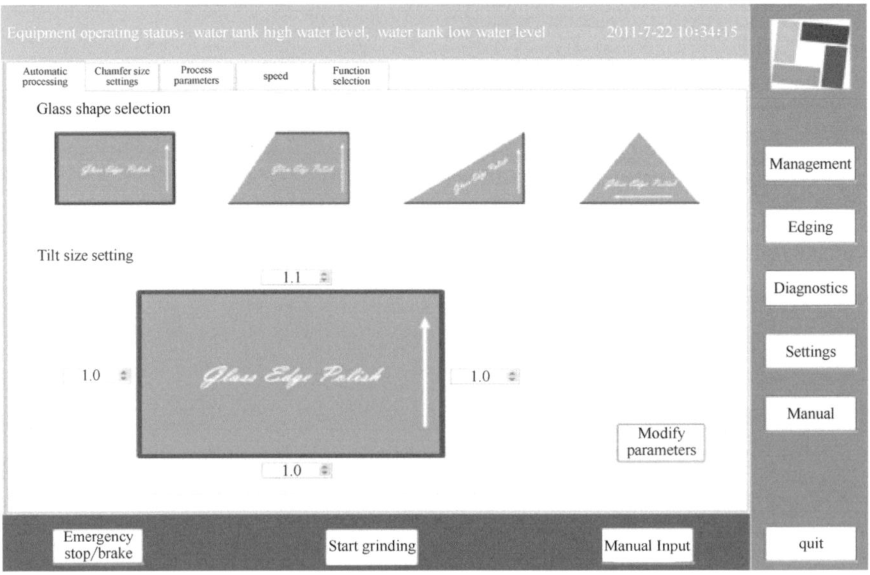

Fig. 7.37 Automatic vertical glass edging machine control system

References

1. Meng Guangjun, Xie Fuchun, Luo Xueke. Application of UMAC Motion Controller in Automatic Gluing Machine[J]. Mechanical Research & Application, 2007(06):66–69.
2. Wang Qiang, Luo Xueke, Xie Fuchun. Application of Open CNC System Based on PMAC in Automatic Gluing Machine[J]. Mechanical & Electrical Engineering Technology, 2006(02):85–87, 95, 105.
3. Liu Ying, Xie Fuchun. Numerical Control Implementation of Multi-Component Proportion and Glue Discharging Speed Control of Gluing Machine[J]. Machine Tool & Hydraulics, 2008, 36(12):117, 144–146.
4. Gao Wensheng, Cui Jianzhi, He Xianbao. Discussion on Flat Glass Edging Process[C]// Proceedings of the 11th Annual Conference of the Chinese Association for Science and Technology, Chongqing, 2009:1–6.
5. Popov A V. Increasing the Efficiency of Diamond Edging of Flat Glass[J]. Steklo i Keramika, 2009(6):16–17.
6. Popov A V. Increasing the Quality of Diamond Wheels for Edge Grinding Flat Glass[J]. Glass and Ceramics, 2010, 67(7/8):252–254.
7. Xu Honghai, Li Xiaoyang. Optimal Design of Transmission Ratio for Lifting Transmission System of Grinding Wheel Frame in Vertical Glass Edging Machine[J]. Journal of Machine Design, 2013, 30(12):37–41.
8. Zheng Xiaoli. Optimal Design of Conveyor Rollers for Vertical Glass Edging Machine Based on MATLAB[J]. Mechanical Research & Application, 2012(05):89–91.

Uncited References

9. Hu Qiu. Design of Servo Feed System for CNC Machine Tools[J]. Machine Tool & Hydraulics, 2004(6):54–56.
10. Liu Chen, Jing Xinghong, Dong Gang. Brief Discussion on Technical Characteristics and Wide Applications of Internet of Things[J]. Science Consult, 2011(9):86.
11. Wei Xiaoguang. Modern Industrial System Integration Technology[M]. Beijing: Electronic Industry Press, 2016.
12. Wang Xifu, Chen Xiaoran. Smart Community - Future Home in the Era of Internet of Things[M]. Beijing: Electronic Industry Press, 2016.

MIX
Papier aus verantwortungsvollen Quellen
Paper from responsible sources
FSC® C105338

If you have any concerns about our products,
you can contact us on
ProductSafety@springernature.com

In case Publisher is established outside the EU,
the EU authorized representative is:
Springer Nature Customer Service Center GmbH
Europaplatz 3, 69115 Heidelberg, Germany

Printed by Libri Plureos GmbH
in Hamburg, Germany